WITHDRAWN

COAL SCIENCE AND TECHNOLOGY 1

GEOCHEMISTRY OF COAL

Czechoslovak Academy of Sciences

COAL SCIENCE AND TECHNOLOGY

Vol. 1 Geochemistry of Coal (Bouška)
Vol. 2 Fundamentals of Coal Beneficiation and Utilization (Tsai)
Vol. 3 Coal. Typology, Chemistry, Physics and Constitution (Van Krevelen)

COAL SCIENCE AND TECHNOLOGY 1

GEOCHEMISTRY OF COAL

by

VLADIMÍR BOUŠKA

Faculty of Science
Charles University, Prague

ELSEVIER SCIENTIFIC PUBLISHING COMPANY
AMSTERDAM · OXFORD · NEW YORK 1981

CZECHOSLOVAK ACADEMY OF SCIENCES

Scientific Editor Academician Bohuslav Cambel

Scientific Adviser Ing. Gustav Šebor, DrSc.,
Corresponding Member of the Czechoslovak Academy of Sciences

Published in co-edition with ACADEMIA, Publishing House of the Czechoslovak Academy of Sciences, Prague

Distribution of this book is being handled by the following publishers

for the U.S.A. and Canada
Elsevier/North-Holland, Inc.,
52 Vanderbilt Avenue
New York, New York 10017

for the East European Countries, China, Northern Korea, Cuba, Vietnam and Mongolia
ACADEMIA, Publishing House of the Czechoslovak Academy of Sciences, Prague

for all remaining areas
Elsevier Scientific Publishing Company
1 Molenwerf
P.O.Box 211, 1000 AE Amsterdam, The Netherlands

Library of Congress Cataloging in Publication Data

Bouška, Vladimír.
  Geochemistry of coal.
  (Coal science and technology; v. 1)
  Revised and updated translation of: Geochemie uhlí.
  Bibliography: p.
  Includes index.
  1. Coal.   I. Title.   II. Series.
TP325.B7313   1981   662.6"22   81-4728
ISBN 0-444-99738-5 (Vol. 1)
ISBN 0-444-41970-5 (Series)

With 69 Illustrations

© Vladimír Bouška, Prague 1981
Translation © Helena Zárubová, Prague 1981

All rights reserved. No part of this publication may be reproduced, stored in a retrieval system, or transmitted in any form or by any means, electronic, mechanical, photocopying, recording, or otherwise, without the prior written permission of the publishers

Printed in Czechoslovakia

CONTENTS

Preface to the first edition	7
Preface to the second edition	9
Introduction	10
Accumulation of plant material	16
Palaeosalinity of the environment	27
Plants − the source material of coal	38
Chemical composition of plants	43
Comparison of the chemical composition of fossil and recent plants	69
Destruction and decomposition of plant matter	72
The peat stage	75
The post-peatification stage	89
Mineralogical characteristics of peat bogs	91
Coalification	96
Elementary composition and chemical compounds of coal	128
Chemical structure of coal	141
Mineralogy of coal seams	152
Contents of chemical elements in coals and coal ashes	164
Geochemical concentration and distribution of trace elements in coals	189
Mechanism and time scale of the enrichment of coal in trace elements	217
Application of the geochemical distribution of elements in coal seams to the identification, correlation and stratigraphy purposes	225
References	232
Index	260

PREFACE TO THE FIRST EDITION

The recognition of the chemical composition and properties of coal promoted the development of new trends of investigation in this field. The present handbook should give up to date answers to the questions that were previously studied separately by chemistry, mineralogy and geology and that today fall in the sphere of geochemistry. They involve, for example, the factors controlling the formation of coal, geochemical views of the alteration of plant material during coalification under biochemical and geochemical conditions, geochemical principles governing the concentration and circulation of trace elements in coal, and the study of the structural-chemical relations. The book is intended not only for geochemists and undergraduates ("Geochemistry of Coal" is the subject of a lecture course to advanced students at the Faculty of Science, Charles University), but for all geologists and technicians interested in the geochemistry, geology and treatment or usage of coal.

I have endeavoured to use uniform terminology for the coal types and uniform units (ppm or per cent) to express the amounts of elements in ash or coal.

I would like to mention two Czech publications which summarize the information on the chemical properties of coal and have contributed greatly to the development of Czechoslovak geochemical research in this field. The classical "Chemistry of Coal" of F. Pavlíček appeared in 1927 and the other bearing the same title was written by J. Hubáček et al. (1962). A wealth of important data on coal geochemistry are contained in many other Czechoslovak and foreign books dealing with the petrography of coal and geology of seams, and in numerous original papers. The great advance of geochemistry in the last decades has necessarily brought forth a real explosion of publications which, naturally, did not avoid the topic of coal geochemistry. Since it is beyond the range of possibility to cite all the studies, the Bibliography had to be restricted only to the most important works. Special attention has been paid to the Czechoslovak literature.

It is my pleasant duty to thank prof. Ing. Dr. F. Špetl, DrSc., Corresponding Member of the Czechoslovak Academy of Sciences, for his invaluable assistance and stimulating criticism, and to Academician B. Cambel for a thorough revision of the text.

My sincere thanks are due to my colleagues, Assistant professor RNDr. J. Pešek, CSc., who very kindly read the draft of this study and offered many constructive suggestions, to prof. RNDr. J. Staněk, CSc. for useful comments on the passage concerning the organic compounds in coal, to Associated professor RNDr. J. Obrhel, CSc. for critical reading of the chapter "Plants — the source material of coal" and to M. Pačesová, M.Sc. for her help in the preparation of the manuscript.

Last but not least I am obliged to the workers of the Publishing House Academia for their interest and assistance in the production of this volume.

Prague, April 1975

Vladimír Bouška

PREFACE TO THE SECOND EDITION

As the Czech first edition of this handbook was out of print within a year, I have been asked to prepare the second edition, which now appears in English as a co-edition between Academia and the Elsevier Scientific Publishing Company. A large amount of new data accumulated since the first manuscript was completed five years ago. It was therefore necessary to extend the book to include new material, particularly concerning the organic substances in peat and coal (information on which has markedly increased, for example the distribution of n-alkanes), to update the literature, and to add a chapter on the palaeosalinity of the environment. The degree of coalification can now be determined with greater precision using new instrumental methods which give excellently reproducible and comparable results. The section covering this topic had therefore to be expanded and revised to provide the reader with an up to date survey of the geochemistry of coaly caustobioliths.

I wish to express my thanks to the scientific editor, Ing. Gustav Šebor, DrSc., Corresponding Member of the Czechoslovak Academy of Sciences, and to the reviewer Academician B. Cambel for their interest in this second edition of the book. I am much obliged to Ing. Milan Streibl, CSc., for his valuable advice on the subject of organic substances and to M. Pačesová, M.Sc., for all her help in the preparation of this edition.

Prague, December 1980   Vladimír Bouška

INTRODUCTION

The organic rocks occupy an important place among the sedimentary rocks. They are denoted as bioliths (from Greek *bios* = life, *lithos* = stone) and represent a comprehensive group of zoogenic and phytogenic rocks. Sedimentary rocks of organic origin that are combustible were called by Potonié (1910) the caustobioliths (from Greek *caustos* = combustible); they are not of uniform genesis and include not only solid but also liquid and gaseous deposits. The caustobioliths are divisible into two basic groups according to the predominant primary material:

(1) The primary materials are prevalently plant organisms (microorganisms).
(2) The primary materials are dominantly animal organisms and microorganisms.
There are transitional stages between these two groups.

On the basis of genesis the caustobioliths can also be divided into two groups:
(1) Group of coal — peat, brown coal, bituminous coal, anthracite and gases liberated from coal.
(2) Group of natural hydrocarbons — natural gas, oil, asphalt, mineral waxes.

Caustobioliths of the first group have a relatively higher oxygen content, those of the second group contain typical hydrocarbons that recede to the background in the first group. Humic substances as the end product of decomposition are characteristic of the caustobioliths of the first group; those of the second group differ from them chiefly in a higher content of lipids and proteins.

Liptobioliths are also placed among the solid caustobioliths of the coal group; their principal classification feature is the content of waxes, resins, spores, cuticles or suberains (resistant tissues). The true humic coals have a minimum content of these substances. Havlena (1963) gave 20 per cent of wax and resinous matter as the boundary between humite and liptobiolith. The transitional type is denoted as liptohumite. Compared with other components, waxes and resins contain a relatively high amount of hydrogen and carbon. Potonié (1910) grouped also fossil resins with liptobioliths. Sapropelites, including bituminous sediments, form an additional group of solid caustobioliths of the coal series.

The origin of the caustobioliths of the coal series is considered in close connection with the origin of organic matter and the evolution of life, particularly of plants, on the Earth. The oldest primitive organisms are known today to have derived from the Precambrian of South Africa. Barghoorn and Schopf (1966) found in black quartzites of the Fig Tree Formation microfossils for which they erected a new genus — *Eobacterium*. The radiometric age of the rocks has been determined by Rb/Sr method as 3.1×10^9 years, which indicates that the beginning of life falls in the first third of the Earth's history. Abiogenic hydrocarbons produced by geothermal processes furnished energy to the first true organisms, which were probably anaerobic heterotrophs (Echlin, 1970). The Earth's atmosphere being in the early stage of evolution was of reducing character and oxygen-free. Berkner and Marshall (1965) have calculated that oxygen appeared in the atmosphere approximately 3.0×10^9 years ago. The development of the oxygen level was intimately associated with the development of organisms. Oxidizing metabolism which provided the organisms with many times greater energy than the primitive fermentative processes could come into existence only when the level of oxygen had reached one per cent of its content in the present-day atmosphere. Berkner and Marshall (1965) have placed the "first critical phase" at the beginning of the Palaeozoic, i.e. in the period about 600 m.y. ago. The aerobic processes were thus made possible but since the intensity of ultraviolet radiation was still very high, the organic cells must have been protected by an at least 10 metres thick water layer. According to the above mentioned authors the "second critical phase" occurred approximately 420 m.y. ago, at the onset of the Silurian, when the oxygen level attained 10 per cent of its contemporary atmospheric level. The oxygen level was high enough to protect the organisms from the deleterious effects of ultraviolet radiation, and terrestrial flora and fauna began to develop. The source of energy for biological processes was photosynthesis, by which carbon dioxide and water are converted into carbohydrates at the simultaneous release of oxygen. Solar radiation is the source of energy and natural pigments act as photocatalysts. About 100×10^9 tons of CO_2 is photosynthetized into carbohydrates every year. It is thought that during the biological history of the Earth about 10^{20} tons of organic matter, corresponding broadly to the weight of the Earth, has been synthetized by living organisms (Abelson, 1957). The content of organic carbon on the Earth is estimated at 6.4×10^{15} tons (Vinogradov, 1967); 5×10^{15} tons from this amount falls to sedimentary rocks and 1.4×10^{15} tons to the remaining rocks. Geological coal reserves on the Earth, estimated to a depth of 1,200 m for bituminous coal and to 800 m for brown coal are computed at 15×10^{12} tons (Strakhov, 1960). The bulk of organic matter in sedimentary rocks consists of amorphous polymeric kerogen, which is insoluble in common organic solvents (Fig. 1).

Coal masses could have been forming only after the evolution of plants. The

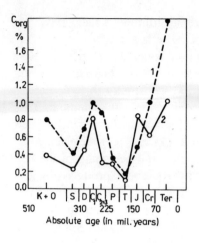

1. Average contents of C_{org} in sedimentary rocks (after Ronov, 1958). 1 — U.S.A., 2 — Russian Platform (K + + O — Cambrian and Ordovician; S — Silurian; D — Devonian; C_1, $C_2 + C_3$ — Carboniferous; P — Permian; T — Triassic; J — Jurassic; Cr — Cretaceous; T — Tertiary).

2. The development of vegetation during the geological periods (Mägdefrau, 1953).

diagram of Mägdefrau (1953) presents a survey of the most important plant groups (Fig. 2) and their qualitative and quantitative importance in individual geological periods.

Brown coals cannot be separated in time from bituminous coals. Brown coal is known to occur in the Lower Carboniferous (e.g. in the Moscow Basin) and in the Jurassic (in the Urals). On the other hand bituminous coal dating from the Upper Cretaceous and Tertiary occur on Sakhalin Island. It is believed that in general coals developed since the beginning of the Cretaceous are predominantly brown coals.

The oldest coal occurrences are the anthracite (shungite) from the Jatulian of Finland, Precambrian coals from Kazakhstan (Egorov, 1948) and some coal deposits in America. Small coal deposits of Cambrian or Silurian age are a special type of algal origin. Of Devonian age are bituminous coals found on ancient shields (Russian Platform, Siberian Shield, Canadian Shield). The oldest exploitable coal seams in the whole world are Upper Silurian 0.3–0.6 m thick coal seams in the area of Tashkent and Kokand (anthracite of Kenagaz, Ak-Kluchungai and Kyzyl Bulak deposits). Also the deposits of the Devonian bituminous coal on the Medvezhi Islands and in the northern part of the Kuznetsk basin at Barzas SE of Tomsk are mineable; several seams are there as much as 3.5 m thick. The coal is interlayered with bands and beds of liptobiolith, termed "sapromyxite", which is composed of tightly packed cuticles of the psilophyta genus *Orestovia*.

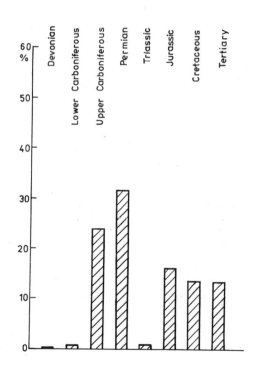

3. Relative distribution of coal in the geological systems, as related to total world reserves.

The first large sources of plant matter giving rise to enormous coal reserves did not appear before the Carboniferous and in this system the first important coal deposits have been preserved. A large amount of coal also developed in the Tertiary (Fig. 3). A coal seam of the greatest thickness is known from this period. It is a ca. 60 m thick seam of Eocene coal in the area about 180 km NE of Vancouver in British Columbia, with reserves estimated at more than 10^{10} tons (the area was the coast of Tertiary continental sea, on which deltaic marshes were developing for a long time). The contemporaneous peat bogs and swamps are nearly of the same extent as the original area of the Tertiary coal basins. Even at present there exist conditions, particularly in some tropical regions, in which a vast amount of plant material is accumulating.

As a matter of fact, coal is an enormous accumulator of luminous and thermal solar energy, which in the long geological history was coming to the Earth. Only part of it has been preserved in the form of coal until the present, and we convert it predominantly into thermal (combustion or coking) and chemical (hydrogenation of coal) energy or indirectly into mechanical and electrical energy (through the water vapour, i.e. thermal energy).

In the modern world coal covers about 50 per cent of required energy, but its reserves make up ca. 90 per cent of the total energy. The dynamic growth of demands on the sources of primary energy in the whole world and the uncertainty concerning the supply of oil and natural gas lead to ever increasing interest in coal as a potential source of energy. Of the sources of primary energy only solar, water, wind and geothermal energies are more or less inexhaustible; solid, liquid, gaseous and nuclear fuels are of limited amount.

The geographical distribution of coal deposits is quite favourable when compared, for example, with oil deposits.

The world reserves of coal (in 10^9 tons) amount to:

anthracite	19 (proved)	28 (total estimate)
bituminous coal	1 058	8 093
lignite and brown coal	343	2 633
sum	1 420	10 754

From the total world coal amount of $10{,}754 \times 10^9$ tons 13.10 per cent is proved and 5.5 per cent exploitable. Bituminous coal makes up three quarters of the amount, the proportion of anthracite is very small (0.1–0.2 %). The largest reserves are further north than N38° lat., mainly in the U.S.S.R., U.S.A. and Ch.P.R., each of which has more than $1{,}000 \times 10^9$ tons of reserves.

With the world production and consumption of coal as they were in 1974, i.e. approximately 3×10^9 tons per year, the total reserves will suffice for 3,600 years, the proved reserves for 470 years (Vrtal, 1978). Actually the reserves which can be

4. Open pit Jiří in the Sokolov Basin. Photo by F. Tvrz.

mined with present day technique and are usable with the present day technology, are estimated for 200 years approximately.

In addition, coal is also used as a material for coke, synthetic substances and mineral fertilizers and its application in the manufacture of heavy chemicals will certainly increase in the future. The importance of coal has come recently to the fore not only as a result of energetic crisis but also of the progressive decline of oil and natural gas reserves. Coal has so far been carbonized to produce coke; gasifying of coal for the production of energetic and technological gases, especially low- or high-caloric, awaits its realization. New processes for the production of synthetic natural gas (SNG) are given a trial in the U.S.A. The conversion of coal into liquid fuel is also a permanent theme of specialist treatises. The possibility of the manufacture of heavy fuel oils, medium to light distillates and even synthetic oil is considered. The consumption of coal in the chemical industry shows a rising tendency. Olefins (ethylene, propylene) and aromatics (benzene, toluene, xylene) are being produced.

The importance of coal also increases with regard to the concentration of rare elements such as U, Ge, V, Mo, Ga necessary in industry. Coal has recently become a much-sought-for and more valued industrial material, which requires a most rational management.

ACCUMULATION OF PLANT MATERIAL

The causes of accumulation of coal-forming plant material and the characterization of the environment where coal is forming are dealt with comprehensively in the books of Moore (1940), Francis (1961), van Krevelen (1961) and others.

The cumulation of particular plant (less faunal) associations gives rise to peat bogs under certain diastrophic, climatic and morphological conditions. Peat deposits are the geologically youngest deposits of caustobioliths, the coal deposits being their fossil analogues.

Essentially, three types of peat bogs are differentiated:

(a) lowmoor, (b) highmoor, (c) transitional peat bogs.

Lowmoor bogs develop in lowlands by gradual overgrowing of marshes or water basins. They are fed both with surface and ground water. The formation of lowmoor bogs depends on the balance of microbial decay of withered plant material. Where microflora (bacteria, fungi) is capable of decomposing all material accumulated in one year into mineral components, humus cannot heap up and peat cannot develop. Since the lowmoor bogs are fed with ground water, which brings an abundance of dissolved nutritive substances, their vegetation cover differs from that of highmoor (raised) bogs. The vegetation is more varied; it is rooted under the water level in soil generally enriched in nutrients, particularly Ca and K.

In Central European regions lowmoor bogs form by overgrowing of water basins, by drying out of lakes, ponds, pools and dead arms of rivers, or of vanishing springs on tectonic fractures. Water basins are overgrown with hydrophile flora from the banks towards the interior. The water surface is covered with minute algae, stems and leaves of water-lilies, duck-weeds and other plants which provide shelter and food to the plankton. The dying organisms drop to the bottom together with sedimentary material and putrify giving rise to sapropel. Aquatic plants take hold on extinct subaquatic organisms. Sapropel accumulates particularly along the banks where water plants such as reed and *Typha* grow, whose dead bodies contribute to the formation of humus, in which dry-land plants and trees can root. In this way water basins may be completely filled with vegetation and the cycle may repeat itself several times. In general, the sequence of deposits is as follows: 1. sapropel, 2. sapropelic peat, 3. true peat (rush, sedge etc.), 4. woody peat.

5. Lowmoor bog Ruda near Horusice (Bohemia) grown with cotton grass (*Eriophorum vaginatum*). Photo by V. Bouška.

Another theory assumes the overgrowing of water basins by floating vegetation (so-called *plaur*), which grows from the water level downwards. The vegetation "carpet" thickens by every new generation until the basin is filled in. The plaur could form on the surface of both freshwater basins and sea embayments. The vegetation thrives better on freshwater surfaces and grows more rapidly. Řehák (1971) presumed a similar development for the Prokop seam in the Ostrava-Karviná coal district and for other coal seams of the Czechoslovak part of the Upper Silesian Basin.

In fossil lowmoor bogs two types are distinguished on the basis of the mode of plant matter accumulation:

(a) stagnant peat bogs, formed by prolonged overgrowing of water body;
(b) irrigation peat bogs, formed by increasing water-logging of the area.

The development of the two types is often linked up and repeats itself cyclically. It is controlled predominantly by diastrophism and climate.

Highmoor bogs develop at higher elevations under cooler and more humid climatic conditions. They grow on soils poor in nutrients (mainly in calcium) or in basins

6. Highmoor bog "Na Čihadle" (Bohemia) formed in a shallow depression on a watershed. Photo by J. Rubín.

7. Highmoor bog "Jizerka" (Bohemia). Photo by J. Rubín.

with clayey impermeable bottoms, invariably above the ground-water table and are fed almost entirely with atmospheric water. Precipitations bring only few nutrients and the plant species are therefore represented by some grasses and sphagnum moss. The lower parts of sphagnum gradually die away and are transformed into true peat; the upper parts grow continuously and the bog may thus rise up to several metres above the ground-water table. The yearly accretions are small, of about 3—4 cm. In the midst of the bog, where the environment is most humid, the moss grows most rapidly and the bog is upheaved assuming a loaf-shaped form. The development of a highmoor bog is occasionally interrupted by periods of drought, when so-called boundary horizons, rich in inorganic material wind-blown on its surface, are formed.

Transitional peat bogs representing a transition between the preceding types develop where a highmoor bog was forming on the lowmoor substrate. Essentially, progressive accumulation of plant material in a lowmoor bog will produce earthy to pure peat as long as the plant roots can reach the mineral substrate. Afterwards the plant association is replaced by associations which need only a minimum of inorganic matter, i.e. by the genus *Sphagnetalia*. The factors affecting the development stage are climatic, topographical and others. The character of the development

8. Highmoor bog "V reservaci" on a watershed at Boží Dar (Bohemia). Photo by J. Rubín.

of brown and bituminous coal deposits, both of Carboniferous and Tertiary age, was similar to that of present-day peat bogs except for the type of vegetation. The overgrowing of recent water bodies preceding the formation of peat bogs repeats the infilling of water basins in Tertiary times or sea embayments and freshwater basins in the Carboniferous.

Brief characteristics of peat bog types

	Highmoor bog	Lowmoor bog
Origin	depending on precipitation, deterioration of climate	topogenous, depending on terrain configuration
Growth	rises above the ground surface, independently of ground water	grows only in water-logged basins
Peat-forming plants	sphagnum, cotton grass	sedge, reed, trees (willow, alder)
Nutrient content	very poor, lack of calcium (CaO 0.5 %, max. 0.8 %)	medium to rich (2—4 % CaO), $CaCO_3$ sometimes present
Reaction	acid, pH = 5—5.6 (may be even 3)	neutral or slightly alkaline, pH = 7—7.8 (sometimes also acid, pH > 4)
Vegetation	sparse, poor in species	luxuriant, rich in species
Ash	1—3 %	5—15 %

The nature of transitional peat bogs is inferable from the above types and depends on the development stage.

The study of the hydrogen ions concentration in peat of Borkovická blata in southern Bohemia (Bouška, 1959) has shown that the most acidic reaction is not in the peat bog itself but in the acid and bushy meadows in its close neighbourhood (Fig. 9), where pH amounts to 4—4.5. The pH values increase both inwards the peat deposit and into the adjacent fields. The average pH value is 5.5 in a lowmoor bog and 5 in the highmoor bog. The relatively high acidity in the lowmoor type is surprising. It is very probably due to the fact that at the time of measurement the peat bog was drained by a system of deep canals, which disturbed the primary water circulation and led to a total equalization of acidity, irrespective of the primary types of deposits.

The deposits just discussed are called autochthonous, since they formed by accumulation of humus developed from plants growing in situ. They are characterized by a large areal extent and fairly regular deposition with gradual horizontal transitions. Undisturbed tree trunks are occasionally preserved.

Svoboda and Beneš (1955) compare the section through a peat bog with fossil deposits of caustobioliths as follows:

"In the deepest part of a developing peat bog sapropelites originate; their fossil forms are called in coal terminology cannels and bogheads. In the higher part, i.e. in the section of subaquatic plants, true peat is deposited, whose fossil analogues are brown coals with a low content of xylitic component up to banded brown coals. The Carboniferous equivalents are clarains, such as microspore clarain, megaspore clarain, cuticle clarain. At the level of the ground water, i.e. at the boundary between aquatic and dry-land vegetation, there is the facies of shrub and tree growth, which gives rise to woody peat. This corresponds to Tertiary hemixylites, xylites and more coalified metaxylites, and to Carboniferous vitrains or clarains. The marginal facies of recent peat bogs is represented by dry-land flora, the organic substrate of which is known as humus. The fossil analogue is liptobiolith developed in rather dry environment, and some sorts of durain formed at wetter places."

The coal seams usually show an alternation of these types both in vertical and horizontal directions. The individual layers or bands do not retain the same thickness laterally. Their wedging out, splitting and setting in is accounted for by

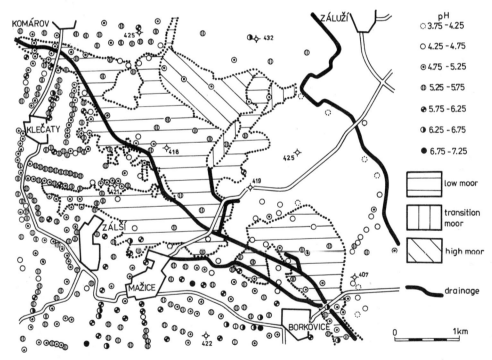

9. A chart showing pH values in the Borkovice peat bog near Veselí nad Lužnicí (Bohemia) (Bouška, 1959).

10. A dying-off pine forest on the "Mrtvý luh" highmoor bog in the Šumava Mts. Photo by O. Leiský.

the sedimentary cycle, which was caused by the variation of vegetation optimum of the individual facies and particularly by morphological changes of the basin bottom and fluctuations of the water level. Diastrophism may also be responsible for the formation of partings.

Deposits which originated from the accumulation of plant debris transported from other places are called allochthonous. They generally contain abundant clastic material and their extent and mode of deposition are varying. In the mouths of great rivers (e.g. the Lena or the Amazon) thick masses of tree trunks and plant fragments and debris are accumulated by floodwaters. This material may be drifted into the open sea and deposited then in an embayment. Such accumulations of tree trunks and wood transported by ocean currents have been described by Potonié (1910) from the bay of Amsterdam Island in the Spitsbergen.

A special instance is the origin of some drift-type seams as a result of a natural disaster, such as hurricane, tsunami or rock slide (Velikowsky, 1956). On the southern coast of the New Siberian Islands up to 90 m thick masses of tree stems interlayered with sand beds containing bones of mammals (mammoth, bison and others) were deposited. In 1958 the earthquake in south-eastern Alaska heaped up large masses of ice in Lituya Bay and immense waves thus provoked destroyed the forest around the bay. Accumulation of dead trees at the mouth of the bay represents potential material for the formation of coal.

Plant material can be accumulated in marine, brackish and especially in freshwater (continental) environments. Marine environment does not provide favourable conditions for the origin of coal. Plant material deposited on the bottoms of bays is often derived from dry land. In its lower course the Mississippi carries a great load of plant material (stems and branches, plant detritus), which it deposits into the ocean at its mouth. Thick masses of plant material are not deposited either in coastal lagoons (for example the Gulf Coast in Texas with prolific growth of *Thalassia*) owing to the transitional character of the majority of lagoons. Occasionally, sapropel may originate in a shallow open sea. The coal deposits thus formed have a high ash content and relatively constant thickness over large areas. Of this type are the kukersite deposits in Estonia and combustible shales in the Volga-river basin. Brackish environments of bays, deltas, lagoons and coastal swamps are more favourable to the accumulation of coal-forming material than marine environment. Large accumulations of palustrine plants form, for example, in the upper parts of estuaries, such as Chesapeake Bay and swamps of Virginia and Maryland. In the Florida Everglades plants and mangroves cumulate together with terrestrial plant material brought by flood-water and streams. A recent representative of such paralic basin is e.g. the Dismal Swamp on the coast of the Atlantic Ocean; it spreads only a few feet above sea-level and together with the Everglades it covers about 3,000 km^2. Since the end of the Pliocene the coast there sank and rose three times; at present it is sinking (Moore, 1940). The sea transgression terminates the coal-forming cycle and unproductive beds begin to be deposited. Their thickness depends on the duration of transgression and rate of subsidence. The vegetation cover is renewed in the post-regressive cycle, when the environment has been freshened again and swamps and marshes have come into being. In geological history, such a type is represented by the Ostrava-Karviná

11. Peat bog destroyed by sea. Seehestadt in the Fed. Rep. Germany. Photo by J. Pešek.

coal district in Czechoslovakia, the Donets Basin in the Soviet Union, Devonshire (England), Ruhr Basin (Germany), some basins in Belgium, and others.

The freshwater (continental) environment is most productive and the greater part of coal has been formed in it. The coal deposits formed inland are called limnic. They were deposited in fluviatile (meanders, branched streams, alluvial plains), and lacustrine environments or in intermontane depressions (closed basins, alluvial cones). Coal is at first of sapropelic character and later develops from lowmoor bogs. Coal seams are usually few (1 to 3). Limnic basins are, for example, Central Bohemian Carboniferous Basins, Rosice-Oslavany Basin, North Bohemian Brown-Coal Basins, the Saar Basin (Germany) and Irkutsk Basin (U.S.S.R.).

Lowmoor bogs and open eutrophic lakes provide the most suitable environment for coal formation. Marshy areas are rich in forest vegetation, savannah flora, reeds, algae and others. The pH value varies between low alkaline and low acidic and Eh between slightly positive and moderately negative. The Mg/Ca ratio of eutrophic peat bogs is $0.5-1$, of oligotrophic >1. The resulting coal contains spores, pollen and cuticles which are resistant to decay.

Where a forest gets over the margin of swamp or eutrophic lake, a terrestrial facies develops, bordering the aquatic facies. The facies may migrate both laterally and vertically as a result of cyclic climatic changes. In the Latrobe Valley (Australia), for example, three layers of brown coal with a total thickness of some 400 m have been exposed. The presence of remains of conifers, angiosperms, fungal fibers and hyphae and sclerotia suggests that the plants important for the formation of coal had grown above water level in the swamps. The cycles of coal formation may last several hundred thousand years up to many million years.

Brown coals in the Rhenish Basins are estimated to have been deposited over more than 50 million years, from the Eocene to the Pliocene. During this period the climate changed from subtropical evidenced by the growth of palms (*Liquidambar*), ferns (*Mohria*) and club-mosses (*Lycopodium*) to temperate, represented by coniferous and deciduous trees, such as oak, alder, birch and chestnut (Swain, 1970).

Cyclic sedimentation is known to have existed in various geological periods. It is thought at present that cyclic sedimentation of coal-forming material is caused by diastrophic movements. The subsidence of a plain with coal-forming peat bogs is followed by sea incursion. A coastal plain borders a sea on one side and the elevated land on the remaining ones, which may grade into mountain ranges It disappears when owing to intensive orogenic movements the sea regresses and maritime conditions are replaced by continental.

Regional subsidence leads to the evolution of four types of basins:

(1) *Basins of geosynclinal systems*. The profile exhibits a regular periodical alternation of marine and continental sediments. A great thickness is typical of the complex. The basins of this type develop in the marginal parts of subsiding sectors of the Earth's crust (foredeeps) or in extensive coastal plains. They may be exemplified by the Ostrava-Karviná Basin (Czechoslovakia), Donets Basin (U.S.S.R.),

Ruhr Basin (Germany), basins of Belgium, a considerable part of Chinese deposits, the Appalachian Basin in Pennsylvania, and others.

(2) *Basins on the platforms*. They are of larger extent but smaller thickness and develop in more stable parts of the Earth's crust. The coal seams are less frequent but attain a fairly great thickness. Examples of this type are the Central Bohemian Carboniferous Basins or the Moscow Basin.

(3) *Basins of intermontane depressions* (and grabens). The material is deposited during the subsidence of the bottoms of depressions. Of this type are, for example, the North Bohemian Brown-coal Basins (Czechoslovakia), Chelyabinsk Basin (U.S.S.R.), Midland Valley (England) and Lowlands (Scotland).

(4) *Basins of transitional regions* comprise segments of various types (e.g. the Kuznetsk Basin in the U.S.S.R.).

The origin of cyclic sedimentation is also explained as due to climatic changes. Every glaciation represents a stage of extensive alluviation and formation of lakes and swamps. In the following interglacial stage the sea may spread over such coal-forming basin and deposit there clastic material with drifted plant remains and later on the marine limestones may develop. During the subsequent glaciation, the resumed sedimentation favourable to coal formation may begin again with the deposition of clastic sediments.

One of many other more or less plausible theories presumes that barriers separating the bays from open sea were recurrently broken and built again (Miocene marine and freshwater sediments separated by coal seams alternate, e.g. on Bataan Island, the Philippines).

Fossil peat bogs from which coal seams have developed were lowmoor bogs of tropical nature. Such tropical swamps exist even at present, for example in Sumatra, where in impenetrable jungles layers of organic matter attain a thickness of about four metres. A similar formation is the Dismal Swamp in Virginia and North Carolina with *Taxodia* virginal forests. The filling of the swamp, six or more metres thick, consists of common palustrine plants and decayed parts of cypresses, poplars and cedars. Virginal forests also exist on the seashore of Sri Lanca and elsewhere suggesting a paralic development of the basins.

Coal deposits corresponding to present-day highmoor bogs have not yet been found. Highmoor bogs with their growth of sphagnum, moss, various grasses and heath occur in cool northern and mountainous regions. They are known from western Wales and northern and central England.

Virginal forests in the states Oregon and Washington provide another example of terrestrial accumulation of plants. However, the predominating positive oxidation-reduction conditions in such places affect very intensively the decomposition process and so hinder the accumulation of plant material.

The accumulation of plant material depends on many factors which are basically controlled by diastrophism and climate. Diastrophic movements produce subsidence of some parts of the Earth's crust and upheaval of others, thus giving rise to source areas and sedimentary basins. The thickness of the basin filling and the cyclic structure of the basin are also affected by diastrophism. The climate governs the evolution and composition of vegetation, the ground-water conditions being one of the major accessory factors. Environments suitable for the accumulation of plant material developed only at some places of the Earth, depending on the course of mountain ranges, climatic zone and on the evolution stage of fossil flora (Havlena, 1965). Migration of the Earth's axis created favourable conditions for the formation of coal deposits even in the areas that are glaciated at present.

The climate is controlled by cosmic agents such as the intensity of solar radiation (Rukhin, 1962) and by geographical factors as, for example, distribution of oceans and continents, direction of mountain ranges, elevation of land above sea level and circulation of water (Brooks, 1950).

The coal deposits provide evidence of a warm and humid climate at the time of their formation. The zone of tropical climate is governed predominantly by cosmic agents and its breadth is therefore relatively constant. The same circumstance can also be presumed for the tropical zones in the past geological periods and applied to the solution of the problem concerning the origin of coal deposits in various parts of the Earth and in various time intervals.

Dohnal et al. (1965) estimate that the world reserves of peat are divided as follows: 61% the U.S.S.R., 9% Finland and Canada each, 5.3% the U.S.A., fractions of one per cent to 4% the remaining countries. Czechoslovakia has got only 0.03% of the world reserves.

PALAEOSALINITY OF THE ENVIRONMENT

The knowledge of the character of depositional environment, particularly of palaeosalinity, is of importance not only in palaeogeographic studies or as an indirect method of stratigraphical correlation, but in many paralic coal basins also as a source of information about the occurrence or absence of coal seams. Geochemical methods help in assessing only the substantial differences in the palaeosalinity of sedimentary basins. They make it possible to readily distinguish between *freshwater* and *marine* sediments, but the *brackish* environments are more difficult to define. The information, however, may be relevant for the recognition of aridization of climate in certain periods, which is clearly reflected in the character of sediments, in pH and Eh of the environment and the formation of minerals.

The determination of soluble or sorbed cations in Carboniferous claystones proved to be an important method for distinguishing between freshwater and marine environments (Spears, 1974). In relation to freshwater and brackish sediments, the marine deposits have higher soluble Ca^{2+} and Mg^{2+} and lower Na^{1+}

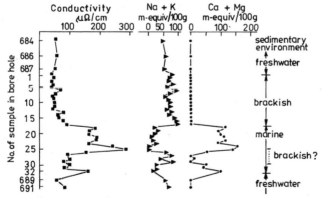

12. Conductivity and soluble cations in Carboniferous claystones, depending on the environment character (after Spears, 1974).

and K^{1+}. These differences are obviously due to the differences in the primary salinity of the environment, although the influence of subsequent diagenesis must be taken into consideration. The total concentration of leached ions is higher in sediments of marine origin. On the basis of this finding Tomšík (1959) proposed the measurement of water extracts from clayey sediments as a method for distinguishing between marine and freshwater bands in the Carboniferous claystones and siltstones of the Ostrava-Karviná coalfield. The method was later verified by Spears (1974) (see Fig. 12).

The differentiation of freshwater and marine, mainly pelitic sediments on the basis of geochemical distribution of the elements has been studied by Ernst et al. (1958), Keith and Degens (1959), Potter et al. (1963), Tomšík (1963), Ernst and Werner (1964), Degens (1965), Reynolds (1965), Eagar and Spears (1966), Michalíček (1967), Walker (1968), Keller (1970), Bohor and Gluskoter (1973), Bouška et al. (1975), Bouška and Pešek (1976), Bouška et al. (1979), and others. Their observations have revealed that the most suitable element for the determination of palaeosalinity is boron (Reynolds, 1972; Walker, 1972).

Goldschmidt and Peters (1932) have drawn attention to the role of boron in marine sedimentation. The sea water contains on the average 4.6 ppm B, which is generally present in the form of undissociated boric acid and its Na-salt. The average content of boron in most continental waters is only about 0.01 ppm or is absent altogether.

During sedimentation boron is extracted from sea water. Small amounts probably take part in the formation of syngenetic or authigenic tourmaline, but the predominant process is the bonding of boron to clay minerals, particularly potassium clay minerals — illite and glauconite. The clay minerals were thought to sorb boron in amounts proportional to its amount in the solution, but experiments have shown that the reaction is affected not only by concentration but also by the reaction time and temperature of aqueous solutions. The adsorption of boron on the surface of clay minerals is regulated by Langmuir's adsorption isotherm.

The sorption capacity of illite, montmorillonite and glauconite is similar and much higher than that of kaolinite. Frederickson and Reynolds (1960) and Walker (1968) assumed that boron enters tetrahedral sites during illite authigenesis. Evidently, Al—B diadochy in outer tetrahedral layers of clay minerals is involved, although B—Si diadochy has also been corroborated (Stubican and Roy in Walker, 1968). Boron is namely firmly held in clay minerals and cannot be leached out even by hot mineral acid.

In the sedimentary rocks boron is distributed as follows: 20 ppm in carbonates (bound to a mineral from the illite group) and 35 ppm in sandstones (depending on the proportion of detrital, syngenetic or authigenic tourmaline).

In Buzzards Bay (Massachusetts), for example, Moore (1963) determined B concentrations of 65 ppm and locally up to 168 ppm in beach sands. The boron content of this mechanical concentration is related to tidal current energy and the coefficient of sorting. When the tidal energy decreases, light minerals are transported preferentially and the residual environment becomes enriched in tourmaline.

Evaporites contain a larger amount of salts and in extreme cases also hydrated borates of Ca, Na and Mg (colemanite, borax, kernite and others). The presence of these minerals contributes to an unusually high boron content in the rock.

According to literature data (e.g. Reynolds, 1972), the boron content in marine claystones and shales is normally 100—200 ppm, whilst it amounts only to 10 to 50 ppm in freshwater sediments. Claystones deposited in brackish environment contain 80—110 ppm B, but the values cannot be determined precisely. The relations differ somewhat in individual basins, since many factors are known to be involved.

Redeposited mica minerals derived originally from metamorphosed or non-marine sedimentary rocks usually possess a low B concentration. It can be said that a sediment continues to develop even under the conditions of diagenesis. Metamorphic events generally change the pattern of geochemical facies to a high degree or completely. New minerals appear and important migration of elements occurs. According to Ernst et al. (1958), the boron content in epimetamorphosed rocks changes considerably. In originally marine or brackish sediments the content decreases as much as to a half: from 0.040 to 0.018 % in marine clayey rocks and from 0.030 to 0.013 % B_2O_3 in brackish ones. The facies distribution, particularly of trace elements, is still more obscured in sediments that suffered a higher-grade metamorphism, but this problem does not usually refer to coal basins.

The boron content is also increased by the presence of volcanogenic material (Cody, 1970). The enrichment in boron may be observed along tectonic lines, at the margins of basins, in the proximity of thermal springs or exhalations (e.g. in New Zealand). Boron contents as high as 2,000 ppm have been measured in illites associated with evaporites.

Reynolds (1965) used the relation between the boron content in illite and in sea water for the determination of boron concentrations in Precambrian seas. He computed the relative palaeosalinity from the distribution coefficient and B content in the illite fraction of the rock.

Boron is usually determined in the pelitic fraction of sedimentary rocks (<2 μm). The silt and sand components and carbonates present in a clayey rock essentially diminish the boron content. Clayey rocks containing kaolinite usually contain less boron than the illitic rocks. The environments in which illite forms and accumulates are evidently most suitable for taking up and bonding of boron. Walker and Price (1963) correlated the boron content with the content of potassium as an indicator of the illite content. Reynolds (1965) gave the following relation to be used for the correction of the total boron to boron bound in the illite fraction:

$$B_{illite} = B_{in\ sample} \times \frac{7.7}{\%\ K_2O\ in\ sample}.$$

He separated the clay mineral by washing and then analysed the sample for boron and potassium. Minimal contamination of detrital tourmaline, micas and feldspar should be ensured. The presence of other K-containing minerals than illite can be determined by X-ray diffraction. The value 7.7 represents the theoretical content of K_2O in pure illite.

The values of equivalent boron bound to illite are given as 300–400 ppm in marine pelitic sediments, less than 200 ppm for fresh-water pelites and 200 to 300 ppm for the brackish environment. It appears, however, that the relationship between the boron content (both equivalent and total) and palaeosalinity is not so simple (Dewis et al., 1972) and that B contents depend to a high degree on the character of rocks of the source areas.

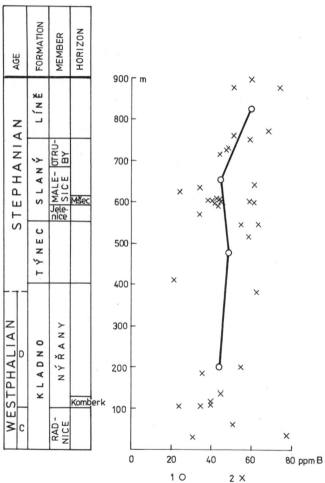

13. Boron contents in Carboniferous sediments of the Central Bohemian Coal Basins. 1 — average content of each formation; 2 — individual samples (Bouška and Pešek, 1976).

Bouška et al. (1975) and Bouška and Pešek (1976) have studied boron contents in the claystones, siltstones and tuffogenic rocks in the coal basins of central Bohemia (Westphalian C to Upper Stephanian). The average boron value of 48.4 ppm has confirmed sedimentation in freshwater environment as has been presumed. A moderate increase in concentration has been found in the Týnec and Líně Formations, which had been deposited during the initiating aridization accompanied by an increased concentration of salts in water (Fig. 13).

The study of marine and freshwater bands in the Ostrava-Karviná coalfield (mostly Namurian) (Bouška et al., 1979) has provided evidence that the content of boron and amino acids in claystones can be used as the indicator of depositional

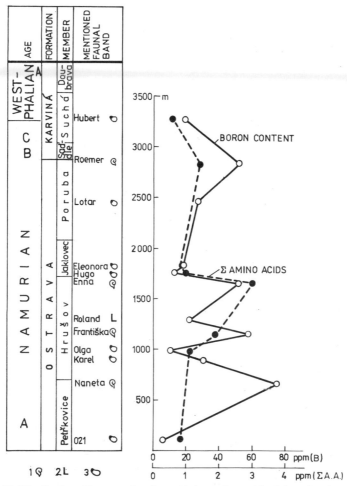

14. Distribution of boron and amino acids in marine and freshwater bands of the Ostrava-Karviná coalfield. 1 — marine bands, 2 — linguloid fauna, 3 — freshwater bands (Bouška et al., 1979).

salinity (Fig. 14). However, the marine sedimentation (evidenced palaeontologically) did not occur in a highly saline environment, as is inferable from the low boron values — an average of 52.0 ppm in claystones of the marine bands and 25.0 ppm in freshwater claystones (see also Polický et al., 1976). Compared with other world basins all these contents are low, but it can hardly be stated that some precisely limited values exist for freshwater, brackish or marine environments in the world basins.

The distribution of boron in individual marine bands (in Fig. 14) shows a gradual but distinct decrease of this element upwards, which is most probably due to a progressive freshening of the basin towards the younger stratigraphical units. If, on the contrary, the average or representative boron contents in the freshwater bands are linked up, its content slightly increases towards the overlying beds. This may be accounted for by general increase in temperature during the Late Carboniferous; the same tendency is observable in younger sediments of the Central Bohemian Carboniferous Basins (see Fig. 13, Bouška and Pešek, 1976).

The boron contents in coal ash are usually definitely higher than in claystones in the close neighbourhood of the coal seam. Somasekar (1971) studied this relationship in one geological unit of the same region.

Adams and Weaver (1958) proposed to apply the Th/U ratio as an indicator of the sedimentary environment. This interesting method is based on the assumption that owing to the different characters of Th and U their ratio changes in the course of the sedimentary cycle. According to the authors it ranges from 0.02 to 21.

Particularly characteristic is the high Th/U ratio in oxidic continental sediments (above 7), whilst that of marine sediments is much less than 7. Using this method it is possible, for example, to discriminate between marine and continental sediments of red-beds type. The idea stems from the different behaviour of the two elements in the weathering process: uranium is leached out and dissolved, whereas thorium being less mobile remains in residual fractions (detritus) or is sorbed by hydroxides and clays. The authors mention many examples of the application of this method to stratigraphical correlation (Fig. 15). They also claim that it is easy to recognize unconformities shown by the occurrence of weathered fossil soils, porous and dolomite zones, and others.

The influence of post-sedimentary processes (diagenesis) on the Th/U ratio is still unclear. The ratio may be modified considerably by the changes of redox potential in the course of diagenesis.

Marine clayey sediments may be additionally richer in chromium, copper, nickel and vanadium than freshwater clayey rocks. The barium, lithium and strontium contents also rise with increasing salinity. Rubidium amounts to 100 – 1,000 ppm in marine clayey deposits and to 70 – 650 ppm in freshwater ones. The available observations have confirmed that freshwater clayey rocks are enriched rather in lead, gallium and thorium. It has been likewise confirmed that the increased content of trace elements in clayey rocks correlates with the organic matter, be it of humite or bituminous character. The rocks richer in bitumen component are usually richer in vanadium and molybdenum, and the org. C/org. N ratio is lower than 10; in clays containing humite matter this is higher than 10. The C/N ratio can also be used for distinguishing marine sediments from those of freshwater origin (Bloxam, 1974).

A suitable tool for discriminating between freshwater and marine environments is the study of the content of trace elements in pyrite. Pyrite in freshwater clayey rocks is generally richer in Co, As, Ag and Cu than pyrite of marine clayey rocks. The concentration of trace elements in pyrite, however, cannot be compared directly with their enrichment in clayey materials, because different processes were probably involved in each case.

Important for the estimation of depositional salinity are the data on the total content of amino acids in claystones (Degens – Bajor, 1962), and of arginine (Degens – Bajor, 1962), and on the ratio of glycine and alanine content to glutamine and asparagine (Degens – Mopper, 1975). The content of amino acids is known to rise with the salinity of waters and does not depend closely on temperature and, consequently, on climatic oscillations. According to Swain (1970), the content of amino acids decreases with the stratigraphical depth of deposits,

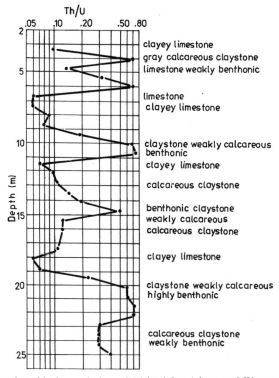

15. Stratigraphical correlation of Th/U (after Adams and Weaver, 1958).

Table 1

Content of amino acids in freshwater (f.b.) and marine (m.b.) bands (in ppm) in the Ostrava-Karviná coalfield (Bouška et al., 1979)

	1 – f.b. Hubert	2 – m.b. Roemer	3 – f.b. Eleonora	4 – f.b. Hugo	5 – m.b. Enna	6 – m.b. Enna	7 – m.b. Františka	8 – m.b. Františka	9 – m.b. Františka	10 – m.b. Františka	11 – f.b. Olga	12 – f.b. seam 02
LYS	0.065	0.20	0.108	0.095	0.300	0.272	0.160	0.220	0.198	0.150	0.095	0.067
HIS	0.020	tr.	0.022	0.027	0.050	0.062	–	–	0.060	0.070	0.015	0.019
ARG	0.030	0.60	0.078	0.085	0.300	0.104	0.160	0.050	0.080	0.300	0.060	0.032
ASP	0.010	0.036	0.013	0.015	0.050	0.052	0.045	0.050	0.080	0.060	0.030	0.022
THR	0.010	0.037	0.022	0.025	0.050	0.156	0.055	0.075	0.040	0.040	0.030	0.032
SE	0.042	0.60	0.072	0.085	0.265	0.406	0.175	0.280	0.236	0.160	0.102	0.047
GLU	0.035	0.088	0.060	0.070	0.125	0.372	0.100	0.175	0.120	0.050	0.071	0.072
PRO	0.075	0.80	0.128	0.142	0.150	0.598	0.250	0.370	0.201	0.110	0.125	0.083
GLY	0.025	0.30	0.035	0.040	0.190	0.190	0.125	0.140	0.120	0.100	0.025	0.037
ALA	0.075	0.40	0.083	0.095	0.275	0.310	0.165	0.255	0.237	0.140	0.155	0.131
VAL	0.090	0.00	0.115	0.110	0.150	0.452	0.160	0.230	0.110	0.070	0.130	0.110
ILE	0.060	0.080	0.073	0.078	0.175	0.222	0.110	0.155	0.120	0.060	0.095	0.066
LEU	0.075	0.30	0.110	0.107	0.200	0.414	0.215	0.315	0.156	0.090	0.135	0.102
TYR	tr.	0.002	tr.	–	0.005	–	0.025	0.020	0.003	0.002	tr.	tr.
PHE	0.030	0.076	0.030	0.030	0.100	0.092	0.075	0.110	0.118	0.070	0.060	0.067
LYS + ARG	0.095	0.280	0.186	0.180	0.600	0.376	0.320	0.270	0.278	0.450	0.155	0.099
GLY + ALA	0.100	0.270	0.118	0.135	0.465	0.500	0.290	0.395	0.357	0.240	0.180	0.168
GLU + ASP	0.045	0.127	0.073	0.085	0.175	0.424	0.145	0.225	0.200	0.110	0.101	0.094
	0.640	1.439	0.949	1.004	2.385	3.728	1.820	2.445	1.879	1.472	1.128	0.887

Explanation: LYS – lysine; HIS – histidine; ARG – arginine; ASP – asparagine; THR – threonine; SE – serine; GLU – glutamine; PRO – proline; GLY – glycine; ALA – alanine; VAL – valine; ILE – isoleucine; LEU – leucine; TYR – tyrosine; PHE – phenylalanine.

because the decay of organic components is more advanced in older sedimentary rocks.

The experiences are virtually based only on the works of Degens and Bajor (1960, 1962), who maintain that amino acids can be used for discrimination between marine and freshwater environments when the rocks were consolidated under diagenetic conditions.

In the early diagenetic stage the organic material is usually strongly reworked by microorganisms, but without selection (Degens – Mopper, 1975), so that the origin of individual components can still be recognized. The authors have found out that in amino acids originated by decomposition of plankton glycine and alanine highly predominate over asparagine and glutamine. The planktonic association of amino acids is also characterized by a higher arginine content and a low concentration of amino sugars. In amino acids derived from continental sources glutamine and asparagine generally prevail over glycine and alanine, arginine is lacking and the content of amino sugars is substantially higher (Parson – Tinsley, 1975). In the Ostrava-Karviná coalfield (Bouška et al., 1979) glycine and alanine predominated over asparagine and glutamine in all samples examined (Table 1). This prevalence is not so prominent as given for marine sediments by Degens and Mopper (1975). Combined with the low values of the total concentration of amino acids it suggests littoral sedimentation. Glycine and alanine also predominate over glutamine and asparagine in all samples taken from freshwater bands.

Table 1 and Fig. 14 clearly show the difference in the total contents of amino acids between freshwater and marine bands in the Ostrava-Karviná coalfield. The trends of this relationship are in good agreement with the distribution of boron, which corroborates the applicability of amino acids content to the determination of depositional palaeosalinity.

The differences are revealed even in glycine + alanine, glutamine + asparagine and lysine + arginine sums. Of the amino acids, it is glycine, asparagine, threonine and histidine that discriminate well between marine and freshwater bands, although

Table 2
Glycine contents in freshwater and marine bands in the Ostrava-Karviná coalfield
(Bouška et al., 1979)

Sample	Freshwater bands		Sample	Marine bands	
1	Hubert	0.025 ppm	2	Roemer	0.130 ppm
3	Eleonora	0.035 ppm	5	Enna	0.190 ppm
4	Hugo	0.040 ppm	6	Enna	0.190 ppm
11	Olga	0.025 ppm	7	Františka	0.125 ppm
12	021	0.037 ppm	8	Františka	0.140 ppm
			9	Františka	0.120 ppm
			10	Františka	0.100 ppm

histidine has not been identified in samples 7 and 8 and in sample 2 it occurred only in traces. As is seen from Table 2, in claystones of the Ostrava-Karviná coalfield glycine is an adequate indicator of the sedimentary environment.

The average glycine content in freshwater claystone samples is 0.032 ppm and 0.142 ppm in marine claystones.

The distribution of *n*-alkanes is another discriminative parameter of palaeo-salinity of the environment (e.g. Bloxam, 1974; Powell et al., 1976). Powell et al. (1976) found out that marine coal shales from East Midlands and Northumberland provide a higher extract of organic carbon than those of non-marine origin. The distribution of *n*-alkanes in samples of marine shales shows the peak between C_{17}—C_{20}, and the maximum content of heavy hydrocarbons in non-marine shales ranges from C_{23} to C_{27}. The content of hydrocarbons depends on the degree of diagenesis, on the source of organic matter (Allen – Douglas, 1977) and on the conditions of organic material deposition.

Bloxam (1974) has studied shales belonging to the *Gastrioceras subcrenatum* marine band. The *n*-alkane distribution and C/N ratios have shown a large proportion of the organic material in the offshore shales to be non-marine in character.

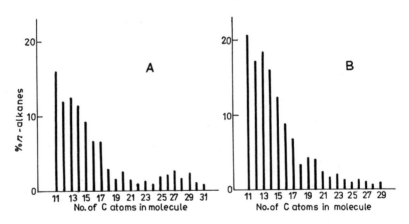

16. Relative representation of higher *n*-alkanes in the rocks of the Upper Silesian Basin (Carboniferous – Namurian). A – siltstone, lacustrine environment, Anticlinal Seams Member; B – siltstone, marine environment, the main Gaebler's band. (After Kříbek, personal communication, 1978.)

The analysis of higher *n*-alkanes (Dr. B. Kříbek, Dept. of Economic Geology, Charles University, 1978, personal communication) from the rocks of the Karviná Formation in the Upper Silesian Basin (Namurian-Westphalian A) has shown considerable differences in the distribution of individual hydrocarbons (Fig. 16). In contrast to marine sedimentary rocks, the continental sediments possess increased amounts of hydrocarbons having 25—29 C in a molecule. These are obviously derived from waxes of continental macroflora, which are characterized by long hydrocarbon chains.

The high pristane/phytane ratios have been used to characterize organic matter of non-marine origin (Brooks et al., 1969; Powell – McKirdy, 1973; Connan, 1974). This is based on the assumption that pristane in coal is derived in part from phytanic acid by decarboxylation; phytanic acid may be produced from phytol or dihydrophytol by oxidation and this could be expected to occur in a terrestrial environment where there is a plentiful supply of free oxygen. The n-alkane distribution and pristane/phytane ratios reflect the relative proportions of organic matter derived from higher plants and that of presumably algal origin.

Palaeosalinity can also be established by determination of Eh and pH values in squeezed-out porous water (including sorbed water) from the clayey sediments and by the method of isotope analysis of some light elements. Ocean water exhibits a very narrow range in isotopic composition, less than 1% for δD and 1‰ for $\delta^{18}O$. Evaporation strongly affects the isotope composition, since it causes a preferential depletion in the lighter isotopes, which become enriched in the vapour phase. Consequently, the remaining water will be heavier. Highly saline waters generally have the greatest D and ^{18}O contents. Low salinities, which are caused by a dilution of fresh waters, correlate with low D and ^{18}O concentrations (Hoefs, 1973). It seems quite certain that ocean water had a constant isotopic composition from the Palaeozoic to the Recent. Greater variations of several per mil and even more might have been possible in the Precambrian.

PLANTS – THE SOURCE MATERIAL OF COAL

In addition to diastrophic movements, formation of suitable basins and the climatic control of vegetation, the production of coal requires that the accumulated plant remains were provided with resistant tissue. This condition is fulfilled only by land plants which, consequently, are the principal coal-forming material. The land plants appeared during the Cambrian as is evidenced by the finds of casts and spores. In that period the level of atmospheric oxygen was not yet high enough to fully secure the existence of organisms on dry land. The first thin coal seams formed in the Late Silurian and important seams in the Devonian. The intensive development of terrestrial plants in the Late Carboniferous, in the Permian and Late Tertiary has been ascribed by some authors to the high content of CO_2 in the atmosphere, produced by volcanic activity. The abrupt development of angiosperms at the Early/Late Cretaceous boundary is explained in the same way. Recent investigations have not confirmed this hypothesis to a full extent, particularly as concerns the Late Carboniferous, since the main effusive volcanism of Upper Palaeozoic occurred as late as the Permian, when the material of the essential part of the Upper Palaeozoic coal resources had long been produced.

More plausible seems to be the theory of Brooks (1950) who also connected the prolific growth of vegetation with increase in temperature but did not postulate an excessive increase of CO_2 content in the atmosphere (after all, the CO_2 level in atmosphere has been since the Silurian broadly constant, the varying enrichment not exceeding the factor of 4, whilst in pre-Silurian times the CO_2 content in the atmosphere was 10 to 100 times higher). His assumption of a large extension of the temperate zone in the Carboniferous, Jurassic, Cretaceous and Tertiary almost as far as the poles, implies the suppression or expiring of polar climate.

The difference in the type of flora in the individual geological periods is evident. The plants developed progressively from lower to more accomplished forms. If, as a result of metabolic process, the organs of a plant individual developed to the same purpose, they were and are in a high degree of the same chemical composition.

Their excretion products were also of the same quality so that in fossil peat bogs similar biochemical decay processes and similar decay products to those of recent peat bogs may be assumed. The differences are most probably quantitative. Essential and qualitative changes are only sporadic. The Carboniferous plants differed chemically from the Tertiary flora in being deficient in resinous substances, which could have affected the coal-forming process.

It is believed that the Upper Palaeozoic cryptogamous flora was represented by trees having thick bark and thin wood. Arborescent gymnosperms of Mesozoic times had already thick wood and thin bark and were rich in resin. Tertiary conifers had also highly resinous wood.

The following paragraphs will discuss the plant species which contributed to the formation of coal seams.

Silurian: herbaceous plants (*Psilophyta*) grew only along the sea coasts. Continental vegetation lived in muddy habitats. Coal seams 0.6 m thick at the most are known from the lagoonal and continental sediments in the area of Tashkent and Kokand in the U.S.S.R.

Devonian: Lower and Middle Devonian land plants are of herbaceous, shrubby and low-tree types (*Psilophyta, Lycopodiales*) and primitive representatives of other sporiferous plants – *Sphenopsida* and *Pteridopsida*. The decayed remains of psilophytes gave rise to the unworkable seam near Daun in Germany and small exploitable seams near Barzas in the Kuznetsk Basin.

In the Upper Devonian vegetation predominated club-mosses of *Cyclostigma* type and arborescent *Archaeopteris*, which in the form of leaves and the mode of reproduction resembled ferns, but their wood was already similar to that of gymnosperms. *Psilophyta* had almost disappeared, whilst arborescent club-mosses, horse-tails and ferns (not in coal-forming environments) were abundant. Trees from the group of gymnospersm complemented the picture of vegetation, which already approached the Carboniferous flora. *Pteridospermae* made their first appearance in the Late Devonian. Workable seams of bituminous coal are on the Medvezhi Islands.

Lower Carboniferous: club-mosses, particularly *Lepidodendropsis*, horse-tails with key species *Asterocalamites* and many *Aspidium* species were prolific and the occurrence of pteridosperms became more frequent. Lower Carboniferous workable seams developed in the Minusinsk and Karaganda Basins in the U.S.S.R. In these basins, however, the flora was different than in other parts of the world, as in some regions of the Earth climatic differentiation began as early as the Lower Carboniferous. Edge coals in Scotland, anthracites of Pennsylvania, coals of the Spitsbergen, Nova Scotia and New Brunswick are also Lower Carboniferous in age.

Upper Carboniferous and Permian: typical of the Westphalian are dendriform club-mosses (*Lepidodendron* and *Sigillaria*), horse-tails (*Calamites*) and ferns (*Pecopteris*). *Pteridospermae* were represented by the genera *Mariopteris*, *Neuropteris* and *Lyginodendron*. These plants produced coal in many European basins (e.g. the

Donets Basin in the U.S.S.R.) and in the east of the U.S.A. During the Westphalian subtropical to temperate and humid climate existed over the whole Earth, except for small cold regions around the poles.

The climate in Carboniferous times was humid and the absence of annual rings from tree trunks suggests a fairly constant temperature, without major oscillation within a year. Forests covered extensive areas with but little variation in the composition of flora. The basins had the character either of forest swamps or peat swamps.

Beginning with the Stephanian the climate became differentiated. The Euroamerican and Katasian provinces developed in the tropical and subtropical zones and the Angara and Gondwana provinces in the temperate humid zone. In the Euroamerican province arborescent ferns (*Pecopteris*) and gymnosperms (*Cordaites* and *Walchia*) flourished. The flora of the Lower Permian was xerophilous and characterized by *Sphenophyllum thoni*, ferns of the genus *Pecopteris*, the pteridosperm *Callipteris* and gymnosperm *Walchia*.

The Katasian province is developed in the Stephanian and Permian of China, Korea and the U.S.A. The main representatives are the equisetum *Lobatannularia* and pteridosperm *Gigantopteris*.

The Angara province appears in Siberia (Pechora, Taimir and Tunguzka Basins extending as far as Chukotka) and in the Kuznetsk Basin. The plant association consists of mosses, low tree-like equiseta *Schizoneura* and *Phyllotheca* and *Cordaites*. In the Gondwana province the flora was similar and enriched by herbaceous genera *Glossopteris* and *Gangamopteris* (rarely penetrating into the Angara province).

Triassic: the Lower Triassic flora of Europe was relatively monotonous and xerophilous, with prominent genera *Pleuromeia* and the arborescent gymnosperm *Voltzia*. The Gondwana genera *Schizoneura* and *Phyllotheca* expanded in the Triassic also to Europe.

The Upper Triassic flora comprised fern-like genera *Camptopteris*, *Dictyophyllum* and *Danaeopsis*, and abundant ginkgoes and cycads. A characteristic pteridospermous species of Europe and Greenland was *Lepidopteris ottonis*. The climate was temperate and humid. Workable seams are developed in the Chelyabinsk Basin (U.S.S.R), the Richmond Basin (U.S.A.) and the Quangyen Basin (V.P.R.).

Jurassic: the temperate climatic zone extended to Greenland, Siberia and Central Asia, covering the so-called Siberian province. Coal deposits are known from Siberia and very rich ones from Central Asia. The flora was of the taiga type with, for example, ginkgoes (*Czekanowskia*) and conifers such as *Elatides* and *Brachyphyllum*.

In the tropical to subtropical climatic zone there was an abundance of ferns (*Cladophlebis*, *Phlebopteris*) and cycads (e.g. *Nilssonia*). This zone occupied Europe, India and China (which has coal deposits) and is termed the Indo-European province.

Cretaceous: In the Early Cretaceous the subtropical to temperate climatic zone spread still to high latitudes, almost to the poles. Ferns of *Onychiopsis* and *Weichselia* genera were abundant but cycads receded to the background. Ginkgoes retreated to the Siberian province. Coniferous trees experienced rich development. The most important Lower Cretaceous coal deposits are in eastern Siberia and in Canada.

The distribution of climatic provinces persisted nearly unchanged into the Late Cretaceous, only the Siberian province expanded into northern Canada. Fairly uniform, subtropical to temperate humid climate existed broadly all over the world. It supported prolific growth of dendriform flora. Trees of angiosperm class, which grow in warm temperate to subtropical climate today, were typical components of the vegetation. The genera *Magnolia*, *Ficus*, *Gredneria* and *Comptonia* predominated. The proportion of conifers was small; branches of *Sequoia reichenbachi* are found most frequently.

Tertiary: in the temperature humid zone deciduous trees were thriving, such as *Populus*, *Fagus*, *Alnus*, *Ulmus* and *Platanus*. Conifers were represented by *Sequoia*, *Taxodium* and *Glyptostrobus*. Palms (genus *Sabal*), *Laurus*, *Ficus* and rubber plants predominated in the subtropical and tropical zones.

17. Arborescent ferns (*Cyathea*). Gran Piedra, Cuba. Photo by V. Bouška.

In the Late Tertiary, deterioration of the climate proceeded from the poles and thermophilous flora migrated southwards. In Europe, the U.S.A. and Central Asia the genera *Populus*, *Fagus* and *Acer* appeared and conifers were abundant. Cooling of the climate and migration of the flora continues until the end of the Tertiary.

Although *Angiospermae* were richly developed and contributed doubtless largely to the formation of lignites, they succumbed to decay and alteration more readily than conifers and the ancient gymnosperms. The woody structure of coal, as far as it is preserved, belongs mostly to the wood of conifers or primitive woods of similar structure. The algae and plankton, which gave rise to cannel or boghead coals, found analogous living conditions in both Carboniferous and Miocene. They lived in open water and the resulting coal types are similar, differing only in the degree of coalification and age.

A detailed description of floras and their distribution in individual geological periods is given in the publications of Němejc (1959, 1963, 1968).

CHEMICAL COMPOSITION OF PLANTS

The chemical composition of plants was changing along with their evolution since the time of their origin until the present day. Since, however, the function of the plant organs has remained essentially unchanged, no great differences are presumed.

The plant accepts nutrients in assimilating CO_2, H_2O and mineral substances from soil. The external process thus becomes an internal one. Carbon dioxide enters the plants through stomata. Gases diffuse well through membranes into the assimilation parenchyma; they dissolve in protoplasm.

The mineral nutrients are accepted from soil in the form of highly diluted solutions by their roots, mainly by root filaments. The circulation of elements in soil is controlled by the solubility of their compounds. Soil solutions contain major amounts of the ions of bases, acids and salts Na^{1+}, Mg^{2+}, Ca^{2+}, Mn^{2+}, Fe^{2+}, Cl^{1-}, SO_4^{2-}, $H_2BO_3^{1-}$, CO_3^{2-} and NO_3^{1-}. The solution is transported by vessels and evaporates at the sites of maximum transpiration, i.e. in the leaves. Elements including the trace elements are extracted from the substratum of a large volume. The plant rebuilds chemically the substances taken from the environment into living matter and storage substances. In dissimilating the plant liberates energy from the assimilates and lives, i.e. grows and develops.

The plant and its roots possess a certain selective power in absorbing the substances from soil solutions. In contrast to earlier opinions, it has been found out that this power is not sufficient enough to prevent the access of harmful substances. The physico-chemical conditions of soil and of plant juices obviously affect the intake of cations and anions. In this context the Gasparian strip built up of endodermal cells is often mentioned. The transfer of ions is mediated exclusively by the living protoplasm of endodermal cells. Plant juices thus regulate the supply of ions, enable some to accumulate but they probably cannot fully prevent the intake of others (in small amounts) which abound in the soil. Consequently, there are elements in the plant organisms whose function and significance are unknown. The plants are capable to adapt to some poisonous substances and to compensate their effects, so that these elements may reach higher than normally fatal concentrations.

The absorption of an element by the plant depends not only on the total content of this element in soil but also on the associated elements, since antagonistic relations exist between some of them (Zýka, 1971). These relations manifest themselves at the intake of elements by the plant, because the individual elements show the tendency of occupying the same sites on the sorption surface of the cell protoplasm. The elements that penetrated into the protoplasm promptly influence the physico-chemical properties of biocolloids and thus directly control the permeability of protoplasm for the ions of other elements. The intake of elements is also affected by the HCO_3^{1-} content, soil reaction (pH), oxidation-reduction potential, water content, activity of microorganism and other factors. At a high pH, for example, the sorption of cations is greater than at low pH.

Some obnoxious elements, such as poisonous lead, are precipitated on the surface of roots, but most of them, particularly in resistant plant and tree types, accumulate at the terminal station of the transpiration process, i.e. in the leaves. The plants dispose of the useless or obnoxious elements annually in shedding the leaves.

The obnoxious and useless substances in trees are generally cumulated in wood and bark which gradually die away (the wood outwards, and bark inwards) and act as storage-room for waste material the harmful effect of which is thus cancelled.

In the leaves, wood, bark and fruit coats these substances and elements are accumulated in a higher concentration but proportionate to their content in soil. On withering they come back into the top soil layer. The elements in the tree roots reaching to a great depth (e.g. poplar to 12 m, beech as deep as 25 m) provide the picture of soil chemistry at this level; this circumstance is applied in prospecting for mineral deposits (Sýkora, 1959).

As the intake of anions is more important for plants with regard to the supply of necessary energy, they accept them more easily than cations. This exchange is important for the regular function of cells. Many elements have a direct importance for the plant and occur in it in a higher concentration provided that their concentration in soil is sufficient.

The biochemistry of plants has been studied in great detail by Devis et al. (1966), and geochemistry of organic substances by Manskaya and Drozdova (1968).

The composition of the plant organism may be considered from two points of view: according to (1) chemical compounds and (2) elements. In a plant organic substances predominate over mineral components. A schematic division is as follows:

(1) Chemical compounds: (a) organic compounds,
                                       (b) inorganic compounds.

(2) Chemical elements: (a) elements of organic matter,
                                       (b) elements contained in ash.

(1a) Organic compounds in a plant body (Fig. 18) are very numerous and many of them so complex that they have not been fully recognized as yet. Of primary importance are hydrophilic substances, i.e. those which dissolve in water. They involve all saccharides (glycides), i.e. monosaccharides ($C_nH_{2n}O_n$) [hexoses ($C_6H_{12}O_6$) –D-glucose, D-fructose; pentoses ($C_5H_{10}O_5$)], disaccharides [saccharose ($C_{12}H_{22}O_{11}$), maltose ($C_{12}H_{22}O_{11}$)], tri- and polysaccharides ($C_6H_{10}O_5$)$_n$ [cellulose, hemicellulose, starch, inulin, xylans, mannans, galactans, arabans, etc.] and allied chitin.

18. Structural units and components of fundamental plant materials.

Of the saccharides, cellulose and starch having a similar chemical composition ($C_6H_{10}O_5$)$_n$ predominate in plant organisms. Cellulose is above all the structural material and starch is deposited in fruits, seeds and stems as storage food. When needed, starch is hydrolysed to glucose, which in turn alters into cellulose which forms the framework of walls. Monosaccharides, glucose and fructose are reserve substances.

D-glucose (Fig. 19 left) is the commonest plant monosaccharide and one of the most stable. It is often found in sediments containing plant debris. Of other monosaccharides, xylose, rhamnose, arabinose are also highly stable in sediments; ribose, α-D-glukose   β-D-galactose   19.

mannose and galactose have a somewhat lower stability. In peat vegetation monosaccharides are represented mainly by glucose and xylose in the angiosperms, and galactose (Fig. 19 right), glucose, mannose and xylose in the bryophytes. Galactose and mannose are not very abundant in deeper parts of a peat bog, where glucose and xylose remain the principal saccharides thanks to their stability (Swain, 1970). Chitin forms the cellular membranes in most fungi and their spores. It belongs to nitrogenous polysaccharides $(C_8H_{12}O_5N)_n$.

In the membranes of older cells there are lignin and polymers of saccharides. Lignin is the matrix or cementing matter which binds up the cellulose fibres into bundles. The origin and chemistry of this amorphous aromatic compound are not known precisely but it is thought that the fundamental structural unit of lignin is phenylpropane. This compound is related to coniferyl alcohol, whose dimer — dihydrodiconiferyl alcohol (Fig. 20) occurs in old spruce wood (Fraser, 1955). The composition of lignin in individual plant species differs (van Krevelen, 1961). Lignin of deciduous trees, for example, is derived rather (at least in part) from sinapyl alcohol, which has one more methoxyl group (OCH_3), than from coniferyl alcohol. The geochemistry of lignin has been studied in detail by Manskaya and Kodina (1975).

20. Dihydrodiconiferyl alcohol.

Important structural materials of plant bodies are organic acids. The most widespread are oleic acid $C_{17}H_{33}.COOH$, linoleic acid $C_{17}H_{31}.COOH$, linolenic acid $C_{17}H_{29}.COOH$, palmitic acid $C_{15}H_{31}.COOH$, stearic acid $C_{17}H_{35}.COOH$, and lactic, glutamic, ascorbic, malic, formic, oxalic and citric acids. The α-lipoic acid of chloroplasts acts as an important oxidation-reduction agent in photosynthesis. Unsaturated fatty acids, scarcely isolated from geological samples, are typical of algae.

Organic compounds with nitrogen, particularly proteins, are further structural components of plant bodies. Amino acids are the basic structural units of proteins. The nitrogenous substances involve in addition to amino acids, amines, amides, purines, nucleic acids and nucleoproteins.

Of importance are also lipids*, i.e. waxy and fatty substances including hydrocarbons and some organic pigments — fats, oils, waxes (esters of higher fatty acids and higher aliphatic monohydric alcohols) and allied impregnating compounds such as cutin forming the leaf epidermis and protective tissues of organs lacking periderm. Chemically, cutin is a complex compound of glycerol esters and higher fatty acids. The cutinic acid $C_{26}H_{50}O_6$ and cutininic acid $C_{13}H_{12}O_5$ are its essential components.

21. Betulin.

Fats accumulate in plant seeds or in the tree bark (mainly in birch, pine and lime), and also in fungi, algae and bacteria. They contain 74–78 % C, 10–13 %H and 9–16 % O. Waxes are composed of 80–82 % C, 13–14 % H and 4–6.5 % O.

Lipids constitute predominantly protective waxes of trunks, leaves and needles. They are partly aliphatic and partly of terpenoidic nature. Terpenoids are represented by such compounds as is, for example, triterpene betulin (Fig. 21), which occurs in a large amount in the bark of *Betula alba (verrucosa)*. Swain (1970) records other lipid substances. Waxes of present-day higher plants show a definite prevalence of odd-numbered *n*-alkanes over even ones and a tendency to a higher molecular weight of the components (n-C_{27}, n-C_{29}, n-C_{31}, n-C_{33}). These *n*-alkanes are ubiquitous components of geological materials. Douglas and Eglinton (1966) informed about the isolation of iso-(2-methyl)-alkanes and anteiso-(3-methyl)--alkanes. Their presence in geological samples is accounted for by their content in plant waxes.

22. a — α-pinene,
    b — p-cymene,
    c — p-menthane,
    d — terpene hydrate.

* According to Bloor (in Bergmann, 1963), the term lipids is used for aliphatic fatty acids, their esters, alcohols and hydrocarbons, and terpenoids and steroids. All these water-insoluble substances have a low O/H ratio, which makes them geologically more stable than are, for example, saccharides. The molecule of lipids remains practically intact even after the elimination of all water.

Skrigan (1951) reported on the presence of α-pinene (Fig. 22a), p-cymene (Fig. 22b) and p-menthane (Fig. 22c) in pine trees. Guild (1922) isolated the two last-mentioned substances and terpene hydrate (flagstaffite) from decaying roots and trunks of pine buried in peat. Cork matter consists chiefly of phellogen, suberin and the phellonic, phellogenic, phloionic, phloionolic acids. Terpenes are generally represented by monoterpenes, sesquiterpenes and mainly by diterpenes — phytol ($C_{20}H_{40}O$), abietic acid ($C_{20}H_{30}O_2$), and others. Diterpenes are contained especially in resin of the conifers. The abietic acid (Fig. 23a) of buried pines is obviously converted into fichtelite ($C_{19}H_{34}$) (Fig. 23b) and retene (Fig. 23c), which by themselves are not known as plant products. In this context it should be mentioned that *Flavobacterium resinovarum* isolated from the soil of a pine forest can under aerobic conditions use the diterpene fraction of pine wood as a sole source of carbon. Microorganisms living both in aerobic and anaerobic environments may play an important role in the early stages of the diagenesis of cyclic terpenes.

23. a — Abietic acid, b — fichtelite, c — retene.

The transformation of α-pinene into p-cymene and of the abietic acid into retene in the roots of fossil pines suggests that the dehydrogenation of aliphatic compounds to aromatic ones is a relevant geochemical reaction. With regard to the gradual conversion abietic acid — fichtelite — retene it is to be remarked that the aromatic hydrocarbon dehydroabietane was isolated from the soil of a pine forest (Anderson et al., 1969). The structural similarity of dehydroabietane with fichtelite and retene indicates that dehydroabietane may be an interstage in the diagenesis of the abietic acid.

Excretion substances (some of them have been mentioned above) do not re-enter the circulation of nutrients in the plant body. They comprise glycosides, tannins, essential oils, alkaloids (e.g. cocaine, atropine, scopolamine) and resins. Alkaloids, purines and chitin, form a special group contained in the leaves of some plants; chitin is a component of the sclerotium and hyphae of fungi.

Natural resins are mixtures of mono-, sesqui-, di-, and triterpenoids; they also contain some terpenoic acids such as resinoic or abietic acids. According to Swain (1970) the basic structural unit of terpenoids is isoprene (C_5H_8). Geranyl pyrophosphate formed by the connection of chains of isopentenyl pyrophosphate and dimethylalyl pyrophosphate is the initial compound of most plant terpenes. The variation in the condensation process contributed to the multiplication of natural terpenoids. The mono-, sesqui- and some diterpene hydrocarbons are the principal volatile fractions of resins; diterpenic and triterpenic carboxylic acids are non-volatile. The remaining constituents may be alcohols, aldehydes and esters. During coalification the predominant part of the volatile fraction of resins disappears, but a small part may be preserved.

The resins are produced by about ten per cent of plants, two thirds of them being the tropical plants. All conifers produce resins but only *Pinaceae* and *Araucariaceae* in a large quantity. The tropical *Leguminosae* and *Dipterocarpaceae* synthesize most resin of all *Angiospermae*. *Gymnospermae* were the earliest plants synthesizing resin. The extinct genus *Cordaites* contained dark resinous material in wood and parenchyma, which can be related to true resins. During the Eocene and Oligocene the *Araucaria* trees of the Baltic region were great resin producers. The large amount of Baltic amber called forth the hypothesis that the trees may have been attacked by a disease ("succinosis"), which caused an excessive production of resin. This assumption, however, disregards the fact that contemporaneous tropical plants of the classes *Gymnospermae* and *Angiospermae* also synthesize enormous amounts of resins. The causes of resin formation and its function both in modern and fossil plants are not safely known. It is tentatively assumed that they have a physiological function in metabolic processes or serve as protection against pests, particularly insects.

Infrared absorption spectroscopy and other physical methods have contributed to the recognition of the structure of the fossil resins (Streibl et al., 1976) and amber. Recent resins and ambers (Gouch and Mills, 1972) have similar spectra in the region of 2.5 to 8 µm (4,000–1,250 cm^{-1}). Absorption bands between 8 and 10 µm (1,250 to 625 cm^{-1}) of individual resins and individual ambers are varying. The differences are probably due to C—O bonds but it is difficult to assign them a specific structure form. The sharply limited band at 11.3 µm (885 cm^{-1}) may belong to the terminal double bond. Some acids, as the agathene-dicarboxylic and copalic acids, which were isolated from recent resins (Swain, 1970), show such a band; it was also found in fossil resins. The analysis of soluble diterpene fractions of several fossil resins has determined among the components the agathene-dicarboxylic acid (Fig. 24a), the sandaracopimaric acid (Fig. 24b) and sandaracopimarinol (Fig. 24b) (Thomas, 1970). These components are also present in the recent resin of *Agathis australis* (Thomas, 1966).

The plants also contain substances that are important for their existence but are present only in a small quantity. They are called ergones and include chlorophylls

24. a — Agathene-dicarboxylic acid, b — sandaracopimaric acid (R = CO_2H) and sandaracopimarinol (R = CH_2OH).

(a and b), enzymes, vitamins (thiamine containing the pyrimidine and thiazol nuclei, riboflavin, pyridoxin, axerophtol, niacin, giberellic acid, tocopherols, etc.) and pigments such as carotenoids (orange β-carotene and yellow xantophylls), flavoproteins, anthocyans, N-heterocyclic substances and others. The walls of pollen grains and spores are other resistant organic materials of biological origin. Their resistance makes it possible to study their morphology and microstructure even after long geological periods. Chemically, they are labelled by the collective term sporopollenins (formed by polymerization of carotenoids). Enzymes support fermentation and other chemical processes and metabolism in organisms. They are complex proteins and act as organic catalysts, which even in a very small amount accelerate all processes in plants. The activity of chlorophyll and vitamins is generally known. Chlorophyll is the principal factor in assimilation and dissimilation.

The plants contain predominantly chlorophyll a ($C_{55}H_{72}MgN_4O_5H_2O$, after Swain, 1970). Its basic structural unit consists of four pyrrole (C_4H_5N) rings connected into the so-called porphyrin system. The pyrrole rings are bonded to a central Mg atom; one double bond of the pyrrole ring is saturated with two hydrogen atoms (dihydroporphyrin) and carries a long chain of phytol ($C_{20}H_{39}OH$).

Chlorophyll is an example of a metal-organic complex, a so-called chelate. It plays an important role in organic geochemistry and particularly in biological processes. In forming metal-organic complexes chelates effect "de-poisoning" e.g. of hydrosphere, in which metals such as Pb, Ni, Mn and Hg would otherwise accumulate. The chelate-type bond on organic compounds is highly probable for Fe, Mn, Cu, Zn, B, Co and Ni. It has, however, not yet been confirmed that metal-organic compounds occur invariably in the form of chelates.

Methyl pheophorbide and other green pigments determined in the Eocene lignite in Geisel (Dilcher, 1967) provide evidence that under moderate alteration conditions the porphyrin system in chlorophyll may remain long unchanged in sediments. This finding reveals the plausible course of earliest diagenesis or low-grade metamorphism. Hydrolysis and elimination of metals are evidently the first reactions to take place.

In summary, it can be said that the dry matter of continental plants is composed essentially of cellulose, lignin, wax, resin and pectins and additionally of proteins, fats and oils, especially in aquatic plants. In regarding the flora as a coal-forming agent, the saccharides and lignin are its most significant components (Table 3). Proteins, which as the basic units of protoplasm are necessary for the life of a plant, are subordinate in the resistant plant fibres. They appear in small concentrations in the coal material. Cellulose makes up 10—63 % of plant material and lignin, which stiffens the cell walls, 0—40 %. The remaining percentage of dry plant matter is composed of some of the above-mentioned substances. Wood contains 8—10 % pentosans and the hard wood as much as 20 %. Part of pentosans is ranked with cellulose. The wood matter contains 2.5—3.5 % methyl pentosans,

Table 3

Chemical substances and average elementary composition of plant tissues (Francis, 1954)

Parts of plant body	Chemical substances	Portion of substances (in %)*	Average composition (in %)**		
			C	H	O
wood	saccharides (only cellulose)	45–65	44.4	6.2	49.4
	lignin	20–45	63.2	6.1	30.7
	proteins	12–16	53.5	7.0	22.0
	fats + waxes	0.2–4.0	82.0	14.2	3.8
	resins	0.5–15.0	80.0	10.0	10.0
bark	saccharides	small			
	lignin	20–50	49.0	6.0	45.0
	proteins	absent			
	fats + waxes except suberin	3–15	82.0	14.2	3.8
	suberin	30–40	72.0	10.5	17.5
	resins	absent			
cuticles, spore exines	saccharides (only cellulose)	10–20	44.4	6.2	49.4
	saccharides (except cellulose)	5–15	68.0	9.5	22.5
	lignin	absent			
	proteins	small			
	fats + waxes (except cutin + sporopollenin)	10–40	82.0	14.2	3.8
	cutin + sporopollenin	25–75	72.0	10.5	17.5

* In dry matter.
** S and N (important in some proteins and resins) are not included.

3—6% hexosans and 0.5—1.5% galactans. Hard wood has about 1% wax and resin, the conifers may contain up to 10% resin in wood. Stumps of coniferous trees have as much as 30% resin. Tannins occur mainly in the tree bark.

The following data (Hubáček, 1948) show the cellulose and lignin contents:

	Cellulose	Lignin
Pine	53.3%	28.2%
Birch	55.5	28.2
Beech	54.5	39.1
Oak	39.5	34.3
Fir	57.0	26.9
Poplar	62.8	20.8

(1b) The principal **inorganic constituents** of a plant are water, which forms the bulk of their bodies, and ash matter. The term ash matter must not be identified with ash. The former include incombustible structural storage or admixed mineral substances represented by compounds of various elements, especially those with low atomic weights. These elements are bound to some inorganic or complex organic acids, or are attached by sorption forces only. The ash is the residue left after combustion of a plant body, which contains the structural elements of plants but not in the original compounds. During combustion at a particular temperature part of volatiles escape and part of elements oxidize (Fe^{2+}, Mn^{2+}) forming new compounds, predominantly oxides, carbonates, sulphates, phosphates and to a minor extent chlorides. The plant ash is usually greyish-white or of various colours, such as yellow to rusty red tinted by iron, or bluish-green by copper. The ash provides only a relative picture of the plant chemistry. The principal constituents of ashes are oxides, silicates, sulphates and chlorides of K, Ca, Mg, Na, Mn, Al and Fe.

As an example is here given the composition of the ash of horse-tails which were analysed by Němec et al. (1936):

	Equisetum palustre	*Equisetum arvense*
SiO_2	61.20%	50.22%
Al_2O_3	1.87	4.35
Fe_2O_3	1.19	3.00
CaO	8.05	7.73
MgO	2.05	2.72
K_2O	15.01	19.70
Na_2O	0.75	0.80
P_2O_5	1.60	2.32
SO_3	2.64	3.60
Cl	5.42	5.30
Total	99.78%	99.74%

The potassium occurs in the ash of plants as K_2CO_3. Its content drops only exceptionally in the ash of wood below 5%, usually varying between 10 and 20%;

it may amount up to 40%. Na_2O amount ranges only from 0.5 to 2%, its higher concentration being only sporadic. The CaO amount is about 20%, MgO 5−10%, and those of Fe_2O_3, Mn_2O_3 and Al_2O_3 are below 1%. The ashes of some conifers are rich in manganese and aluminium. The wood ashes contain 2−30% P_2O_5. Chlorides are generally present in very small amounts or in traces; concentration of ca. 3% Cl has been found scarcely in wood ashes. The content of SO_3 rarely exceeds 5%, and that of SiO_2 is usually 1−3%. For a better apprehension it should be remarked that the ash content in wood is about 0.5−1.5%, the values over 1% being quite rare. The tree bark provides more ashes, as much as several per cent, since the plant deposits unnecessary substances in it. The largest amount of ash is in leaves, where the mineral substances are translocated with solutions and after evaporation of water the mineral salts are accumulating. The content of inorganic material in plants is generally below two per cent and varies in the different parts as well as in different species of plants.

The walls of *Gramineae* are silicate-impregnated, *Characeae* have calciferous sheaths, and *Lycopodiaceae* display a high concentration of Al_2O_3 (26−57% in ash). The tree wood has a lower ash content (about 0.5−1.5%) than the leaves (2.5−3%), and coals derived from wood have therefore a smaller amount of ash than coals originated from leaves or a mixed debris. A high ash content has been determined in the bark of old trees (10−20%). The ash content in cryptogams (*Lycopodium*, *Equisetum*, *Cyathea* and others) ranges from 3 to 12%. Iron is concentrated in leaves and the coals containing a large proportion of cuticles thus contain more iron compounds than those derived from woody parts of plants, and have also more CaO, SiO_2 and Al_2O_3.

Inorganic substances actually present in plants are primarily the acids H_2CO_3, H_4SiO_4, which stiffen the skeleton, H_3PO_4 affecting favourably the breakdown of saccharides and formation of fats, and the structural constituents such as $CaCO_3$ and SiO_2.

Water content is not the same in all plants. In fresh wood it makes about half the weight, wood dried in air contains ca. 20% water after evaporation, some woods even as little as 10%. Green leaves have 70% water on the average, some fungi (*Coprinus*) almost 98% water. Spores and seeds are very poor in water.

(2a) **The elements of typical organic matter** are not very numerous but they prevail quantitatively in the composition of plants. The four most important elements − C, H, O and N − are not present in ashes, as they burn up. Their average amount in wood is given by Hubáček (1948) as follows:

	C %	H %	N %	O %	Ash %	Water %
Wood dried in air	39.60	4.80	0.20	34.60	0.80	20.00
Waterless wood	49.50	6.00	0.30	43.30	1.00	−
Volatile matter of wood	50.00	6.00	0.20	43.80	−	−

The volatile matter of leaves contains on the average 46.5% C, 7% H, about 2% N and the remaining percentage goes to oxygen. G. W. Hawes in Moore (1940) listed the chemical elements and ash content in some significant coal-forming plants (in weight %):

	C	H	O	N	Ash content
Lycopodium dendroideum	47.11	6.39	41.85	1.40	3.25
Lycopodium complanatum	45.78	6.25	40.66	1.84	5.47
Equisetum hyemale	41.94	5.89	39.23	1.12	11.82
Asidium marginale	44.77	5.99	41.97	2.08	5.19
Cyathea caniculata	45.39	6.11	39.82	1.12	7.56
Wood of Pinaceae	50.31	6.20	43.08	0.04	0.37

Hubáček et al. (1962) published the following composition of some plant components (percentage of inflammable matter):

	C %	H %	N %	S %	O %
Woody matter	49–50	5.8–6.2	0.04–0.20	traces	43–45
Cellulose	44.1	6.2	–	–	49.4
Lignin	63.1	5.9	–	–	31.0
Waxes	80.3–81.6	13.1–14.1	–	–	0–11.1
Resins of conifers	76.8–83.6	9.7–12.9	–	–	0–11.1

As concerns the isotopic composition of carbon, the terrestrial plants are poorer in ^{13}C than the marine plants. Craig (1953) determined $\delta^{13}C = -22.2\,°/_{oo}$ in the recent freshwater plant *Rhizoclonium* sp. Silverman and Epstein (1958) reported that the $\delta^{13}C$ values in wood of terrestrial plants range from $-22\,°/_{oo}$ to $-27.4\,°/_{oo}$ (the average value being $-24.9\,°/_{oo}$) in case the values in leaves are $\delta^{13}C = -23.1$ to $-28.6\,°/_{oo}$ ($-25.8\,°/_{oo}$ on the average). The greater the negative value, the less ^{13}C is present relative to the standard value (Fig. 25). Park and Epstein (1960) explain the difference in the ^{13}C content in terrestrial and marine plants by the fact that the terrestrial plants consume more atmospheric CO_2, which is poorer in ^{13}C than HCO_3^- from oceanic water. It seems, indeed, that the use of these two carbon forms by the vegetation contributes to the differences observed in freshwater and marine sediments (water of streams and lakes contain CO_2 at lower pH). Sackett et al. (1965) related that lipids have by as much as $7\,°/_{oo}$ more ^{13}C than the whole plant. The average $\delta^{13}C$ of terrestrial plants and trees is determined by the total effect of fractionation which occurs (a) as the absorption of CO_2 into the leaf cytoplasm, (b) during photosynthetic fixation, (c) during partial conversion to fat. The organic carbon in most terrestrial plants is about 20% lighter than atmospheric CO_2. The isotopic depletion of ^{13}C in plants is due mainly to cyclic depletion of that isotope in the air environment of plants.

Similarly, the water in arborescent plants is richer in deuterium than the ground water under the trees; the leaves are richest in heavy water. This is accounted for by isotopic fractionation during transpiration, which is of highest degree just in the leaves.

Phosphorus and sulphur are necessary mainly for the building up of proteins. According to Hubáček (1948), the wood contains 0.03 % to 0.04 % sulphur. The content of phosphorus varies and cannot be precisely stated, since its greater part is present in the plant body in the form of phosphoric acid or phosphates, which belong to inorganic components of plants. Sulphur is also partly present as sulphates.

(2b) In addition to typical organogenic elements C, O, H, and N, which compose the bulk of the organic matter, the building up, nutrition, growth and other life processes in plants need necessarily many elements, which remain, often in a large amount, in the incombustible portion of organic matter, i.e. in ashes. The further elements present invariably in plant bodies are P, S, Ca, K, Mg and Fe.

Together with the preceding elements they are called the biogenic elements. The average content of nitrogen in plants is about 3 %, the C/N ratio being 15. Organic nitrogenous compounds are degraded by bacteria to ammonia after the death of plants; it may be utilized again by plants but is for the most part oxidized to nitrite and eventually to nitrate by soil bacteria. Nitrogen is indispensable for the formation of proteins (which contain 15–20 % N) and is a component of chlorophyll, alkaloids and amines. Sulphur, which is less necessary in reactive compounds is usually bound to α-biotin and in some amino acids, mainly in cystine and cysteine. Phosphorus is needed for the formation of cell nuclei and ripening of seeds. It is most abundant in flowers (1.5 % of dry matter); pollen grains contain 2–3 % and seeds 0.5–1 % phosphorus. It is an important component of reactive compounds of protoplasm and cell nucleus (di- and triphosphopyridin nucleoids) and is present in nucleic acids and associated proteins and in lipids. The phosphorus compounds mediate the transfer of energy. Potassium influences the quality of fruits, functions as a catalyst and is the principal element of cytoplasm. It is most abundant in the plants that cumulate large reserves of saccharides or starch (in sugar-beet or potato tubers). Calcium is important for the building up of tissues and proper maturing of wood. It is deposited in extinct cells. In combination with

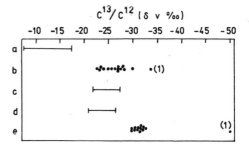

25. Values of $\delta^{13}C$ for various organic materials (Silverman and Epstein, 1958). a — marine organisms, b — organic carbon of marine sediments, c — terrestrial plants, d — coal, e — organic carbon in non-marine sediments, (1) — natural gas.

potassium it controls the intake and movement of water. Potassium increases the intake of water and expanding of protoplasm; calcium acts as a dehydration agent. Magnesium is needed for the formation of chlorophyll. It is contained in proteins of the cell nucleus, in phytin, as enzyme hexokinase, and abundantly in seeds. Iron is not present in chlorophyll but takes part in its formation. It is usually bound in porphyrin rings and functions as a catalyst in respiration enzymes; the plant does not need much iron for its life. All biogenic elements necessary to a prosperous development of plants should be availabe in adequate amounts, otherwise their development is controlled by the least abundant element. It is of interest that neither chemically closely related elements can substitute one another (e.g. iron — manganese). The elements Na, Si, Al, Mn, B, Ba and Sr, which are also important and useful for the plant, occur only in very small to trace amounts. They are designated as microelements or as microbiogenic or oligobiogenic elements. Their functions in plant bodies are not well understood; it is most probable that they function as catalysts. The presence of some of them has been safely determined but their relations to plants are difficult to decide upon. Sodium together with K, Ca and Mg, for example, regulate the transfer of saccharides. The elements B, Mn, Cu, Zn and Mo are considered to be important for higher plants (their number is most likely not final) and Co for lower plants. Some microelements support the growth of plants. Lounamaa (1956) mentioned Li, Be, F, Ti, V, Ni, Br, Sn, I and Pb. Manganese, iron, zinc, copper (cobalt, nickel) take part in oxidation-reduction processes. Aquatic, palustrine plants and conifers (wood and bark) contain more manganese than other plants. It has also been observed that some plants prefer soils rich in a particular element, which phenomenon has been used successfuly in geochemical prospecting for minerals (for references see Sýkora, 1959). Němec and Babička (1934) found out that the increase in toxicity of elements for plant organisms has the following trend: Mg, Al, Mn, Zn, Cd, Tl, Fe, Co, Ni, Pb, Bi, As, Sb, Sn, Cu, Hg, Mg, Pd, Pt and Au, depending on their potential to hydrogen ions and their valence. The higher the valence the stronger is the toxicity. Virtually, all lithospheric elements may be present in plants which they need (even in the least concentration) for the life processes or which they cannot prevent from entering their bodies. The chemistry of plant juices can only affect or partly limit their access. The selective sorption of ions, such as Al by *Lycopodium*, I and K by algae and Au by horse-tails, has not been borne out. The content of elements in plants changes rather with the concentration of elements in the soil of their habitat.

It is of interest that the quantity of elements having a low atomic weight predominates greatly in both the plant body and ash; there is probably a relationship between the permeability of roots and the atomic weight of sorbed elements.

A survey of elements forming part of plant bodies is given in many papers. Detailed data may be found in the summarizing studies of Linstow (1929), Fearon (1947), Lounamaa (1956), Ruhland (1958), and Zýka (1971). The average contents of elements in terrestrial plants and their ashes are listed here in Tables 4 and 5.

Table 4
Average content of elements in terrestrial plants (after Zýka, 1971)

Element	ppm (in dry matter)	Element	ppm (in dry matter)	Element	ppm (in dry matter)
Ca	4 500	Th	6	Sn	0.5
K	3 000	Rb	5.5	As	0.5
Si	1 500	B	5	Te	0.455
Mg	700	Zr	4	Ge	0.422
P	700	Cu	4	I	0.35
S	500	Ti	3.5	Co	0.25
Na	200	V	2.5	Li	0.1
Cl	200	Br	2.5	Ag	0.09
Fe	100	F	2.2	Cd	0.078
Al	35	Y	2.2	Se	0.063
Mn	30	Pb	1.75	Hg	0.019
Ba	30	Cr	1.7	U	0.0033
Cs	22	Nb	1.171	Ra	0.000 000 068
Sr	20	Ni	0.9		
Zn	18	Mo	0.9		

Table 5
Average content of elements in the ash of terrestrial plants (after Zýka, 1971)

Element	ppm	Element	ppm	Element	ppm
Si	150 000	Br	210	Ga	13
Mg	70 000	Rb	182	Yb	13
P	70 000	Cu	152	Li	11
S	50 000	La	130	Co	11
Ca	30 000	Zr	65	F	10
K	30 000	I	50	As	5
Na	20 000	Nb	50	Be	4.1
Al	14 100	V	45	Cd	3.7
Fe	10 000	Y	45	Ag	1.8
Mn	4 500	Ni	42	Se	1.1
Ti	2 100	Cr	40	Hg	0.79
Zn	680	Ge	35	U	0.7
Sr	630	Pb	35	Au	0.007
Ba	430	Sn	25	W	0.00X
B	385	Th	25	Re	0.000 00X
Cl	X00	Mo	19	Ra	0.000 000 2

The following survey presents only the most important data which provide sufficient information on the representation of elements in the plant bodies or ashes. Most records relate to the species that are considered as potential sources of coal material or are remarkable from another point of view.

Ag − silver

The function of silver in the biological sphere is practically unknown. It has been found in plants only in traces. Lounamaa (1956) found small amounts of Ag in the moss group and 1−10 ppm in ashes of conifers. Sporadic higher concentrations of 100−200 ppm were determined in the young growth of *Pinus silvestris* at the Suppuvaara locality (in ashes). The highest concentration in deciduous trees was 10 ppm in ashes.

Al − aluminium

Aluminium is an essential constituent of all soils but since its soluble salts are toxic, the plants resist its penetration. The extractable aluminium occurs in some soils and peat bogs in quite large amounts (such soils are said to exhibit aluminium toxicity). Only *Lycopodium* and the wood and bark of some deciduous trees concentrate Al in larger amounts; the cryptogams tolerate the relatively highest concentrations. Babička (1937) determined the following contents in ashes of lycopods collected in southern Bohemia:

Lycopodium clavatum   −   27.45 % Al_2O_3
Lycopodium annotinum   −   32.72 % Al_2O_3

In his paper there are also further results and a comprehensive bibliography related to this problem. Hutchinson (1945) claims that Al is important only in the biochemistry of woody plants; about 10 % Al_2O_3 has been found in the wood of beech. The whole plants before drying contain only ca. 0.002 % Al (Hutchinson, 1945). Chenery (1948) ascertained accumulated aluminium in 1821 plants, which had more than 0.1 % Al in dry matter of leaves. It probably influences the change of the tree leaves colour. Some plants probably need Al to their growth (they contain about 1 % Al) and therefore prefer more acid soils. Some trees precipitate aluminous salts of succinic acid (indigenous oak in Australia, Lovering, 1959). The sap of Al-accumulating trees shows acid reaction, from 3.9 to 5.2 pH. The tree *Orites excelsa* contains basic aluminium succinate in wood fibres.

In moss ashes there is 4.2 % Al on the average (Shacklette, 1965). Aluminium concentration in swamp waters may be as high as 1 mg/l. Primitive plants consume more aluminium. The presence of Al in humates suggests that the fossilized plant remains in coal may contain considerable amounts of this element.

Alkali metals

According to Bowen (1966) the average amount of lithium in plants is 0.1 ppm, and Li substitutes for K^{1+} and Na^{1+} in metabolic reactions. Sodium and potassium are typical biophile elements. Vernadskiĭ (1925) put forward the idea that the plants exhibit the capacity to separate the isotopes, particularly of potassium, during the vegetation interval. Němec (1947) recorded 22.53 % K_2O in yew ash (the total ash 3.54 %). The growth is impossible without potassium, which is replaced in plant bodies partly by rubidium. The K/Rb ratio in plants is of the order of 1,000. Rubidium has rather toxic effects and concentrates in reproducing organs like potassium. There is about 1,200 ppm Na in plant dry matter. Sodium is one of the elements whose concentration in plants may be higher than in soil solutions. In the ashes of young branches and leaves of trees growing on pit heaps on Kaňk hill near Kutná Hora as much as 400 ppm Rb has been determined (Zýka, 1971).

As − arsenic

Arsenic as a toxic element is little absorbed by plants, but a small amount of it is found commonly in recent plants. Those growing in muddy habitats are richer in As. Also some grasses growing in the neighbourhood of smelting works on arsenic ores have a higher As content. In Anaconda (U.S.A.) 1,550 ppm As was determined in grass dry matter (Linstow, 1929), but the arsenic compounds derived from industrial smoke may have got stuck only on the plant surface. Minguzzi and Naldoni (1950) wrote that the As concentration in tree wood is usually very low, ranging from 11 to 5 ppm. Generally, it does not surpass 2 ppm in plants.

Au — gold

Gold has been revealed in horse-tail ashes. The cumulation of gold in plants has been studied, for example, by Němec, Babička and Oborský (1936). They found 0.061 % Au in the ash of *Equisetum palustre* and 0.0063 % Au in *Equisetum arvense*. *Equisetum palustre* was collected in August 1935 near Oslany in Slovakia, where it grows on soil formed on disintegrated andesite and rhyolite. *Equisetum arvense* came from the locality Hornia Ves, where the soil (on rewashed andesite, tuff and limestone) contains 0.1 ppm Au. In the paper of the above authors (1937) the Au content of 0.11 % is recorded from *Clematis vitalba* and traces of gold from *Salix capna*. The intake of gold by plants is influenced by the composition of soil and, in a high degree, by the SiO_2 content (op. cit.).

Data on the gold content in other plants than horse-tails are given in the studies of Sýkora (1959), Zýka (1971) and Kašpar et al. (1972). Traces of gold have also been identified in the ash of beeches, hornbeams and other trees.

B — boron

Boron is important for the nutrition of plants. Small amounts of boron are present in all plants, even in those growing on soils extremely poor in it. With boron deficiency the shoot apices die away, the fruits do not ripe, and the germination of pollen grains of many plant types is decreased. Lounamaa (1956) recorded the results of analyses made on ashes of a number of plants: mosses contained 300 to 600 ppm B, *Tortella tortuosa* up to 2,000 ppm (dependence on subsoil has not been found); some ferns 300—1,000 ppm; conifers 300—3,000 ppm in the ash of needles and roots. He found the highest B content of 6,000 ppm in *Pinus silvestris* growing on serpentinite at the locality Laajaniemi. The ashes of deciduous trees (poplar, willow, birch, alder) gave 300—3,000 ppm B on the average. Boron is significant for carbohydrate metabolism and the synthesis of ribonucleic acids.

Ba — barium

Barium often appears in plants, although it is slightly poisonous. The physiological function of barium in plants is still unknown. Its largest amount equal to 0.97—1.2 % BaO has been determined in beech wood (Němec, 1948). Goldschmidt (1954) recorded 0.4 % Ba in dry leaves of Brazil-nut tree. Ashes of pine, oak, birch and beech and other deciduous trees gave 0.8—1.0 % Ba. The biochemistry of barium was studied by Bowen (1966) and Puchelt (1967). Bowen (1966) found the average amount of 14 ppm Ba in the dry matter of angiosperms; ferns had 8 ppm and horse-tails 30—4,500 ppm in ashes. The ash of horse-tail was also analysed by Cannon et al. (1968) and Borovik-Romanova (1939), who determined 70—4,500 ppm Ba; this amount corresponded to its concentration in the substratum. Bloss and Steiner (1960) reported a relatively high Ba content (up to 2.30 %) from the ash of oak branches. The ash of conifers yields 10—100 ppm Ba according to Puchelt (1967), and 500—6,200 ppm according to Lotspeich and Markward (1963).

Be — beryllium

The biological role of beryllium is unknown; it appears to accompany aluminium in some plants (Fearon, 1933). It accumulates in plants growing on soils and rocks containing beryl or other beryllium minerals. Its concentration in plants may be four times that in the soil. Dudykina and Semenov (1957) recorded 250 ppm Be in the ash of *Vaccinium myrtillus* from the Khibiny Mts.

Bi — bismuth

Nothing is known of its biological role. It is highly toxic for plants, in which it occurs only in traces.

Ca — calcium

Calcium is an important nutrient of plants. The Ca-deficiency in soil is known to be unfavourable to plant growth. Calcium improves the properties of soil in retaining there the nutrients (particularly the phosphoric acid) and neutralizes various organic acids to form salts (calcium oxalate). The Ca content in plants increases with the age of their organs. According to Rankama and Sahama (1950) the Ca^{2+} ion counterbalances the toxic effects of K^{1+}, Na^{1+} and Mg^{2+}. Calcium is an important component of cell membranes of leaves and vessels. The deficiency of calcium causes the interruption of root growth and yellowing of leaves. Together with potassium Ca mediates the transfer of saccharides in the plant. It removes the acids in neutralizing them to insoluble compounds, which are stored in the plant body. Kletečka (1933) found out that even a very small amount of $CaCO_3$ impairs the growth of sphagnum, whereas $CaSO_4$ does not interfere with it. The author believed that the damaging effect of calcium is paralysed by SO_4^{2-} ion. Němec (1947) recorded 44.72 % CaO in the ash (ash content 3.54 %) of yew tree.

Cd — cadmium

Lounamaa (1956) determined small Cd amounts (1 — 10 ppm in ash) in mosses. Only at the outcrop of an ore vein (containing sphalerite) near Vittinkii (Orijärvi ore district) did *Rhacomitrium lanuginosum* yield 100 ppm Cd. Conifers had a higher Cd content in shoots, maximum 100 — 300 ppm in ash; the amounts below 10 ppm were most frequent. The role of cadmium in plants is unknown.

Cl — chlorine

A very small amount of chlorine is generally present in woods; in the ash of plants 1 — 2 % Cl was found exceptionally. Němec (1948) recorded 6.05 % Cl in the wood of *Arsculus hippocastanum*, and 4.67 % in the wood of *Morus alba*. The content of chlorine in plants varies; it is higher in the vegetation period and in the autumn. Saukov (1950) determined 10 % NaCl in *Salsola hali*.

Co — cobalt

Fearon (1947) gives the Co content in recent plants as 0.02 — 2 ppm. According to Němec and Babička (1934), Co produces toxic effects on plants, causing chlorosis. In its presence the amount of accepted iron in plants decreases. Minute amounts of cobalt have been determined in the wood of oak and in mosses. Lounamaa (1956) recorded maximum Co content in mosses as 60 ppm, in the roots of conifers up to 300 ppm, but in needles only 3 — 100 ppm. In plants on ultrabasic rocks the values are higher. A relatively high concentration was found in plants growing on the nickel deposits in the Urals (mainly in *Pulsatilla patens* whose teratological nickelous forms had more than 0.001 % Co in ash). The Ni/Co ratio in plants usually ranges from 5 to 10.

Cr — chromium

The role of chromium in plants has not yet been fully explained. Pirschle (1939) thinks that plants can contain 0.005 % Cr. Lounamaa (1956) recorded maximum content of Cr in mosses growing on outcrops of ultrabasic rocks as 300 ppm and only 10–30 ppm in those collected on limestone substratum. The same author also gives the following Cr values: up to 1,000 ppm in the roots of ferns, 1–60 ppm in conifers (contents in those growing on ultrabasic rocks are higher), 10 ppm in deciduous trees and 100–300 ppm in those on basic rocks.

Cu — copper

Copper is commonly present in recent plants. For details see the studies of Fearon (1947) and Vogth and Braadlie (1942). Copper supports the growth of plants and has catalytic effects but is toxic in larger amounts. Despite that, some plants are able to concentrate it. Extremely high Cu contents were reported by Prát and Komárek (1934) for *Agrostis alba* (stalks with withered leaves), amounting to 5.2 % in 18.1 % ash, and for *Melandrium silvestre* (green leaves) up to 1.02 % in 18.1 % ash. The plants were growing on pit heaps of the copper mines near the village Piesky in the valley Špania Dolina (Slovakia). The toxicity of copper is probably counterbalanced by the presence of some salts. As a constituent of chlorophyll copper takes part in oxidation and reduction processes; it also plays an important role in the synthesis of the porphyrin nucleus of plant pigments. The poor growth of vegetation on peats and podzol soils is due to copper deficiency (Němec, 1948). According to Prát and Komárek (1934) grasses and caryophyllaceous plants are most tolerant of copper. Němec (1948) recorded 0.06 % Cu in the ash of oak wood. According to Czapek (in Němec, 1948), finely dispersed metallic copper can be found in tree wood. Lounamaa (1956) reported the following results: the average Cu content in mosses 100 to 300 ppm; in *Tortella tortuosa* growing on the residue of rapakivi granite 3,000 ppm and in *Rhacomitrium lanuginosum* from an outcrop in the Orijärvi ore district — up to 6,000 ppm. Mosses on limestone substratum yielded the least values. Ferns — 30 to 300 ppm Cu, some conifers — up to 3,000 ppm; trees and plants growing on soils on ultrabasic rocks had the smallest amount of copper.

F — fluorine

The content of fluorine in plants is generally low, except for tea plants, in which it amounts to 400 ppm. Leaves and needles contain more F than wood and fruits. According to Bowen (1966), the F content in plants varies between 0.5 and 40 ppm, only in industrial areas where there are high fluorine emissions does its content rise 2 to 250 times.

Fe — iron

Iron is an important biophile element supporting the growth of plants. In green plants it co-acts in the formation of chloroplasts, is indispensable for the origin of chlorophyll and induces transfer of oxygen in plant respiration. Babička (1937) determined about 10 % Fe_2O_3 in *Lycopodium* and Němec (1947) 2.18 % Fe_2O_3 in the ash of yew tree (ash content 3.54 %). As far as is known, iron concentrates mainly in the leaves or needles of trees. Terrestrial plants contain 0.3 % Fe in dry matter on the average (Oborn, 1960).

Ga − gallium

Goldschmidt (1954) placed gallium in the association with aluminium as concerns the biochemical processes in plants (probably in view of Hutchinson's studies, 1945). Pelíšek (1940) maintains that gallium is an ubiquitous element in soils and plants, although present in a very small or trace amounts. Lounamaa (1956) stated 3−30 ppm Ga in mosses growing on siliceous rocks; maximum 10 ppm in conifers; and 60 ppm in the ash of the lichen *Stereocaulon pascale*. No gallium has been found in plants growing on soils on the ultrabasic and limestone bedrock. According to Zahradník et al. (1959), gallium is often present in moulds (e.g. *Aspergillus niger*) and in some palustrine plants. The terrestrial plants show a tendency to concentrate Ga in larger amounts than is its content in soil (Shacklette, 1965).

Ge − germanium

The role of germanium in plant life is not yet well understood. Geilmann and Brünger (1935) have shown experimentally that barley, oat, mustard and buckwheat readily absorb germanium from the soil, particularly oat. Germanium was contained both in shoots and roots of the plants. The Ge percentage in plant ashes is relatively high, in oat up to 1 %. The plant extracted 75 % Ge from very diluted soil solutions and about 50 % from concentrated solutions. Higher contents of germanium had toxic effects. With Ge concentration of 5 ppm in soil, about 50 % accumulates in the plant. Kudělásek (1960) proved experimentally that already 10 ppm Ge in plant is fatal for *Equisetum arvense*, *Equisetum hyemale*, *Picea excelsa* and *Secale cereale*.

On the ash heaps near Hannover some grasses and flowers contained 1−4 mg Ge in 5 g of dry matter and poplar leaves 1−2 mg Ge. It is possible that germanium was absorbed by plants as a result of the analogous properties of Ge and Si, which is usually abundant in horse-tails (Rotter, 1952).

All plants do not take up Ge to the same extent. The study of Švasta et al. (1955) gave the following results:

(1) plants are capable of absorbing Ge from soil;
(2) at a low Ge concentration the plants may cumulate 50−70 % of it;
(3) higher concentrations of Ge are harmful to plants, causing chlorosis;
(4) plants are capable of taking Ge also from coal ash but in very small quantities;
(5) some plants, particularly lower and aquatic plants and moulds, have a higher capacity of absorbing Ge and tolerate higher concentrations.

It can be said in general that the amounts of germanium in coal cannot be derived from plants or only from plants, but are the result of later processes. Zýka (1971) recorded maximum 200 ppm Ge in the ash of plants from the ore district in Saxony (G.D.R.).

Hf − hafnium

Zýka (1971) found 0.92 ppm Hf in the ash of young shoots of heather growing on pit heaps at Kaňk near Kutná Hora, and only 0.18 ppm Hf in the ash of birch leaves from Turkaňk. The shrubby birch grew on a stone dump.

Hg − mercury

The role of mercury in plant biology is little known. It probably acts as a catalyst and is sorbed by many plants from the soil. If the soil contains less than 1 ppm Hg, all vegetation concentrates it; if, however, the soil contains tens or hundreds ppm Hg, its contents in plants are lower than in the soil.

The ash of some plants growing on polymetallic or mercury deposits may yield up to 100 ppm Hg. Plants have been found on Hg-rich soils whose organs (e.g. leaves) contained drops of metallic mercury (Rankama and Sahama, 1952).

I — iodine

Iodine is a frequent component of plant organism; it is concentrated primarily in marine and then in palustrine vegetation; the accumulation is rather passive than functional. Iodine is dissolved in cell sap, particularly of young plants, supporting the formation of green matter (Stoklasa and Bareš, 1926). Iodine is present chiefly in sea-weeds (e.g. *Laminaria digitata*). Marine algae concentrate both bromine and iodine. Zýka (1971) found 360–410 ppm iodine in the dry matter of *Sphagnum sp.*

Mg — magnesium

Magnesium is common in plants, forming the essential part of chlorophyll; it is also dissolved in nucleoplasm. The ash of chlorophyll contained 4.5 % MgO (Prát, 1932). It is known to act as a catalyst in photosynthesis. Němec (1947) determined 10.34 % MgO in the ash of yew (total ash content 3.54 %).

Mn — manganese

Manganese is a common element in plants, which absorb it most easily in an environment having pH < 6. Fearon (1947) gave 0.0001–0.02 % as absolute values for Mn in recent plants. Manganese is important for the growth of plants, has a share in reduction of nitrates and in the catalysis of some enzymatic processes, but is toxic in larger amounts. The presence of manganese in conifers was studied by Němec (1948). He determined 0.71–0.102 % Mn in the seeds of pines, 0.067–0.0919 % Mn in the seeds of spruce and 0.07 % Mn in *Pinus strobus* (growing on acid soil), 0.019 % Mn in the seeds of *Pinus nigra* growing on calcareous soil, and 0.016–0.039 % Mn in fir seeds. On the whole, the content of Mn is higher in coniferous than in deciduous trees and in wood higher than in seeds. Němec cited the highest Mn amount of 10–18.36 % Mn_3O_4 in the ash of birch wood and 5.2 % Mn_3O_4 in oak wood. A strikingly high Mn amount was found in hornbeam wood. Moore (1940) recorded 25.53 % Mn_3O_4 in the ash from needles of the Norwegian spruce and 41.23 % Mn_3O_4 in the ash from the bark of the same tree. There is an interrelationship between manganese and calcium: the higher Ca content, the smaller is the amount of manganese and vice versa. Lounamaa (1956) reported the following observation results: mosses growing on granite residual soil contained up to 6,000 ppm Mn; the roots of ferns as much as 30,000 ppm Mn; the Mn content in conifers varied between 1,000 and 30,000 ppm, being higher in needles than in roots; deciduous trees yielded 1,000–30,000 ppm Mn and the amounts were particularly high in leaves. The plants growing on calcareous soils contained least manganese.

The function of Mn is not known precisely; its deficiency causes yellow spots on leaves.

Mo — molybdenum

The presence of molybdenum in plant bodies is common; it is more abundant in some species of rushes, fungi and particularly in leguminous plants. Fearon (1947) gave the average Mo content for all plants as high as 9 ppm. According to Zázvorka (1947), molybdenum has favourable effects on the development of soil microflora. It is concentrated mainly in bacteroid tubercles on the roots of leguminous plants. In affecting the soil bacteria that assimilate nitrogen, it indirectly influences the higher plants.

Molybdenum functions most probably as catalyst but is toxic in larger amounts. Vinogradova (1954) recorded higher contents of molybdenum in the seeds of leguminous plants. She determined 4 ppm Mo in the seeds of woody plants. Lounamaa (1956) reported 10–30 ppm Mo in mosses, sporadically up to 200 ppm; less than 10 ppm Mo in conifers, sporadically up to 60 ppm; in the resins of conifers up to 800 ppm, which corresponds broadly to the Mo content in amber (700 ppm).

Ni — nickel

The plants contain very small amounts of nickel. Fearon (1947) determined its average content in plants as 1.5–3.3 ppm. Němec and Babička (1934) noted a high toxicity of Ni for plants when present in larger amounts. Its function in plant organism is unknown. A high Ni content, nearly 0.04 % in ash, was found in *Pulsatilla patens*. Němec (1948) assessed traces of this element in oak wood. Lounamaa (1956) determined 30–100 ppm Ni in mosses growing in a granite terrain and 1,000–6,000 ppm in mosses on ultrabasic rocks (*Campylium polygonum* had the largest Ni amount). As much as 10,000 ppm Ni was found in the ash of the fern *Asplenium viride*. Analogous values have been reported from the ashes of some plants growing on ultrabasic rocks in the Southern Urals (Zýka, 1971). Lounamaa found more than 1,000 ppm Ni in conifers, especially in their needles, and remarked that plants growing on calcareous substratum contain less Ni than those in granitoid or serpentinite terrains. Zýka quoted (1971) the Ni contents of 36.0 ppm in the ashes of *Pinus silvestris* growing on podzol at the locality Ostankino, and 620 ppm at loc. Tyulenevski Mine (U.S.S.R.).

P — phosphorus

Phosphorus is a very important biogenic element; it is a component of cytoplasm and influences the growth of plants. Němec (1947) determined 12.85 % P_2O_5 in the ash of yew (total ash 3.54 %) and 10,124 ppm in the ash of *Equisetum arvense* from Oslany (Němec et al., 1936). Phosphorus is generally present in plant ash in the form of calcium phosphate. Moore (1940) reported a high P content in spores (0.078–0.0228 %) and 0.288 % P in the pollen grains of *Ceratozamia mexicana*.

Pb — lead

Lead has been identified in ashes of many terrestrial and marine plants. It concentrates in plants wherever it is excessive in soil although it is poisonous, but usually precipitates right in the roots. Vernadskii (1929) assumed that plants might contain up to 0.001 % Pb. Švasta et al. (1955) refer to the paper of Emmerling and Kolkowitz (without any further citation), in which the following values are recorded: 0.48 % Pb from the trunk of *Salix viminalis* and 1.73 % Pb from its roots, and 0.13 % Pb from spruce wood. These plants were growing on soil containing galena. Lounamaa (1956) found 100 to 300 ppm Pb in mosses growing on siliceous rocks and on the average 100–600 ppm Pb in conifers. The young shoots of conifers, however, contained up to 3,000 ppm Pb. In the roots of willows growing on pit heaps of the lead mines in the Harz Mts. 1.73 % Pb was found (in ash) and in grasses from the same place 3.56 % Pb (in ashes). Lead is also present in discernible amounts in wood, e.g. in the ash of spruce 0.13 % was determined (Sýkora, 1959). In general, the plants do not absorb much of this poisonous metal, which most frequently concentrates on the surface of roots in soils rather rich in lead.

Re — rhenium

The function of rhenium in the organic cycle is little known. Myers and Hamilton (1961) found 300 ppm Re in ashes of plants growing on U- and Mo-rich soils. Shacklette (1965) analysed many samples of cryptogams but determined Re in a single sample in the amount of 1,500 ppm.

S — sulphur

Sulphur is concentrated mainly by sulphur bacteria (e.g. *Baggiatoa, Thiothrix*) and is involved in metabolic processes (e.g. in the genus *Dessulfovibrium*). Sulphur is a constituent of proteins and is thus invariably present in plants. Němec (1947) assessed 2.29 % SO_3 in the ash of yew tree (total ash content 3.54 %). High S contents occur in horse-tail ashes. Zýka (1971) reported 36,380 ppm in *Equisetum telmateia* (Ehrh.) and 27,560 ppm in *Equisetum arvense*.

Sb — antimony

The biological role of antimony has not yet been recognized. Trace amounts have been found in some plants. Shacklette (1965) related up to 50 ppm Sb in the ash from the branches of *Betula resinifera* rooted in a mineralized zone.

Sc — scandium

Scandium is concentrated in plants only locally and in very small to trace amounts. Dobrolyubskiĭ (1962) wrote that Sc takes part in oxidation-reduction processes in plants. Maximum Sc values in ash of plants are in the range of 10–15 ppm. Zýka (1971) reported 0.97–3.4 ppm Sc in the ash of young branches and leaves of the trees growing on pit heaps at Kaňk near Kutná Hora.

Se — selenium

Selenium has been found in some plants (e.g. *Astragalus*) on seleniferous soils of the Western United States and Canada. Some plants tolerate Se in soil to a certain degree and act as Se colectors. Vinogradov (1965) recorded up to 0.31 % Se in a dry plant. Zýka (1971) reported 70 ppm Se in the ash of young branches and leaves of birch and lime growing on the pit heaps at Kaňk near Kutná Hora.

Si — silicon

Silicon occurs in plants mainly in the form of silicic acid, which impregnates the cell membranes. It makes the essential part of the ash of horse-tails and related plants; it may amount to more than 80 %. Grasses also contain an increased content of SiO_2. Němec (1947) reported 5.08 % SiO_2 in the ash of yew tree (total ash content 3.54 %). Opal was found in the fibres of many *Tracheophytes*. Moore (1940) related the amount of up to 12 % SiO_2 in the stems of horse-tails. According to Lovering (1959), tropical plants accumulate the largest amount of SiO_2. Organic Si complexes have not yet been identified.

Sn — tin

Our knowledge of the biological function of tin is meagre. The average Sn content in recent plants ranges between 0.001 and 0.1 % (Pirschle, 1939). Lounamaa (1956) determined 10–60 ppm Sn in mosses, in largest concentrations in the Vittinki ore district. Only some samples of conifers yielded tin. Fresh shoots of *Pinus silvestris* and *Picea alba* contained 10–100 ppm Sn. Sarosiek and Klys (1962) found as much as 46 ppm Sn in the ash of some plants.

Sr — strontium

Strontium is present almost in every plant. According to Rankama and Sahama (1950), it is associated in plants with calcium and has no specific physiological function. The Sr amount in plants rises with its content in soil. Linstow (1929) reported that beet grown on fields fertilized with Sr-containing lime yielded up to 170 ppm Sr in dry matter (0.0206 % in ash) and clover even 10,000 ppm in dry matter. The plants growing on celestite-bearing marls also contained increased strontium.

Ta — tantalum

The contents of tantalum in plants are very low. The ash of fern from the area of a tantalum ore deposit gave 8 ppm Ta and *Equisetum silvaticum* hundredths of one per cent in ash (Zýka, 1971).

Tb — terbium

Robinson et al. (1957) found 18.2 ppm Tb in the dry matter of leaves of the hickory tree (*Carya alba*).

Th — thorium

The function of this element in plants is unknown. The Th content in the ash of tree branches and leaves is in the range of 30–60 ppm. Exceptionally, as much as 1,000 ppm Th was found in the ash of *Populus tremula* growing above a pegmatite deposit rich in REE and Th in eastern Siberia (cited by Zýka, 1971).

Ti — titanium

Titanium being a common element in most rocks also appears in the ash of almost all plants. Němec and Kás (?) studied the function of Ti in plant life; they came to the conclusion that Ti concentrates in assimilation organs and it can replace iron as oxidation catalyst in plant cells. A direct proportionality exists between the Ti content in soil and in plant ash. Babička (1937) proved high Ti contents in lycopods: 1.20 % TiO_2 in *Lycopodium clavatum* and 1.18 % TiO_2 in *Lycopodium annotinum*. In the book "Gmelin Handbuch", part Titanium, the Ti content in recent plants is given as 0,001 %. Němec et al. (1936) found Ti especially in *Leucoium aestivum* in the amount of 4.55 % TiO_2 in ash.

Tl — thallium

Thallium was found in the ash of terrestrial plants (0.5–10 ppm) but its function has not been recognized.

Tm — thulium

Robinson et al. (1957) determined 6.4 ppm Tm in the dry matter of hickory leaves.

U — uranium

The plants contain 12–32.9 ppm U (Rankama and Sahama, 1950). Sýkora (1959) determined 6 ppm U in the ash of birch wood. It seeems that uranium is not indispensable for the life of plants. Zýka (1971) cited the results of analyses for U in various plants (300–600 ppm in ash). The plant roots generally contain more uranium than other plant organs. The ash of leaves and branches of trees growing on pit heaps of the Kaňk deposit near Kutná Hora contained up to 31 ppm U (Zýka, 1971).

V — vanadium

Although the V contents in recent plants are often quite considerable, it is not reliably known whether it is necessary for the life of plants. Vanadium is most probably assimilated by plants in the form of VO_4^{3-}, which serves as a source of energy. Its concentration depends on its content in soil. Zázvorka (1947) related vanadium to be favourable to the development of soil microflora. In affecting N assimilating bacteria it exerts indirect influence on the development of higher plants. The optimum content for plants is relatively low, in larger amounts it is toxic.

W — tungsten

The biochemical function of tungsten is practically unknown. Dekatès (1967) reported that plants growing on mineralized substratum had 2 to 18 times the W amount of the substratum (2.7 ppm). The increased W contents in coal (see in further text) may thus be interpreted in terms of concentration by palustrine vegetation or of sorption on decayed plant material.

Y — yttrium and rare earth elements

Lounamaa (1956) recorded the following values of Y contents in mosses: *Tortella tortuosa* growing on pegmatitic rapakivi granite, sensitive to Y, contained 300 ppm of this element; mosses on ultrabasic rocks had up to 10 ppm Y, which was present also in plants on calcareous rocks. In conifers he found rarely small amounts of yttrium (10–60 ppm) and less than 10 ppm in deciduous trees. Robinson et al. (1957) identified in the hickory leaves Y, La, Ce, Pr, Nd, Sm, Eu, Gd, Tb, Dy, Ho, Er, Tm, Yb and Lu. Forty to ≈ 60 ppm $\sum Y + La-Lu$ was found in the tree trunks and 981 ppm $\sum Y + La-Lu$ in leaves of a hickory tree. Relative to the contents in soil, a hickory tree does not induce a marked fractionation of lanthanides. Pappas et al. (1963) found 30–150 ppm $\sum Y + La-Lu$ in the opium ash of various derivation. The origin of opium is determined from the distribution diagram of REE in opium ashes.

Zn — zinc

Zinc is an important element of the biosphere; in small amounts it supports the growth of plants, but is poisonous in larger amounts. Rankama and Sahama (1950) recorded 16 % Zn in the ash of *Thlaspi calaminarium*. In Sýkora (1959), the following results are cited: *Viola calaminaria* up to 13.12 % Zn in the ash, *Thlaspi calaminarium* (the whole plant) 21.30 %, *Silene inflata* and *Armeria hallerii* as much as 13 % ZnO. Photoassimilation of CO_2 is hindered by zinc deficiency. Zázvorka (1947) stated defects in the formation of chlorophyll due to the lack of zinc. Lounamaa (1956) reported 1,000–3,000 ppm Zn in mosses, in ore fields up to 10,000 ppm Zn; 100–600 ppm in ferns, in the needles of *Pinus silvestris* even 6,000 ppm Zn. Of the cultivated plants pea contains considerable amounts of zinc in the ash. In the plant dry matter zinc contents are generally in the range of 50 to 1,000 ppm. Zinc-organic complexes have been observed in plants but they have not been characterized as yet. Zýka (1971) found maximum 14,000 ppm Zn in the ash of leaves and young branches of trees growing on old pit heaps at Kaňk near Kutná Hora.

Zr — zirconium

The biological function of zirconium is not known satisfactorily. According to Pirschle (1939), Zr occurs in plants in amounts not exceeding 0.001 %. Lounamaa (1956) determined an average content of 10–30 ppm Zr in mosses, with the least amounts in moss growing on ultrabasic rocks. Conifers were found to contain 10–100 ppm Zr. Only the samples of moss (*Rhacomitrium lanuginosum*) growing on the deposit outcrop near Orijärvi and Vittinki gave 60–100 ppm Zr.

Some other elements are known to be present in plants but there are no detailed data available concerning their function; in most cases they were only qualitatively determined. From the above survey some general conclusions can be drawn. The plants can concentrate a great variety of elements in amounts varying from traces to several per cent. The presence and quantity of elements in plants are mainly controlled by their amounts in soils which provide the nutrients to the vegetation. Some plants, however, require a particular element, which they concentrate even where it is scarce in the soil. The deficiency in such element may cause a disease or decay of plants. It has been ascertained that in the course of time a plant may get used to higher concentrations of an element which is normally toxic in larger amounts.

The elements that are present in plants may be expected to occur also in coal ash, even if we know that also other factors took part in the concentration of elements in coal. Since, however, the coal derived from plant material, the chemistry of vegetation must be regarded as a relevant parameter in coal composition.

COMPARISON OF THE CHEMICAL COMPOSITION OF FOSSIL AND RECENT PLANTS

The comparison of fossil and recent plants reveals some differences in the content of organic substances.

The alkane fraction, for example, which is present in the fossil *Voltzia brongniarti* (Triassic) in the amount 0.01 %, contains *n*-octacosane. Although *Voltzia* is extinct and cannot be correlated with the recent form, the essential component of the cuticular wax of leaves of the recent *Taxodium distichum* Rich. is *n*-octacosanol (Knoche et al., 1968). The hydrocarbon n-$C_{28}H_{58}$ in the fossil *Voltzia* formed most probably by the reduction of primary alcohols n-$C_{27}H_{55}$.CH_2OH. Knoche and Ourisson (1967) compared the distribution of *n*-alkanes in extinct horse-tails (Triassic *Equisetum brongniarti*) with that in present-day common *Equisetum silvaticum* and identified the same odd-numbered alkanes n-$C_{23}H_{48}$, n-$C_{25}H_{52}$, n-$C_{27}H_{56}$ and n-$C_{29}H_{60}$ in both of them. It is remarkable that the determined hydrocarbons have remained unchanged over the period of some 200 million years.

The study of saccharides in Palaeozoic plants (Swain et al., 1967, 1968) has shown that the principal saccharides of the plants examined were glucose and galactose. They were derived allegedly from cellulose or starch. Xylose and small amounts of mannose, rhamnose and arabinose have also been identified. Arabinose, which is an important constituent of present ferns, has not been found in their fossil forms, but in the genus *Cordaites* (gymnosperm). Glucose is the principal saccharide in most Palaeozoic plants and in all recent ones, with a few exceptions. The predominant saccharide in the Upper Devonian primitive *Archaeopteris* (*Callixylon sp.*) was galactose, which may imply that galactans were the main constituents of the cell walls of this plant. The types and ratios of monosaccharides in the fossil *Psilophytales* and *Lycopodiales* (Devonian – Carboniferous) resemble those in modern ferns.

The fossil *Calamites* and some *Equisetales* contained more galactose than glucose, which again suggests that galactans were probably the main structural polysaccharides in those primitive forms in contrast to modern horse-tails having glucose as the principal saccharide. The major amount of galactose in fossil than in present horse-tails implies an important change from galactan polysaccharides to glucans of recent horse-tails. Galactans were more likely the principal structural polysaccharides of fossil forms than cellulose, which is the dominant polysaccharide of recent horse-tails (Swain, 1970). Some forms of the *Lepidodendron* and *Lycopodiales* also contained rather galactans than glucans in important amounts. The samples of the pteridosperm *Alethopteris* and the horse-tail *Annularia* from Radstock in England (Upper Carboniferous) contain a sufficient amount of galactose and mannose (Swain et al., 1969) to allow us to presume that galactans and probably mannans were constituents of the polysaccharides in the cell walls of these extinct plants. The shift from galactan structural types of polysaccharides (typical of algae) to cellulose was evidently caused by evolutionary changes such as are known in fossil *Cordaites* to have occurred between the Devonian and Late Carboniferous. The saccharides of Palaeozoic plants resembled at least structurally the polysaccharides of recent plants. The differences in their proportionate amounts are still more conspicuous if we realize that the higher forms of present terrestrial plants are also rich in xylans, which have not been identified in the Palaeozoic plants, except for small amounts of xylose (Swain et al., 1967). In three species of the Palaeozoic plants small amounts of linear α-1$\rightarrow$4D-glucopyranose and linear β-1$\rightarrow$4D-glucopyranose and β-1$\rightarrow$3 linked polysaccharide of laminaribiose type have been identified (Swain et al., 1969). In assessing the saccharide content in fossil plant remains it must not be forgotten that after the death of plants and during the fossilization process the total amount of saccharides is decreased depending on the character of the plant material and of the environment. The neutral to slightly alkaline medium is more suitable for their preservation, despite the fact that individual saccharides are less stable in alkaline than in acid solutions. In polymeric form, however, the saccharides in sediments are somewhat more stable in alkaline than in acid conditions. The negative Eh values characteristic of sediments showing an alkaline reaction are responsible for producing the reducing conditions which contribute to the preservation of saccharides.

It can be said that in the genetic series of plants the contents of cellulose and lignin increase gradually and the amount of soluble saccharides, pectins and proteins decreases. Bacteria and algae are richest in proteins. Microbes and algae have a high content of fats; waxes are relatively widely distributed and the true resins appear only in the highest plants. This change in the composition of the primary organic substances also depends on the stratigraphic age and facies changes that occurred during the formation of coal. The oldest natural caustobioliths formed from the decayed remains of extinct primitive organisms and consist predominantly of kerogen and bitumen. Only towards the end of the Devonian had the terrestrial sporiferous plants developed to such an extent that coal seams of a great thickness could have formed.

The chemistry of lignin was changing with the evolution from lower to higher plants (see Manskaya and Drozdova, 1964; Manskaya and Kodina, 1975). Lignin of the present-day lycopods and horse-tails can not be compared to lignin of their Palaeozoic ancestors. Lignin of sphagnum differs in composition from that of higher plants in the so-called methoxyl number or the n-hydroxyphenol substances. Lignin in gymnosperms contains $14-16\%$ methoxyl but in angiosperms $20-22\%$. There is also a difference in plant cutins. Fossil plant cutins are widely distributed in sediments and their morphology is often preserved, but their chemical reaction differs essentially from the reaction of recent plant cutins. The latter are hydrolyzable to hydroxy acids in alkaline conditions, whilst the former are stable in alkaline hydrolysis even if effective means are used. There is also a difference in their infrared spectra. The infrared spectra of recent plant cutins exhibit absorption of a carboxyl group in ester, which is lacking in spectra of fossil cutins.

DESTRUCTION AND DECOMPOSITION OF PLANT MATTER

The processes to which the plants are exposed after death are divisible into four groups: (a) rotting, (b) mouldering, (c) putrefaction, (d) peatification.

All of them occur at the participation of aerobic or anaerobic bacteria, fungi, larvae, worms, various enzymes and under partial action of water, mineral admixtures and gases (the processes can be compared broadly to fermentation). Aerobic microorganisms are able to exist only at the presence of free oxygen. Anaerobic microorganisms take the necessary oxygen from chemical compounds. Hydrogen sulphide is toxic for most microbes — saprophytes; they thrive in an alkaline medium produced by ammonia, which was derived from proteins.

The decay of plant material (destruction of biochemical macromolecules) usually occurs first by the action of aerobic agents at the access of air. Subsequently, anaerobic microorganisms (mainly bacteria) continue decomposition, particularly when the access of air was limited or interrupted altogether. The burial of material leads to the change of Eh from $+0.05$ to -0.5 (as a result of anaerobic decomposition) within the depth range of 0 to 130 m, which contributes to the change in the quality of coal.

In free air the organic plant remains decompose completely. This process is actually perfect oxidation and is called rotting. The main end products are CO_2 and water. Only resins and waxes remain lying unchanged on the surface. Their accumulation and concentration may give rise to a special genetic group of caustobioliths — liptobioliths.

Mouldering occurs under insufficient air access, where the plant material was saturated with water and compressed. The main end products are CO_2 and water. The organic matter is not fully decomposed and a solid carbon-rich residue remains. It may be found accumulated in humid forests as a dark humic matter. The plant remains that did not succumb to decay are, in addition to mineral components of the plant body, resins, waxes, suberin, cutin and other more resistant substances. A similar process to mouldering but shorter and more intensive, is imperfect burning, e.g. coalification of plant tissues in forest fires (fusinitisation).

Humus is the product of a slow decay of organic substances in humid areas. At the presence of the organic substances of humic soils a fractionated extraction takes place removing all soluble compounds, particularly the salts of alkali metals, Mg and Ca salts and anions of some strong acids. Trace elements are trapped in the humus layer, probably by their bonding on tannin and humic acids (Fuchs, 1935). In this way some elements concentrate in the soil of old forests and are so toxic for plant organisms that the undergrowth with shallow roots cannot vegetate. This holds true particularly for arsenic. According to Olsen and Bray (in Baumeister, 1952) the average sorption capacity of humus is higher than that of natural colloids of the montmorillonite group. The numerous trace elements that occur in humus also appear in coal. The circulation of elements in soil is controlled by the solubility of their compounds. The soil solutions contain a major amount of the ions of strong bases and acids: Na^{1+}, Mg^{2+}, Ca^{2+}, Mn^{2+}, Fe^{2+}, Cl^{1-}, SO_4^{2-}, $H_2BO_3^{1-}$, CO_3^{2-} and NO_3^{1-}.

Putrefaction is a predominantly anaerobic process which decomposes dead bodies of plant and faunal organisms in the presence of water, without air. The environment generally has negative Eh and pH $7-9$. In the moving water the upper 20 to 30 cm of sediment has Eh + and pH <7. The alkalinity is due to the accumulation of ammonia from decomposing proteins. This process involves decomposition of nitrogen-free substances, fermentation, and decomposition of nitrogenous substances (proteins), which is putrefaction in the stricter sense. Putrefaction occurs in stagnant, insufficiently aerated waters. Although the access of oxygen is inhibited or strongly restricted, the organic substances undergo interior decomposition using their own oxygen; this process is analogous to slow distillation. The main gas products are methane CH_4, ammonia NH_3, N_2O, N_2, H_2S, and CO_2. Methane forms chiefly by decomposition of the saccharides of plant tissues. The remaining part consists of liquid or solid bituminous substances, relatively rich in hydrogen. Potonié (1910) termed this entire decomposing process the bitumenization. In addition to bitumen soluble in normal organic solvents also kerogen, insoluble organic matter, is formed. Putrefaction is a process to which succumb mainly aquatic organisms such as dead plankton, microplankton, algae, small crustaceans, insects, drifted spores and pollen. They are overlaid by organic ooze — sapropel, formed by putrefaction of proteins. The zone of sapropel formation is often called the H_2S zone. Sapropel is rich in C as is the peat but in contrast to the true peat it contains a relatively large percentage of H and N. The recent sapropel showing nearly no structural organic remains under a microscope is called saprocol. The sapropelic sediments form a separate group of caustobioliths called sapropelites, which may be rich in spores and pollen grains (cannels) or oleaginous algae (bogheads). The Tertiary solid saprodil and Carboniferous solid sapanthracon are further members of the sapropelic series. The C/N ratio in recent sapropelic sediments is lower than 3 and amounts to $20-30$ in fossil sapropelites. The activity of bacteria depends on the C/N ratio. At the boundary of the nil redox potential sapropel cumulates uranium and lead; Zn is cumulated at lower and Cu, Ag, V, Mo and Ni at higher negative values of redox potential. Sapropelic sediments have often an increased amount of Sn.

At the base of peat bogs and their lateral transitions appear sediments that are a mixture of peat, sapropel (saprocol) and inorganic substances. Sediments containing 20–50% organic matter are called peat-clays or sapropel-(saprocol-)clay. Gyttja is an internationally used Swedish term for clay containing 25% sapropel. The C/N ratio of recent gyttja is <8, that of the fossil analogue 50 to 350. Manganese concentrates at a higher Eh. An increased chromium content is another characteristic feature of gyttja.

Of interest are also the ratios of some elements as given by Krejci-Graf (1972). The V/Mo ratio is >5 in sapropel and <5 in gyttja. The V/Cr ratio in gyttja is about 1, in sapropel >1. The V/Ni ratio is about 1 in gyttja and 2–10 in sapropel. The Cr/Ni ratio in gyttja is >1 and <1 in sapropelites. The Ni/Co ratio is near to 1 in gyttja.

Putrefaction is very often associated with pyritization of overlying and/or underlying beds. The escaping H_2S reacts with infiltrating Fe^{2+}-bearing solutions to form FeS_2. The formation of H_2S indicates a strongly reducing environment; ferric ions are reduced in peaty medium by bacterial activity to ferrous ions. Pyrite is probably not formed directly; Berner's (1964) experiments have shown that at 20 °C and pH 7–9 a poorly crystallized black ferrous sulphide appears first. It crystallizes in time until it is discernible by X-rays. The black sulphide is either mackinawite (FeS) or greigite (Fe_3S_4). With increasing depth of burial it recrystallizes to more stable pyrite. The excessive H_2S is absorbed partly by water which percolates into underlying beds, where it produces the same reaction as the gaseous hydrogen sulphide in the overlying beds.

The difference between humic and sapropelic substances is well seen from the above facts. Sapropel forms in stagnant water, humus in mud or on the ground. Sapropel is derived almost exclusively from aquatic organisms, humus from terrestrial plants; whereas sapropel is the product of putrefaction, humus is produced by mouldering and peatification.

The last type of biochemical decomposition of the plant material is peatification, which is a coal-forming process. It occurs below the water level in the region of aquatic and subaquatic growths, i.e. in places with a partial access of air. Palustrine plants, for example, are rooted in water. If during their withering the access of air is not interrupted, they undergo the mouldering process. The younger generation that grows immediately on the dead plants prevents the access of air to the atrophied plant parts until the medium is prepared for the following process — putrefaction. The gaseous products of decomposition are CO_2, H_2O, CH_4 and less NH_3.

The absence of oxygen and the newly formed organic acids retard further microbacterial degradation and the organic substances are conserved. A minor product of the humification of plant remains is ammonia, which neutralizes humic acids reducing thus their toxicity. In this way favourable conditions for intensive decomposition are renewed. Neutralization of humic acids by calcium has similar effects.

Generally it holds good that the less oxygen and the larger molecule has the plant, the more resistant it is to decomposition in a peat bog. The chemical stability of substances of the higher plants rises approximately in the following order: saccharides and pectins, starch, hemicellulose, cellulose, lignin, resins, higher fatty acids, betulin and cutin, waxes and sporopollenins. The spore exines are most stable. Resins are stable biochemically but because they contain unsaturated groups they show a tendency to polymerization and condensation.

Peat, mineral components and liptobioliths remain as the undecomposed residue.

Should the peatification result in coal formation, the burial of plant material must proceed rapidly. Therefore it is understandable that the largest coal deposits developed in littoral swampy areas, which meet this precondition. A rapid burial of the organic matter prevented its decomposition by fungi and other aerobic organisms. Phenols, terpenes and the remaining impregnating substances retard a rapid microbiological decay, which leads to the preservation of lignin.

Generally, the mineral compounds of plant bodies and resistant substances of resin and wax character — liptobioliths — are preserved, solidifying during the coalification process. An example of a recent liptobiolith is provided by the South African plant *Sarcocaulon*, which is protected from oppressive heat and parching winds by a wax sheath, which burns when ignited. The wax remains of the dead plants may be transported and heaped up by wind or water and thus give rise to a liptobiolith deposit. On the basis of their nature, liptobioliths are divided into the resinous, suberin, spore, and cuticle groups (Havlena, 1963).

As peatification is the most important process in coal formation, its role in the decomposition of the individual components of plant organisms will be discussed in more detail.

The peat stage

Peatification is above all a biochemical process, involving hydrolysis, oxidation and reduction; the end product is peat.

The number of essential structural substances of the plant body is increased in peatification by a new component — the humic acids. Swain (1970) summed up their formula as $C_{68}H_{53}O_4OCH_3(OH)_4(COOH)_4$ (?) with 57.15% C and 4.43% H. The remainder to 100% falls to oxygen (without ash and moisture). Francis (1954) gave the average composition of humic acids as follows: C = 56.5%, H = 5.5%, O = 36.0% and N = 2.0%. Fungi, microfauna and bacteria are doubtless active in the humification process. Peatification is governed by a number of factors such as the reaction of environment (pH <7), the oxidation-reduction potential (Eh +), access of air, activity of microorganisms and, to a lesser or even negligible extent, temperature and pressure. The destruction processes — putrefaction, mouldering and peatification — combine in different ratios to give humolith as the end product. The character of humolith depends, among others, also on the stage of decomposition and on the type of the most active process.

The movement of water in peat bogs occurs in several ways. During capillary movement the elements leached from the rock substratum are transferred into the upper part of the peat bog. On the contrary, percolating water impoverishes the upper layer of peat in nitrogen and organic products of peatification, which concentrate at the base of the peat bog or in the underlying rocks. In deltaic peats and fens, humic acids are washed out and removed by slowly flowing water. As a result of these processes the medium of eluviation is poor in humic substances, which are deposited in calm parts of the depression as a flocculate precipitate, labelled by the Swedish term "dy". It is clearly a subaqueous deposit, in fact a colloidal humic matter.

During decomposition the principal components of the plant matter (cellulose and lignin) swell in the water environment, lose their cellular structure and are gradually broken down into aggregates of molecules (molecular dispersion). The aggregates may combine to form colloids or separate into molecules, which associate into other aggregates and these in turn into new colloids. The process is usually dependent on the content of dissolved solids in water. Many colloids, however, are unstable at the contact with water and disintegrate into smaller particles called micelles, which subsequently may form a water solution. This process, essentially dissolution of colloids is termed peptization. Molecular dispersion and association are sometimes denoted by a common term gelification. It takes place as in peat bogs so during coalification, and hydrolysis is part of it. The moisture content of the medium, its acidity and dissolved-solids content are very important for the aerobic and anaerobic organisms involved in decomposition. The degree of anaerobic process is also controlled by the level of ground water.

Peats invariably contain besides partly decomposed matter (saccharides, lignin, proteins) and undecomposed structural substances (waxes, resins) also a humic component formed by hydrosols and gels, which were formed mainly by direct chemical processes. The proportion of the humic component controls the rate and type of chemical reactions to some extent. Its chemical effect is most intensive towards the end of the biochemical phase, when the peat is already covered with sedimentary material but the endogenous agents (temperature and pressure) are not yet active.

Cellulose and proteins decompose most readily of all substances present primarily in plant organisms. Cellulose is either hydrolysed by strong mineral acids or decomposed by the aerobic and anaerobic bacteria (*Bacillus cellulosae methanicus* and *Bacillus cellulosae hydrogenicus* are best known), mycobacteria and fungi. Fungi are active mainly in a medium with pH lower than 5.6 and bacteria at pH >5.6. Since cellulose is deficient in nitrogen, the microorganisms must take it from proteins and amino acids. If it is unattainable to organisms, the decomposition of saccharides is retarded or ceases altogether. Under favourable conditions the process is completed within a relatively short time. Wood contains 45–65 % saccharides, and the woody peat only 26 %. The average content of cellulose in peat is about 15–25 %. The portion of lignin does not change appreciably (wood — 20 to 45 %, woody lowmoor peat ca. 23 %). In addition, the microorganisms need K, S, P and other elements for building their cells. Phosphorus forms up to 2.5 % of ash matter of the microbial cells (Hadač, 1953). The product of cellulose decomposition is the incompletely decomposed cellulose rich in carbon, which is the principal structural constituent of the humolithic rocks. It is brown to black amorphous matter, insoluble in water, ethanol, benzene or in boiling concentrated NaOH and KOH solutions. In various environmental conditions arise intermediates such as pyruvic acid CH_3COCO_2H, lactic acid $CH_3CH(OH)CO_2H$, acetic acid CH_3CO_2H, butyric acid $CH_3CH_2CH_2CO_2H$ and acetaldehyde CH_3CHO. The end products are, for example, CO_2, H_2O, H_2 and CH_4. Enders (1943) claims that methyl glyoxal was an important intermediate in degradation of saccharides in predominantly anaerobic conditions, less favourable for the metabolism of microorganisms. In aerobic conditions the normal end product of decomposition of saccharides would be CO_2; in anaerobic conditions methane can also be produced by the activity of some organisms

$$C_6H_{12}O_6 \xrightarrow{\text{Methanosareina}} 3CO_2 + 3CH_4 .$$

The plants of lowmoors usually contain 20–30 % cellulose and lowmoor peats themselves 0–6 %. In highmoor peats the decomposition is not so rapid and the difference in the content of cellulose between the plants and peat is not so great. The lowmoor bogs are more favourable to decomposition because of their higher content of nitrogenous substances, subneutral reaction of water and higher temperature.

Hemicelluloses are hydrolysed to simple saccharides, hexoses and pentoses, and decomposition continues as desribed above. Owing to the activity of microorganisms further hemicelluloses are produced as components of microbial cells or form thin films. Polysaccharides undergo a similar process, the decomposition also occurs only in part of the cycles.

Monosaccharides, disaccharides and trisaccharides dissolve readily in water and are easily destroyed by microorganisms. Alcohols and organic acids form as intermediate products. The end products of the processes are CO_2, H_2O and the cell substances of microorganisms. At the presence of fungi the saccharides decompose into citric acid $HOC(CH_2CO_2H)_2CO_2H$, oxalic acid HO_2CCO_2H, lactic acid, butyric acid, ethyl alcohol CH_3CH_2OH, acetone CH_3COCH_3 and others. At the presence of oxygen the saccharide molecule splits into lactic and pyruvic acids, CO_2 and H_2O. There is less saccharide in humoliths than in the primary plant material. Under specific conditions the cellulose and polysaccharides may be decomposed even to humic acids. They are amorphous yellow-brown plastic substances of colloidal character, insoluble in benzene. However, they may be dissolved in boiling solutions of alkaline carbonates or in alkaline lyes, from which they can be precipitated by mineral acids. These processes have not been fully recognized, but even if they prove true, the amount of humic acids thus formed would be small, as cellulose is more easily decomposed by simpler processes.

The decomposition of starch is very rapid. It is hydrolysed by acids or enzymes — amylase and maltase — into glucose; in the process some intermediates are produced.

The main portion of humic acids in peat and coal is produced by lignin (Fischer and Schrader, 1921; Fuchs, 1931). Lignin is characterized by the methoxyl group and is of aromatic nature. It is very resistant to decay, being decomposed only by some fungi (*Basidiomycetes, Hymenomycetales* in woods, *Ascomycetes* and *Fungi imperfecti*) and by microorganisms *Polyporus, Lenzites* and others, in anaerobic conditions also by *Flavobacterium* and *Pseudomonas*. Bacterial activity is important in the last stages of decomposition of lignin and other aromatic substances; the aromatic rings of simple phenol compounds which formed by fungal degradation of lignin, are destroyed and aliphatic products originate. In an environment where cellulose undergoes fermentation, the lignin substances are hydrolysed to protohumic substances and these are oxidized to humic acids. The humic acids of this derivation have a benzene nucleus and belong in the phenol group. Lignin may accumulate in humoliths, especially in anaerobic conditions. The scheme showing the conversion of lignin into humic acids in the peatification stage was proposed by Manskaya and Kodina (1975) (Fig. 26).

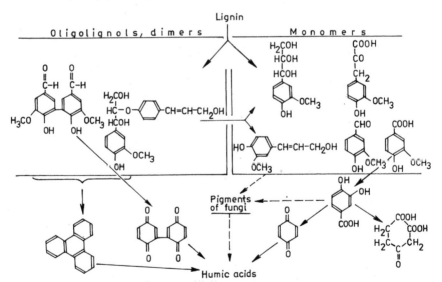

26. Scheme showing the alteration of lignin into humic acids in the peat stage (after Manskaya and Kodina, 1975).

Chitin is also resistant to decomposition.

Tannins are converted into phlobaphenes or the ellagic acid $C_{14}H_6O_8 \cdot 2H_2O$ by oxidation or by diluted mineral acids. These products are relatively stable and have some conservation capability.

Glycosides split by fermentation into a saccharide component and aglycone. In phenol glycosides the aglycone is represented by phenol C_6H_5OH. Of the humolith-forming plants the species of the families *Ericaceae* and *Vacciniaceae* contain most glycosides.

The *p*-hydroxyacetophenone ester of palmitic acid has been isolated from a sphagnum peat from Alfred, Ontario, Canada (Morita, 1975). The finding of this ester is significant because naturally occurring phenolic esters of fatty acids are rare. Fatty acids and phenolic compounds, however, are common components of peats. In this connection it should be noted that *p*-hydroxyacetophenone has been identified in plants as glycoside or the aglycone (Neish, 1964).

Fats degrade into glycerol and higher fatty acids; the latter are degraded into lower fatty acids, CO_2 and H_2O by the activity of fungi and bacteria. The process occurs mainly under aerobic conditions.

The nitrogenous compounds — proteins and amino acids are destroyed by microorganisms, who withdraw from them nitrogen for the construction of their cells. The structural units of proteins are amino acids. It is thus assumed that the end products of decomposition of proteins and amino acids are fatty acids, ammonia and amines. Fatty acids break down into H_2O, CO_2 and CH_4.

In peat bogs the amino acids concentration generally increases downwards. In the Dismal Swamp in Virginia (Swain et al., 1959), which may be taken as a plausible example of coal-forming basin, the highest concentration of amino acids was established in the lower part of the reddish-brown woody peat. At a depth of $0.7-1$ m below the surface amino acids have not been found. Owing to the rapid growth of palustrine vegetation the surface layers probably contain undecomposed proteins or polypeptides and humus. The humic and amino acids concentrate to the base of the woody peat, where they precipitate. They are gradually leached from the upper peat layers immediately upon their formation.

The distribution of amino acid types is not the same in all peat bogs (Fig. 27).

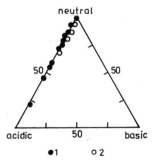

27. Distribution of acidic, neutral and basic amino acids in peat from the Cedar Creek Bog (1) and Dismal Swamp (2) (after Swain, 1961).

According to Swain (1970), neutral and acidic amino acids predominate, for example, in the peat of the swampy area around Lake Ponchartrain and basic amino acids occur e.g. in the Dismal Swamp. The peat from Dismal Swamp shows a more acidic reaction, and under these conditions the amino acids having several carboxyl groups will be more liable to bacterial degradation (decarboxylation). Basic amino acids are more likely to be preserved in such conditions.

In the neutral swampy environment, the neutral and acidic amino acids are preserved approximately in the ratio of $6n:1a$; alkaline muds are suitable for the increase of the proportion of acid amino acids to a ratio of $3n:1a$. Acid muds preserve some basic amino acids, giving rise to the ratios of $75-95n:0-10a$ (Swain, 1970). Such proportions may be expected where a certain degree of natural equilibrium in a closed peaty environment has been achieved. Drainage and other interference disturb the equilibrium resulting in the change of content and types of the amino acids. According to Swain (1961), their total content varies between 2 and 4,000 ppm, with a predominance of neutral amino acids. According to Swain et al. (1959), alanine reaches the highest concentration.

Proteins are of importance for the formation of coal seams only when faunal organisms, plankton and microplankton were present in the coal basin during its development, since proteins occur in plant bodies only in very small amounts relative to lignin and cellulose. Under some conditions caused by the concentration of hydrogen ions, the salinity and acidity of the decomposition environment vary; in such case the nucleoproteins (proteins contained in the cell nuclei) decompose into phosphoric acid, similarly as does chlorophyll. Owing to the acid and reducing medium they break down into organic acids with complex bonded manganese or into esters of these acids.

Waxes, terpenoids and resins are most resistant to chemical and microbiological decomposition. Streibl and Herout (1969) and Wollrab and Streibl (1969) presented a comprehensive information on their behaviour. The so-called montan waxes are known especially from the peats of England and Scotland (yield $3-12\%$ in dry matter).

According to Wollrab and Streibl (1969), the data on the composition of peat resins and waxes are rather sparse. General information on the composition of peat resins is to be found in the work of Kwiatkowski (1965). Gilliland et al. (1960) referred on the presence of the aromatic hydrocarbon perylene.

Waxes are esters of aliphatic acids with a long chain and of higher alcohols. In general, this group encompasses all substances accompanying wax esters and having a long paraffin chain (hydrocarbons, alcohols, carboxyl acids, ketones, etc.). Wax acids have usually a higher number of C atoms ($C_{16}-C_{40}$) than acids derived from fats ($C_{10}-C_{24}$). In the group of fats are also placed plant resins, cutin, waxes of cuticular tissues, suberin of cork tissues, sporopollenin of spores and pollen grains, chitin of fungi, and similar substances. Resins are most resistant of all of them. In humoliths, waxes, fats and resins together with pigments form bitumen, which is very stable and accumulates with time. Peats normally contain 5—15% bitumen. A typical peat bitumen contained 52.5% wax, 20.4% paraffin and 24.6% resin (Kaganovich and Rakuskiĭ, 1958).

Tannins enter the reaction with proteins during decomposition to give a jelly-like precipitate; hydrolysable tannins with a benzene nucleus contribute to the increased content of humic acids.

An important constituent of humoliths are humic acids. Some authors thought that they derived from cellulose or other substances, others considered lignin to be the primary material. Rakovskiĭ (1949) argued against the "lignin" theory, claiming a relationship between humic acids and carbohydrates. Humification (formation of humic acids and humates) is a complicated process, controlled by many factors, and the acids obviously formed from various compounds. Humification is accelerated by a higher nitrogen content, a higher temperature, more alkaline reaction, better aeration, and others. It appears that a greater number of plant

Table 6
Mechanism of humification, according to Enders (van Krevelen, 1961)

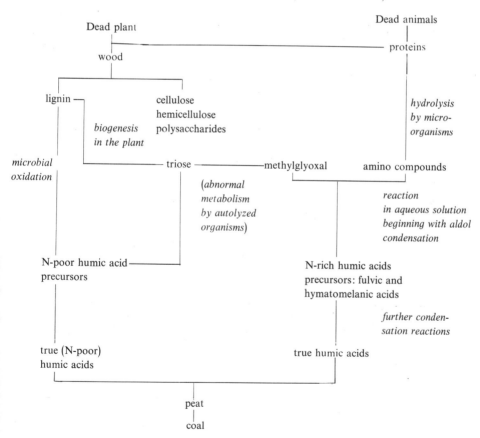

substances take part in the formation of humic acids at a substantial activity of microorganisms. The theory most widely accepted today maintains that besides lignin and cellulose, which seem to be the principal precursors of humic acids, other biochemical compounds or products of protein breakdown may also play a role in their formation (see scheme in Table 6). The intermediate products of this process possess a high reactive ability and may strongly affect the concentration of metallic elements.

Francis (1961) presumed that in the peatifying process lignin is decomposed into complexes of phenol compounds (a series of phenol compounds derived from lignin decomposition by whiterot form of fungus *Polystictus versicolor* was determined by Flaig, 1966; see Fig. 28) and the products of protein breakdown appear. The mechanism of these processes producing humic acids is based on oxidation and polymerization reactions. Breger (1951) inferred the structural identity between lignin and humic acids extracted from peat, from the infrared spectra. However, during anaerobic decay of plant tissues cellulose is decomposed when lignin and other resistant components remain still intact. This suggests that rather cellulose is the material for humic acids in peat, and that lignin conversion into humic compounds begins during peatification but plays a major role in the post-peat stage or in coalification processes.

28. Products of lignin decomposition by the fungus *Polystictus versicolor* (Flaig, 1966). a — R = —CHO p-hydroxybenzaldehyde, R = —COOH p-hydroxybenzoic acid, R = = —CH=CH—COOH p-hydroxycinnamic acid; b — syringic acid; c — guaiacylglycerol-β-coniferyl ether; d — R = —CHO vanillin, R = —COOH vanillic acid, R = —CH=CH—CHO coniferyl aldehyde, R = —CH=CH—COOH ferulic acid, R = —CHOH—CHOH—CH$_2$OH guaiacylglycerol, R = = —CH$_2$—CO—COOH guaiacyl-pyruvic acid; e — dehydrodivanillin.

The ratio of aliphatic/aromatic constituents in humic acids may depend on the chemistry of the primary material, the type of environment and on the character of metabolism of microorganisms. If one arborescent plant forms the original material, which is characteristic only of some coals, the humic acids can have more aromatic nuclei, but if the humic acids are derived from a mixture of plant and animal proteins and lipoidal substances, they are rather of aliphatic character.

The humic constituent of peat consists of humic acids, humates and accompanying substances such as fulvic acids and hymatomelanic acids.

During peatification the humic acids and thus also carbon accumulate (cellulose contains 50 % C, peat 60 % C). Peatification, which is largely due to the desorganization of the lignin-cellulose group and the microbial effects, provides the end product (peat) within a relatively short period, compared with coalification. Organic matter passes through the stage of aerobic degradation, when the unstable substances are destroyed. The accumulation of lignin or cellulose results in a high C/N ratio, <15 in recent deposits and ~50–100 in fossil deposits.

In the end product practically all inorganic plant components are present. Only some of them passed probably into solutions during decomposition and others became bonded to new compounds. With regard to the fact that the content of plant bodies during decay was reduced, they became relatively enriched.

Peatification does not pass gradually into coalification, although peat is universally regarded as one of the stages in the series of alterations from fresh plant material to brown and bituminous coal and eventually to anthracite. Peatification is a qualitatively different process from coalification. This assertion, however, does not deny the possibility of coal formation from peat. The chemical composition of peat is given by the environment and the type of vegetation, which influence each other and control the course of alterations in the atrophied material. Decisive for the degree of decomposition and for the character of humoliths are the processes functioning in the surface "peat-forming" layer. The alterations at a greater depth are immaterial. The content of humic acids is higher in lowmoor bogs than in highmoor bogs.

The humoliths contain the following organic substances: unaltered parts of plants — proteins, pectins, monosaccharides, disaccharides, hemicellulose, cellulose, chitin, lignin, protobituminous substances; products of humification — amino acids and other products of protein decomposition, secondary proteins, volatile acids, soluble saccharides, "humolignins", humic acids and their polymers, humates, fossil bitumen, phenols, ketones, organic acids, hydrocarbons, etc. Free sterols, particularly

29. β-sitosterol (R = C_2H_5).

β-sitosterol (Fig. 29) and β-sitostanol were isolated from peat by Ikan and McLean (1960). Belkevich et al. (1963) reported β-sitosterol from peat wax. McLean et al. (1958) identified triterpenoids from peat. A small amount of chlorophyll was also extracted from peat, and flavinoid pigments have been recorded.

The inorganic substances are represented by earthy particles, water, substances of plant bodies and less of animals, and substances drifted by water or wind, which often contain trace elements.

The chemical characteristics of some lowmoor and highmoor peats of Czechoslovakia are listed in Tables 7 and 8.

Vasiľev and Dyakova (1958) found in one peat sample: 5% of water-soluble

Table 7
Chemical analyses of highmoor peat from Libverda and of lowmoor peat from Mariánské Lázně
(Hadač, 1953)

	Highmoor peat Libverda	Lowmoor peat Mariánské Lázně
Inorganic substances:	%	%
K_2O	0.08	0.20
Na_2O	0.31	0.53
CaO	0.24	3.90
MgO	0.08	0.83
Fe_2O_3	0.55	2.24
Al_2O_3	1.12	3.18
SO_3	0.03	1.41
P_2O_5	0.07	0.10
CO_2 bound	0.11	1.25
SiO_2	2.31	7.70
sulphate sulphur	0.01	0.08
elementary sulphur	0.02	0.02
pyrite	0.00	0.02
Organic substances:		
organic sulphur	0.54	0.65
non-hydrolysed N	0.62	1.07
bitumen	8.71	4.66
pectins, saccharides, proteins — soluble in hot water	5.24	3.93
proteins (N 6.25)	5.01	4.06
hemicelluloses	19.69	11.50
cellulose	11.79	4.50
humic acids	22.22	39.97
substances accompanying humus (fulvic acids)	8.69	7.59
humin + lignin	14.25	9.91
pH of extract	4.1	5.8

component, 41.4 % humic acids, 5.6 % cellulose, 22.5 % hemicellulose, 10.5 % lignin, 12.9 % bitumen and 2.1 % other substances.

The element analysis of several peats from Germany gave the following results (Francis, 1961):

C	51.13 %		58.48 %
H	6.05 %	increasing maturity	5.64 %
N	1.83 %	$\longrightarrow$	2.34 %
O	40.99 %		33.54 %

The concentration of trace elements in peat rises most intensely in the early diagenetic stages, that is in the end phase of peatification and in the post-peatification stage. The concentration depends on the chemical composition of the organic matter, on the physico-chemical properties of the element, and in which conditions the reaction between the element and organic matter occurred. The accumulation of elements may be syngenetic or epigenetic. The ion exchange and absorption processes occurring during the life of plants and during their decay cause the concentration of some elements in peat (B, Ge, As, Bi, Be, Co, Ni, Cd, Pb, Ag, Sc, Ga, Mo and U) and subsequently also in coal. The accumulation of metals attains such degree in some peat bogs that these become ore deposits of economic significance. Zýka (1971) quoted, for example, up to 10 per cent Cu in the peat (dry matter) from Sackville (Canada) and up to 16 per cent Zn and a high Pb in the peat (dry matter) from Manning (U.S.A.). Aľbov and Kostarev (1968) studied Cu contents in peats of the Central Ural; Kovalev and Generalova (1969) examined the migration of iron in the peat bogs of the Belorussian S.S.R. The enrichment of uranium in

Table 8
Contents of principal structural substances in peats (Hadač, 1953)

Type of peat, locality (dominant plants)	Percentage of structural substances and inorganic admixture*					
	saccharides	lignin	proteins	fats + waxes	resins	inorganic admixture
lowmoor peat — Svět near Třeboň (reed, sedge)	14.96	14.41	8.31	9.10	absent	19.07
lowmoor peat — Rájov near Kynžvart (reed, birch, spruce wood)	26.00	23.13	1.50	5.83	absent	12.63
highmoor peat — Jizerská louka near Libverda (sphagnum, cotton grass)	32.43	10.89	4.36	10.78	absent	3.85

* In original state.

peats was investigated by Kochenov et al. (1965). Other data on trace elements in peats were presented, for example, by Saprykin and Sventikhovskaya (1965) and Savchuk (1967) from the area of Novosibirsk and Tomsk, by Tarakanova (1968) from the Urals, and by Szalay and Szilagyi (1969) from Hungary. Peat from the Masugusbyn in Sweden contains up to 3.1 % U in dry matter.

Of interest is the finding of Sillanpää (in Göttlich, 1976) that at the *Sphagnum/Carex* boundary horizon, i.e. ca. 1 m below the peat-forming layer the concentration of microelements is minimum, and the mineral substratum has a higher content. He came to this result by studying the distribution of some elements on two profiles in the Finnish peat bogs (Table 9). On the other hand, Casagrande and Erchull (1976) found that in Okefenokee Swamp (Georgia) the sandy bedrock is not an important contributor to metal concentration in peat.

Recent opinions that the bonding of metals to organic substances may be explained on the basis of chelation have been confirmed experimentally. Chelates contain a central atom or ion (usually metal) attached to radicals of organic molecules, which are donors of electrons (ligands). If the ligands have two or more groups of donors, the whole structure represents a chelate. Many chelate compounds have already been recognized, some of them play a role in weathering processes.

Copper, germanium, vanadium, nickel, cobalt, uranium and some other elements form chelates in peat and coal, particularly with humic substances. Peats contain little Ge, which suggests that it was sorbed from water towards the end of peatification, and mainly in the post-peatification stage, by bonding to humic substances, especially to humic acids (Breger, 1958). Humic acids show a high fixation capacity for many elements. They have a stable polyaromatic skeleton, to which oxidic, hydroxyl and carboxyl groups are attached. Metal ions are attached to humic acids through reversible substitution of hydrogen capable of ionization by cations. The polycarboxylic nature of humic acids makes them an effective chelating agent. Goldschmidt (1954) has shown that soils and oak leaves have 0.5×10^{-3} % GeO_2 in ash, oak and birch humus 7×10^{-3} % GeO_2 in ash, peat 5×10^{-3} % GeO_2 in ash, but coal ash contains as much as $100-3,000 \times 10^{-3}$ % GeO_2. This example indicates that peats do not usually exhibit a higher Ge concentration, which was also confirmed by study of some Czechoslovak peat bogs (Bouška, 1959).

Freshwater and coastal peat bogs contain 0.1 – 10 % of total sulphur (Casagrande et al., 1977). Sulphur precipitates in the early stages; the formation of pyrite depends on the pH of the medium. Generally, the S content increases with depth, particularly in marine peat bogs. A considerable part of sulphur in peats was fixed by plants. Carbon-bound sulphur accounts for 70 % of total sulphur in freshwater peat and 50 % in marine peat. In the marine environment over 15 % of total sulphur was in pyritic combination, as compared with the values one order of magnitude lower in freshwater peats (~1.5 %). An ester-sulphate fraction represents 25 % of total sulphur in both peat types.

Table 9
Content of elements in relation to depth in Finnish peat-bogs I and II after Sillanpää (Helsinki, 1972), in Göttlich (1976)

Depth cm	Ash wt.%	pH	Al %	Fe %	Mn ppm	Ni ppm	Co ppm	Cr ppm	Mo ppm	V ppm	Sn ppm	Zn ppm	Pb ppm	Cu ppm	Sr ppm	Note
20 I	3	4	0.1	0.05	50	2	0.5	5	0.5	3	1	50	10	4	10	layer 0–20 cm (*Sphagnum*)
20 II	2.5	3.5	0.15	0.1	30	3	0.5	4	0.8	3	1	45	20	7	10	
100 I	2	4.1	0.04	0.1	20	2	0.3	2	0.4	1	0.2	20	6	3	8	boundary horizon (*Sphagnum–Carex*)
100 II	2.2	3.6	0.1	0.1	15	3	0.6	2	0.4	1	0.2	10	3	10	10	
250 I	5.5	4.3	0.8	0.3	80	4	2	11	1	20	0.5	15	6	20	20	peat with *Carex*, ca. 50 cm above mineral substratum
150 II	5.5	4.2	0.8	0.25	30	10	2	5	1	5	0.3	8	3	20	15	
300 I	90	5.1	9	5	700	50	30	500	5	200	2	100	40	90	220	mineral substratum
200 II	90	5.1	7	4.8	650	50	25	500	5	170	1.5	60	30	65	300	

The mean sulphur contents in marine and freshwater peats are as follows (in %): H_2S — 0.015 to 0.0007; elementary sulphur — 0.39 to 0.0034; sulphate sulphur — 0.36 to 0.0036; ester-sulphate sulphur (type C—S—O) — 1.22 to 0.046; pyritic sulphur — 0.74 to 0.0026; organic sulphur — 2.44 to 0.127 (Casagrande et al., 1977).

Senesi et al. (1977) used ESR (electron spin resonance method) and Mössbauer spectrum for the determination of the oxidic stage and position of iron bound to humic and fulvic acids. All iron bound to humic material was found to occur in trivalent form: (1) Fe^{3+} fixed firmly in tetrahedral or octahedral coordination; this form is resistant to complexing and reduction. (2) Fe^{3+} adsorbed on the surface, loosely attached octahedrally, easily complexed and reduced.

Casagrande and Erchull (1977) studied 13 elements (Ba, Ca, Cr, Cu, Fe, Hg, K, Mg, Mn, Na, Ni, Pb and Zn) in plants from the top layer (0—6 cm) of the peat bog and from the water of Okefenokee Swamp, Georgia. The plants of both environments concentrated more easily Ca, Na and Mg, and some species also Hg (*Utricularia*), Fe (*Swamp iris*) and Zn (*Nymphaea orontium*). It should be noted that Cr, Cu, Fe and Pb are present in plants in smaller amounts than in peat. The concentration of elements is obviously controlled by the character of environment, by the type of peat development, the type of plants and the plant organ. The metals in peat are largely associated with the cation-exchange sites on humic acid, fulvic acid and humin materials.

Some plants (e.g. *Taxodium*) contain far more organic constituents such as lignin than others (e.g. *Nymphaea*), and since lignin is a chemically stable precursor of humic acids, some plants have a potentially greater capacity for humic acid synthesis and subsequent metal accumulation in the sediment. *Taxodium* grows in an alternately wet/dry environment, which is conducive to microbial attack and subsequent humic acid synthesis, whilst *Nymphaea* grows in a dominantly anaerobic environment unfavourable to the synthesis of humic acids. Humic acids produced from various plants differ in type; some have more carboxyl functions than others.

Plants which serve as sources of organic matter for humic acid synthesis can determine to a certain extent the quantity of the produced humic acids and thus indirectly influence the capability of the sediment to produce the cation-exchange sites necessary to attract metals. The above observation shows clearly the important influence of plants on the distribution of metals in peat and on its variance in the vertical and horizontal direction of a coal seam. The participation of other factors must of course be also considered.

The enrichment of organic matter in trace elements is known from nearly every sedimentary basin. Various pelitic-psammitic rocks very often contain slightly coalified tree stems and branches in which the content of any element is always many times that of the enclosing rock. The increase depends on the original amount of the element in plant, on its content in percolating water, on the volume of the solutions, the size of the sorbing surface of the stem, on the mode of rock deposition, and on pH and Eh of the solutions and medium as well.

The physico-chemical sorption of an element from the percolating or seeping water is an important agent. Krauskopf (1956) proved that 40–99% Zn can be absorbed from the solution by sphagnum and peat matter. Humates can adsorb as much as 10% Zn of their dry weight (Swanson et al., 1966).

The post-peatification stage

The coalification process begins when the peat bog has been overlaid by clastic or other sediments of a small thickness; if it is covered with a clayey impermeable layer, the gases cannot escape. Since neither temperature nor pressure beneath a thin cover did change appreciably from the conditions that existed in the lower layers of the peat not yet concealed, we speak of a post-peatification stage. This stage precedes the coalification process proper, or it may be regarded as its first phase. Biochemical processes die away gradually to be replaced in the initial stage of coalification by endothermic processes at a slight or greater pressure influence. The activity of bacteria, particularly anaerobic, continues and the plant material is being compressed and reduced in volume in result of water loss. It assumes the gel character.

Some subrecent caustobiolith deposits in the post-peatification stage occur in deltaic sediments of major streams (e.g. Mississippi) or in the bottom of closed lagoons. Of this type is the subrecent peat bog derived from interglacial sediments (Riss-Würmian) near Rostov. Havlena (1963) recorded as an example the so-called "diluvial coal" from the upper course of the Iller and Inn in Austria (diluvial Jungbraunkohle, diluvial Schieferkohle — strongly compressed, laminated) and from Chambéry in France. The Alpine "diluvial coal" was identified as a Riss-Würmian interglacial peat of lacustrine origin. An about 3 m thick seam consists of plant matter that had been strongly compressed by about 900 m thick Würmian glacier. The chemical composition is almost the same as that of recent peats, except for the water content ($W^r \sim 50\%$, relative to $\sim 75-85\%$ in recent peats). The water content of this "young" coal can be correlated with water content of brown coal. The content of carbon and oxygen in volatile matter approaches their content in recent peat and differs widely from brown coal (C^{daf} − 59.74%, H^{daf} − 5.47%, O^{daf} − 33.2%, N^{daf} − 1.56%; ash content $A^d = 7-12\%$ in dry material). Havlena (1963) termed the subrecent caustobiolith of peat nature the subhumite. The sapropel (saprocol) in the post-peatification stage would become solid subsapropel (subsaprocol). Subhumite differs from peat in a low water content and in being solidified; it differs from brown or bituminous coal in chemical composition, particularly in low carbon and a still high oxygen. The development of the caustobiolith in the post-peatification stage is chemically nearer to peatification than to coalification.

The transitional types to brown coal are unknown. Although a continuous series of coalification stages (peat — brown coal — bituminous coal) is often cited, there is an interruption of continuity between peat and brown coal, which can be regarded as a qualitative jump.

MINERALOGICAL CHARACTERISTICS OF PEAT BOGS

The peat bogs have a specific but relatively poor mineral paragenesis. The minerals occur either in peat bogs of all types (lowmoor, transitional and highmoor) or are confined to a particular type. The paragenesis may be influenced by the occurrence of mineral springs or other factors.

The principal minerals of lowmoor bogs and transitional bogs near to them are sulphates (gypsum, epsomite, mirabilite, melanterite, copiapite, minerals of alum group), sedimentary bog iron ores (hematite, limonite), pyrite, marcasite, sulphur, soda and occasionally vivianite (most frequently in old lowmoor bogs covered with highmoor bogs). On the other hand, organic minerals such as fichtelite or dopplerite occur in all peat bog types.

Peat bogs generally developed on little permeable clayey substratum. Moreover, the humic acids which intensely affect the rocks adjoining the peat bogs induce podzolization. As a result, hardpan is often formed (Bouška, 1956) and as this is very little permeable, the peat area becomes more or less a separate closed whole.

30. Woody layer in the profile of the "Na kanále" peat bog, Borkovická blata (Bohemia). Photo by V. Bouška.

The lowmoor bogs that are in sufficient contact with the neighbourhood mediated by water show a more varied chemistry. The mineral paragenesis is there richer than in highmoor bogs. True highmoor bogs contain in addition to organic minerals only limonite, gypsum, sometimes opal and clay minerals, chiefly kaolinite, and rarely vivianite.

Fichtelite is a hydrocarbon $C_{19}H_{34}$. It impregnates brownish tree trunks or roots of pines (*Pinus uliginosa* and *Pinus silvestris*) lying in peat. Besides white or greyish-white crystalline crusts, it forms typically domatic crystals, up to 1 cm in size and tabular parallel to the base. Fichtelite is a product of the degradation of resinous substances of pines. At many localities, e.g. at Redwitz in the Fichtelgebirge, at Rosenhein in Upper Bavaria and at Holtegoard in Denmark, fichtelite occurrence is confined to pines (to *Pinus australis* besides the above mentioned). The abietic acid appears to be converted in the buried pine trees into fichtelite and retene, which are otherwise unknown as plant products. They are products of decarboxylation and dehydration or hydration of their common source — the abietic acids (Eglinton, 1968).

Fimmenite formed by accumulation of pollen grains; it is light in colour, is neither plastic nor elastic and when dried is lighter than water. Owing to a high fat content, it burns with a bright flame. The largest and best known occurrence is in the Oldenburg region, where it constitutes an up to 120 cm thick layer (K. Bülow, 1929, Allgemeine Moorgeologie, Berlin).

Dopplerite is not a strictly defined mineral. This calcium salt of humic acids has a varying composition with absorbed Al_2O_3 and Fe_2O_3. The main localities of dopplerite are Aussee and Obbürg in Austria, where other organic substances such as theocretine, phylloretine and ozocerite have been found. Dopplerite is also known to occur in Obbergen, Dachmoos, Gonten, Aurich, New Mexico, Sluggan Bog and other places. It is a black lustrous, elastic mineral with marked conchoidal fracture, and very similar to asphalt; after drying out it becomes hard and brittle. The author found a mineral corresponding to the description of a typical dopplerite in the peat bog near Borkovice and Mažice in southern Bohemia (Bouška, 1959), where it formed black coating on pine woods extracted from peat. When fresh, the samples were honey-yellow, elastic, but in time they turned brown to black, plastic and eventually solid. Meantime they lost ca. two thirds of their weight. The formation of dopplerite may be associated with the origin of humic acids.

Vivianite occurs on the roots of pines, and forms by decomposition of plant and animal organic substances during the peatifying processes. The animals are known to contain phosphorus, but only some plants (e.g. *Scheuchzeria palustris*) have a higher phosphorus content. Phosphorus may be transported into the basin together with iron to separate from it during diagenesis and form vivianite.

All sulphates occur as white or yellowish powdery films or coatings on dug-out peat, usually in mixtures. Copiapite is sometimes pure and in this case is of yellow colour.

31. Whitish and yellow sulphate efflorescences on the dug-out peat. Ruda lowmoor bog near Horusice (Bohemia). Photo by V. Bouška.

The bog ores, limonite and hematite make up elongated lenticular bodies. Limonite may be found as precipitate in bog waters. The bog ores colour rusty the peat waters; they do not precipitate throughout the year but only in certain periods. The formation of recent bog ores is explained today mainly as due to biogeochemical processes (Sýkora, 1959). A typical iron ore is of a deep-red colour and earthy; when dried it is ochreous in colour (hence its name ferric ochre). Bouška (1962) determined in the Ruda peat bog near Horusice hydrogoethite or, better to say, goethite γ-FeOOH with a variable content of freely adsorbed water. Other precipitated minerals are siderite, ferrohydrite — $n\,Fe(OH)_2 . m\,Fe(OH)_3$, and even magnetite, hematite and maghemite. The origin of the last three minerals in a peat bog is connected with the oxidation of siderite and ferrohydrite, mainly in winter when the oxidation of other substances in the peat bog drops to minimum. Lukashev et al. (1971) also reported the occurrence of kerchenite, calcite, rhodochrosite, oxidized phosphates, psilomelane, pyrolusite and rarely chlorite.

Using spectral analysis of ash from the peat of highmoor bog Borkovická blata and fens at Ruda, near Schwarzenberský pond and at Vimperka near Třeboň, Bouška (1959) found the following elements: Ag, Al, B, Ba, Be, Ca, Co, Cr, Cu, Fe, Ga, K, Mg, Mn, Mo, Na, Ni, Pb, Si, Sr, Ti, V, Y, Yb, Zn and Zr. The association did not show such a variety as is found in some kinds of coal. Most elements, except for the commonest ones such as Si, Al and Fe, occurred only in trace amounts, which indicates that the peats studied developed on a poor mineral substratum.

32. Mud pots and sulphate efflorescences on the fen surface with polygonal pattern. Hájek (Soos) near Františkovy Lázně. Photo by J. Rubín.

A well-known Czechoslovak lowmoor bog is the sulpho-ferruginous fen Hájek (Soos) near the spa Františkovy Lázně (Dohnal et al., 1965). The mineral paragenesis of this fen comprises epsomite (efflorescence on dug-out peat), fichtelite, limonite, alum (not defined strictly), marcasite, melanterite, mirabilite, pyrite, gypsum, sulphur, soda and vivianite. Marcasite forms porous nests and fossilizes the plant bodies. The peat also contains pelosiderite nodules, asphalt matter termed cladorite, earthy clayey ochre rich in Fe hydroxides and Fe humates. It was called palliardite at one time.

The highmoor bogs near Mariánské Lázně (Dohnal et al., 1965) have a considerable mineral content with epsomite coatings, fichtelite crystals under the bark of *Pinus uliginosa* (together with retinite), glauberite, ferric ochre, marcasite, melanterite, pyrite, gypsum, sulphur and vivianite.

In addition to organic minerals formed by the decay of organic matter proper, some minerals originated under the co-action of water and the dissolved salts. The decomposition of organic matter creates a favourable environment for the origin of H_2S and of sulphides derived from it, pyrite and marcasite; sporadically the reduction proceeds, through the influence of organic substances, to elemental sulphur. On the other hand, sulphur dioxide released by decomposition reacts with water and the produced sulphate anions form predominantly water-soluble sulphates. After peat has been dug out and deposited at a dry place, sulphates crystallize on its surface as efflorescences and coatings.

In peat there are usually some minute inorganic constituents, which are found in ashes (Table 10) but do not fit in the paragenesis of typical peat minerals. They are either fragments and detritus of common rock-forming minerals supplied by water or wind, or minerals that occurred at the surface of bedrock and during the peat formation were lifted by plants into higher layers. This group of minerals comprises abundant clay minerals, especially kaolinite, quartz, muscovite, feldspars, biotite (in most cases decomposed) and some mafic minerals.

Table 10
Composition of ashes from peats of various types (in weight %)

SiO_2	10–45	K_2O	0.1–2.5
Al_2O_3	1–11	Na_2O	0.2–5
Fe_2O_3	1.5–5.5	P_2O_5	1–3
CaO	2–45	SO_3	5–20
MgO	1–20		
MnO	0.1–0.3		

Lower values generally apply to highmoor bogs.

COALIFICATION

Coalification is the process which leads to the formation of **coal**, i.e. the inflammable organogenic rock. It is an irreversible process, i.e. it cannot be reverted. The end products are caustobioliths both solid (**brown coal, bituminous coal and anthracite**) and gaseous (**natural gas** released from coal).

According to Havlena (1974) the brown coal, bituminous coal and anthracite coalification stages are discriminated:

Brown coal stage comprises three types:

Hemitype (lignite)	$<68\% \ C^{daf}$	V^{daf} content indecisive
Orthotype (lignite and massive brown coal)	$68-73\% \ C^{daf}$	$58-50\% \ V^{daf}$
Metatype (bright brown coal)	$>73\% \ C^{daf}$	$<50\% \ V^{daf}$

Bituminous coal stage includes three types:

Hemitype (subbituminous coal)	$74-82\% \ C^{daf}$	$42-36\% \ V^{daf}$
Orthotype (bituminous gas and fat coals)	$82-87\% \ C^{daf}$	$36-29\% \ V^{daf}$
Metatype (bituminous coking and anthracitic coals)	$87-92\% \ C^{daf}$	$29-10\% \ V^{daf}$

Anthracite has $C^{daf} > 92\%$ and the content of volatile matter $V^{daf} < 10\%$.

Coal is a product of physical and chemical alterations of plant remains that had accumulated during the geological history. In the course of this process only a minor part of carbon escapes into the atmosphere in the form of carbon dioxide and methane. The remaining part is cumulated through the gradual alterations in peat (about 50 % C^{daf}) and coal to anthracite (above 92 % C^{daf}). The beginning of coalification is usually placed in the time interval when a peat layer is covered with another rock.

In the early alteration stages, i.e. in the stages of humic acids and peat, microorganisms and the oxidation-reduction potential of the environment play the principal role. The **biochemical phase** is broadly terminated by the formation of peat. Subsequently, lignite, brown coal, bituminous coal and anthracite are developing as a result of slow "inorganic maturing". On the basis of carbon content, coal is divided into ranks. The higher C content, the higher is the degree of coalification and the more its chemistry differs from the chemical composition of the primary plant material. During coalification two alteration phases occur synchronously: coalification in peat and bitumenization in sapropel. The alterations of substances present in peat and sapropel into the components of coal cannot be observed directly. These processes occurred during the geological epochs either beneath a relatively thin cover or deep under the Earth's surface as dynamochemical and thermochemical processes. In contrast to peatification and post-peatification stages, a geochemical phase was involved.

The interval between the covering of peat and the cessation of microbial activity (biochemical phase) corresponds to the process in the zone of diagenesis. The geochemical phase begins with the termination of microbial degradation; it corresponds to the zone of advanced diagenesis and in part, probably, to the zone of epigenesis. As the diagenesis and epigenesis cannot be strictly discriminated, no clear boundary between the biochemical and geochemical phase can be established. Generally speaking, the biochemical phase ends at depths of hundred--metres order, although the percolating water can transport the microorganisms

33. Outcrop of the main seam in Maxim Gorkii open mine. The Chomutov-Ústí Basin (northern Bohemia). Photo by J. Pešek.

even deeper. The biochemical processes may be fading out to a limited extent even in the brown-coal stage.

In the brown-coal stage, particularly in the initial phase, when the organic matter is settling, the porosity decreases and the bound water is squeezed out (Škuta, 1972). Colloidal precipitates formed by purely chemical processes penetrate the more resistant cell tissues, saturating them. Concurrently, the polymer structural units, which resisted so far to decomposition, are rebuilt by temperature and pressure effects. As a result, the proportion of cellulose, lignin and free humic components decreases. Water becomes an important component of this gradually reconstituting colloidal system. With the decrease of water content the system loses its colloidal character. The compression of coal matter induces the diminishing of micelles and thus also of intermicellar spaces. The bound water was squeezed out of them and its content was gradually reduced despite its elimination from the coal matter.

In the bituminous-coal stage gelification fades out under the steadily decreasing water content. The huminitization process is replaced by vitritization. This physicochemical process attacks not only huminitized substances, including the preserved humic component, but also the as yet intact plant constituents such as waxes and resins.

The principal agents of the geochemical phase controlling the coalification are temperature, pressure, volume and time. The end products are solid coaly caustobioliths (brown coal, bituminous coal and anthracite with transitional types to clayey, silty and sandy sediments) and the natural gas escaping from coal during coalification. According to Zhemchuzhnikov (1952) the end product of coalification depends on the state of organic matter at the end of the biochemical phase and on the coalification degree attained in the geochemical phase.

Temperature which may act for a long time is a very important factor, even if it is not too high. It is assumed that temperatures of $100-150\,°C$ over a long period will appreciably affect the coalification process. The principal source is the Earth's heat. The geothermal gradient of $1\,°C/33\,m$ of depth (ca. $50\,°C$ at a depth of 1,500 m), however, is only schematic, being modified by a number of local conditions such as the proximity of igneous bodies or radioactive effects, which produce the increase in temperature. Francis (1954) calculated the temperature necessary for the formation of brown coal at $10-40\,°C$, $40-100\,°C$ for bituminous coal and $100-150\,°C$ for anthracite, on the basis of changes in the chemical parameters of the Indian coal. Schuhmacher et al. (1960) reported that up to the temperature of $350\,°C$ products comparable to the bituminous types can be obtained from cellulose, lignin and peat, without pressure application or with increasing the pressure to $1-2$ MPa only.

At a temperature above $320\,°C$ and pressure of 180 MPa Gropp and Bode (1932) managed to convert brown coal into bituminous matter. When they compressed plant matter, peat and brown coal by 180 MPa at a temperature of $200\,°C$ they did not observe any change even after a longer period. With the same pressure

34. Antonín seam in Jiří mine near Vintířov, Sokolov Basin (northern Bohemia). Photo by F. Tvrz.

but at ca. 300 °C the plant matter and peat altered into brown coal matter; brown coal remained unchanged.

Szádecky-Kardoss (1952) compared the diagenetic changes in clays and similar rocks with those occurring in coal. From the computation of maximum coalification temperature he derived that the increase in temperature by 1 °C is equivalent to the effect of a pressure ca. 3.3 MPa.

Coalification is an exothermic process. The heat released during coalification decreases with the degree of carbonization to disappear completely in the stage of critical metamorphism (Havlena, 1963).

After the termination of exothermic reactions, coalification is closed or may be interrupted if, for example, the thickness of overburden is reduced. It may, however, be renewed in later geological periods (Günther, 1966). The coal seams may also be directly metamorphosed (formation of some graphites) as a result of heating of the Carboniferous complex by plutonic magmatism in a later period, or of tectonically induced sinking to great depths and burying by younger sediments.

The geothermal gradient in the Carboniferous period is estimated at $12-15$ m/1 °C (Günther, 1966). This lower value relative to the present one is thought to have been caused by exothermic reactions connected with coal formation. The value of the geothermal gradient affects the degree of coalification and, vice versa, the accurately measured degree of coalification is a sensitive indicator of the maximum temperature to which the coal seam was exposed. The maximum depth of the burial of coal sediments can be computed from the degree of coalification and the fossil geothermal gradient, but we must take into consideration such processes as erosion, redeposition of organic matter into newly forming sediments and the effects of contact metamorphism.

The additional coalification may be exemplified by coal seams (coal with $V^{daf} = 35\%$) near Gronningen in the Netherlands, which owing to younger tectonic movements had sunk to a depth of 3,000 m and there underwent further coalification at greatly changed conditions.

The renewal of coal-forming processes in the Ostrava-Karviná coalfield was discussed by Günther (1966). In his paper he writes: "In the Miocene the tectonic activity at the Ostrava-Karviná ridge was followed by the heating of the adjacent complexes owing to a plutonic intrusion, particularly in its north-western and south-eastern parts. The juvenile gases which were released in a large amount during these orogenic processes influenced to a certain extent the coalification stage of coal. The gas produced by these geochemical processes penetrated a considerable part of the ridge. In the Petřvald basin coalification was not renewed because this basin was isolated by crystalline thresholds of the Orlová and Michálkovice faults. The Příbor-Těšín ridge was not affected by young tectonics. The nappe overthrusts exerted only slight influence or none at all."

The study of the coalification degree in the North Bohemian Brown-coal Basin (Zelenka, 1973) has shown that the process was controlled not only by the thickness of overburden but mainly by the unequal rate of growth of the overburden thickness (i.e. of sinking) of individual, even small sectors of the basin immediately after the close of organogenic sedimentation.

In places, heating produced by magmatic intrusions or by oxidation of sulphide ores played an important role. At Handlová (Slovakia) brown coal was converted into anthracitic coal at the contact with the overlying andesite effusions and accompanying hypabyssal bodies. In the Ostrava district the basalt dykes piercing the coal seams induced the conversion of bituminous coal into coke within a few metres from the contact and into definitely anisotropic coal with mineral impregnations at a greater distance.

A similar example was recorded by Kreulen (1935) from Indonesia. The thermal effect of an andesite body on the Pliocene Lematang coal deposit manifested itself by the successive transition from natural coke to anthracite, bituminous and brown coal away from the contact. The whole andesite body was rimmed by the coal types in this succession. The thermal metamorphism of coal by eruptive bodies is quite frequent. Dykes and sills are less effective than major effusions. The direct contact of coal with a volcanic rock is usually rimmed by a layer of natural coke of small thickness and showing columnar jointing, which is oriented at right angles to the surface of the volcanic body. In the Kladno mines, for example, there is a 20 m thick layer of natural coke at the contact of the seam with the volcanics of Vinařická hora.

Sometimes the coal is strongly mineralized at the contact by the accompanying solutions. The coal seams in the Svornost mine in Ohnič (northern Bohemia) are silicified, obviously due to the effects of the overlying basalt sheet. Another example of contact metamorphism of coal seams has been reported by Řehák (1966) from the Těšín area. The effect of a dyke on the properties of the Durham coal is given in Table 11. The influence of the dyke was observable to a considerable distance. Another example is reported by Ivanov (1975) from the Tunguska Basin (Table 12), where the bituminous coal seam was contact-metamorphosed by a vertical dolerite dyke of 3 m thickness.

The contact metamorphism by dyke intrusions is characterized by rapid heating, a high temperature of about 200–750 °C (at 1,000 °C originates graphite) and a relatively short duration, probably of hundreds to a few thousand years, compared with regional metamorphism. The pressure exerted by the overlying dykes is

Table 11
Durham coal in contact with igneous dyke (Northern Coke Research Committee. Prog. Rept., No. 47. Newcastle-upon-Tyne, 1948)

Approx. distance from dyke	Ash % A^d	V^{daf} %	C^{daf} %	H^{daf} %	N^{daf} %	S^{daf} %
	1.1	31.5	86.36	5.29	2.05	1.20
	1.7	30.1	86.68	5.19	1.89	1.69
	1.0	30.0	86.78	5.39	1.86	1.19
	1.2	29.5	87.10	5.24	1.97	1.32
	1.2	28.4	87.38	5.14	1.95	1.29
	2.8	28.7	86.88	5.17	1.85	1.92
	0.7	28.0	87.30	5.16	1.92	1.12
	1.4	27.6	87.46	5.26	1.89	1.32
	1.4	26.9	87.15	5.15	1.98	1.82
	1.3	26.6	87.40	5.07	1.84	1.29
	2.4	24.8	87.42	5.03	1.93	1.81
439 m	4.5	24.8	86.16	4.92	1.95	2.86
	2.9	24.4	87.10	4.96	1.91	2.18
	4.0	23.3	87.28	4.87	1.85	2.37
374 m	2.9	23.3	87.68	4.88	1.85	2.05
	2.5	23.9	87.96	4.78	1.89	1.21
	3.0	21.7	88.61	4.89	1.92	1.59
	2.3	22.7	88.39	4.80	1.90	1.44
	3.1	22.6	88.59	4.78	1.93	1.71
183 m	3.1	21.8	88.04	4.75	1.91	1.91
	6.2	20.2	88.31	4.59	2.00	2.88
	3.8	17.7	88.69	4.38	1.93	1.96
	4.3	14.5	89.52	4.09	1.96	3.15
	5.0	9.4	91.52	3.62	1.88	2.06
80 m	7.2	8.5	91.47	3.17	1.69	3.23

Table 12

Characteristics of contact-metamorphosed coal of the "Srednaya" seam (Kayakskoe mestorozhdenie), Ivanov (1975)

Distance from contact (m)	Water %	Ash %	CO_2 %	V^{daf} %	Calorific value Q kcal†/kg	C^{daf} %	H^{daf} %	ϱ g/cm³	R_{max}
30.0	8.0	2.9	0.3	44.5	7 600	77.6	5.7	1.35	0.46
10.0	8.4	4.2	0.7	44.7	7 600	77.9	5.5	1.35	–
5.0	6.4	6.0	0.5	39.3	7 560	78.7	5.3	1.35	0.72
4.0	3.1	5.5	1.7	33.6	8 130	83.1	5.4	1.32	0.80
3.0	2.1	6.0	1.8	31.2	8 320	84.4	5.6	1.30	0.83
1.65	2.3	8.5	4.6	9.2	8 270	89.3	3.7	1.44	3.21
1.0	2.5	9.2	4.0	9.6	8 250	89.7	3.5	1.45	2.60
0.75	2.4	10.4	7.5	8.8	8 260	89.8	3.4	1.45	–
0.65	1.5	12.8	7.8	9.4	8 370	89.9	3.6	1.44	2.88
0.50	1.2	13.7	8.2	6.6	8 330	93.3	2.8	1.56	5.40
0.25	1.3	21.7	12.5	6.5	8 290	93.7	2.6	1.60	6.09
0.10	1.3	20.0	10.1	4.5	8 140	94.9	2.1	1.66	–
contact*	1.1	22.8	11.4	4.9	8 160	95.0	2.0	1.69	7.77

* Contact of a vertical, 3 m thick dolerite dyke with coal seam.
† 1 kcal = 4 186.8 J.

generally not higher than 40 to 130 MPa. The emanation effect of intrusions, particularly of larger size, should also be considered.

The contact metamorphism of coal seams has been reported from many coalfields, as for example, from the Kuznetsk Basin, from England, Scotland, Australia, from the Ekka deposit in the South African Republic, from the Taimir, where the Permian coals were metamorphosed caustically by basalt trapps, from Ranihandzhi coal seams in India, which are pierced by troctolite and picritic rocks, and from Pennsylvania.

At the contact with volcanics the coal and the neighbouring rocks, tuffogenic or clayey and silty, are occasionally enriched in boron and other elements. Boron may also migrate farther from the contact and distort the original boron contents in the seam and its neighbourhood.

The local increase in temperature caused by the oxidation of sulphidic ores is known, for example, from Saxony, where oxidation of marcasite effected the alteration of coal into coke. Increased coalification can be observed around uraninite, zircons and other minerals that contain radioactive elements. Thermal effects of radioactive sources may influence even major areas.

At present, there is a steady accumulation of evidence that the coalification process may be influenced by the porosity, thermal conductivity and capacity of retaining heat. Porosity regulates the transfer of released gases. Clay minerals in the beds overlying the coal seams may function as catalysts. The lithology of over- and underlying rocks is certainly of importance.

All these examples, though interesting, are only of local significance; the most important thermal factor for the coalification degree and the coal rank remains the geothermal gradient. The depth of coal seam is often decisive for the rank of coal, which increases with depth. For example, Dannenberg (1915) determined the following average carbon contents in seams in the Saar Basin (in descending order): 77.61 %, 79.32 %, 81.16 % and 84.72 % (in volatile matter). This agrees with Hilt's rule (1873), saying that the coalification stage increases with depth, i.e. the content of volatile matter of coal decreases with depth. This relation is not linear and varies in individual localities. Figure 35 shows a nearly ideal instance. The Schürmann rule that water content of coal in mines or boreholes decreases with depth has not been so widely accepted (H. M. E. Schürmann, 1927, Braunkohle 26, 609, 634). Its validity is very approximate, although plausible theoretically under quiet conditions in a sedimentary basin.

The imaginary surface of equal coalification of seams is usually parallel to the surface of the original overburden above the seams. Therefore Hilt's rule is valid above all in cases when temperature and the pressure of the overburden were, besides time, the main coalification agents. Where the influence of temperature and pressure was subordinate, i.e. where the time was the principal factor (age of coal seams), Hilt's rule is not valid (Řehák, 1968). This author reports, e.g. that in the upper coal seams in the Karviná Formation the V^{daf} varies greatly and does

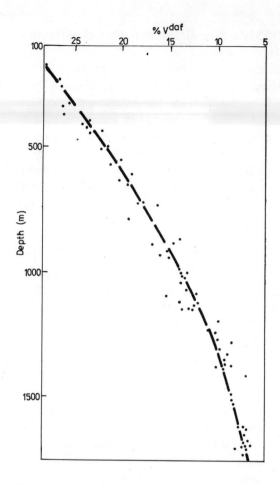

35. Hilt's rule (coalification increases with depth) exemplified by samples from two boreholes — NP 269 and NP 257 (Ostrava-Karviná coalfield) (Weiss, 1976).

not show any relation to their depth under the surface (Řehák, 1971). The Carboniferous sediments were there not deposited above the highest coal seams at such a thickness as to have an appreciable influence on the coalification degree. In the Ostrava Formation Hilt's rule is generally applicable, with the exception of the Poruba coal seams in the Petřvald Basin.

Pressure is an important factor particularly when it produces friction of beds (as is the case in folding). The static pressure itself can induce changes in the composition of coal. Vidavskiĭ and Prokopec (1932) exposed bituminous coal to a pressure of 500 MPa for one and half hour, which caused a change in its chemical composition (in volatile matter):

	C	H	N	O
Before experiment	87.4 %	5.40 %	1.8 %	5.5 %
After experiment	89.8 %	5.25 %	1.8 %	3.3 %

The unconfined pressure in the Earth's crust exerted by the weight of overburden might thus influence the coal-forming process. The effects of pressure and temperature, however, are slowed down and weakened by the water content in the developing coal layer. As the experiments of Vidavskiĭ and Prokopec (1932) have shown, static pressure itself retards the chemical reactions during the alteration of the organic matter, in preventing the removal of the products of the coalification process. The dynamic metamorphism of coal is produced by high orogenic stresses, as has been proved in some intensely folded regions. The coal in strongly folded complexes is also richest in carbon. As a result of loading by overlying beds the pressure rises by about 150 to 250 kPa per 10 m, and thus it is ca. 15 to 25 MPa at a depth of 1,000 m. As a result of pressure the water is squeezed out, the matter densifies and under particular conditions the coal matter becomes plastic. Simultaneously, chemical changes take place, mainly a decrease in volatile components, molecular grouping of atoms and gradual disappearance of the differences between the components. There are many examples demonstrating the influence of orogenic events. In the Ostrava-Karviná coalfield, the folded seams in the west have the rank of bituminous metatypes to anthracite at a depth of 400 m (Havlena, 1963), whereas in the eastern part deformed only by germanotype tectonics, the coal at 1,000 m under the surface is of bituminous orthotype. Similarly, the strongly folded Permian coal seam at Lhotice near České Budějovice (southern Bohemia) is of anthracite type, and the older Kounov seam in the tectonically undeformed Carboniferous of central Bohemia has the rank of bituminous hemitype. In the Donets Basin the folded seams in the south-eastern part are dominantly of anthracitic character, irrespective of the thickness of overburden and age. Towards the north and north-west the anthracite is replaced gradually by bituminous up to brown coal as the intensity of folding declines. The effectiveness of folding stresses was proved by Teichmüller (1951), who found out in the Ruhr Basin that the points of the identical coal rank are located in the identical fold line. In the Ruhr Basin the folding process began towards the end of the Carboniferous, which implies that coalification had been in progress for about 25 million years before this event.

The relationship between the coalification degree and the metamorphic grade of the contiguous rocks has been studied by Quinn and Glass (1958) in the Narragansett Basin. Coal and sedimentary rocks show there progressive metamorphism from north to south and south-west. Both chemical analyses and X-ray study demonstrated the agreement between the rank of coal and the metamorphic grade of the rocks. The coal has a high content of ash (13.8–40.1 %). Coal from Pawtucket, which was determined as meta-anthracite ($C^{daf} > 92.3\%$) petrologically and chemically and as anthracite by X-ray analysis, is associated with the rocks of the muscovite-chlorite subfacies of the greenschist facies (it contains muscovite, paragonite and chlorite). Near Portsmouth, the meta-anthracite ($C^{daf} = 93.7\%$) is associated with rocks of the muscovite-chlorite subfacies (it contains muscovite, chlorite, ilmenite, albite). The coal from the Cranston mine, which is of the highest

Table 13
Scheme of coalification in relation to the thickness of overburden
(modified after Hubáček et al., 1962)

	Depth under the surface (in km)	Temperature (°C)	Hydrostatic pressure (kg/cm²)	Weight of overlying beds (kg/cm²)	Caustobioliths	Volatile matter V^{daf} (in %)
Diagenesis early stage of coalification	0–1.0	0–30	0–100	0–270	peat, sapropelite, young brown coal	50–80
Coalification	1–1.5	30–50	100–150	270–400	brown coal (mainly semi-bright + bright) with a minor amount of hygroscopic water	52
	1.5–2.5	50–75	150–250	400–675	subbituminous coal	35
	2.5–3.5	75–105	250–350	675–945	bituminous gas coal	36–40
	3.5–4.5	105–135	350–450	945–1 215	bituminous fat coal	26–36
	4.5–5.5	135–165	450–550	1 215–1 485	coking coal	18–26
	5.5–6	165–180	550–600	1 485–1 600	baking + slightly baking coal	13–18
	6–8	180–240	600–800	1 600–2 160	anthracitic coal	13
	8–11	240–330	800–1 100	2 160–2 907	anthracite	1–5

The values are not of genera validity and are broadly approximate.

rank, is again meta-anthracite ($C^{daf} = 95.45\%$) associated with the rocks of the biotite-chlorite subfacies.

The principle of coalification in relation to the thickness of the overburden and the resulting temperature and pressure is presented in Table 13 (after Hubáček et al., 1962). In the last column of the Table the V^{daf} values are listed for individual coal types. Since the contents as given by different authors differ considerably, I record here the modified and summarized values from literary data: anthracite $<8\%$, anthracitic coal $8-17\%$, coking and fat bituminous coals $17-33\%$, gas and sub-bituminous coals $33-45\%$, brown coals $45-60\%$, peat $\sim70\%$ and wood $\sim80\%$.

Suggate (1974) reported that the bituminous coals from the Gippsland Basin (Victoria) and the Taranaki Basin (New Zealand) vary widely in rank and type. The original depth of burial is estimated at 5,000 m and temperature at up to 160°C. A systematic correlation exists between the depth, temperature and coal analyses. As these basins are nearly undeformed tectonically, the observation may be applied to terrains where the basins were disturbed or eroded.

The transition from peat to coal is accompanied by the reduction of volume, which is important for the study of the concentration of some elements. The reduction is due partly to the formation of many volatiles (e.g. CO_2, CH_4) during the decay of plant matter, and partly to the weight of overburden. Attempts were made to compute the coefficient of shrinkage of coal. Recently, Hurník (1972) has computed the coefficient of shrinkage for the brown-coal seam in the Most part of the North Bohemian Basin as 6, and approximately 3 for the overlying clays. He defined the coefficient as the ratio of the primary thickness of the accumulated plant material to the thickness of the resulting coal seam.

It may be said that time plays a catalyzing role in the coalification process, which may be achieved even if low temperature and low pressure are active for long periods. The Lower Carboniferous Moscow Basin contains brown coal with a high lignin content beneath an overburden of small thickness. The seams are situated 300 m under the surface at the most and were not affected by tectonic movements. Similarly, in the north-western part of the Donets Basin the Lower Carboniferous coal under a relatively thin overburden and on a little mobile basement matured to the brown-coal orthotype only. Apart from these exceptions, it holds that pre-Cretaceous formation contain bituminous coal and the Cretaceous and younger complexes enclose seams of brown coal.

Under an intensive, particularly regional metamorphism the coal may presumably be converted into graphite. Carbon is often very finely dispersed in some claystones, but it is assumed that high-grade metamorphism creates conditions for catch crystallization, and that the graphite deposits may form in this way. Many authors believe that graphite is the end product of coalification and that graphitization proceeding from meta-anthracite through semigraphite to graphite is the last stage of coalification. Graphitization represents a qualitative jump, when the more or

Table 14
Temperature and time required for the coalification rank in coal series (Francis, 1954)

Rank of coal	Upper carbon content %	Maximum temp. of change °C	Time required to complete change years × 10^6
peat	60	10	67
unconsolidated lignitous coal	68	20	20
lignite	75	30	8.5
subbituminous coal	80	40	4
bituminous coal	84	50	2.5
bituminous coal	87	60	1.6
bituminous coal	89	70	1.0
bituminous coal	91	80	0.9
carbonaceous coal	92	90	0.75
carbonaceous coal	93	100	0.6
anthracite	94.7	150	0.52

less two-dimensional arrangement of crystallites in a colloidal matter (semigraphite, meta-anthracite — crystallite size of 0.01 µm) becomes a continuous three-dimensional structure of a hexagonal mineral. Graphite is incombustible in contrast to meta-anthracite and semigraphite. Anthracite is graphitized at 1,500–2,000 °C and at up to 60 MPa. These different conditions of origin and different properties of the end product place graphitization far from coalification.

During coalification, including anthracitization (the end member is semigraphite), the mineral composition of the neighbouring rocks may also change gradually, particularly in the series: coal – coaly claystone or siltstone – claystone or siltstone. In the metamorphic phase new minerals are formed. Organic substances can be traced even in the zone of greenschists; under the microscope they show a coke-like and inertinite character and partly resemble graphite. Graphite appears in the biotite facies; it may belong to the mesozone, but its characteristic occurrences are in the katazone.

A number of authors have tried to establish a relationship between the coalification degree and the mineral facies. The scheme in Table 15 is based on the V^{daf} and R_{im} (reflectance in immersion) and on the definition of the biochemical and geochemical phases. The biochemical phase is related to the upper part of the zone of diagenesis (depth of burial of hundred-metres order), which implies that it includes the peatification, the post-peatification stage and part of the brown-coal stage. The geochemical phase covers the lower part of the zone of diagenesis and the zone of epigenesis (depth of burial of thousand-metres order). The greater part of the geochemical phase takes part in the formation of bituminous coal and anthracite. There are no sharp boundaries either between phases or zones. Table 15 shows that the geochemical phase may be active synchronously with the biochemical phase in the upper part of the zone of diagenesis.

Table 15

Relationship between mineral facies and the degree of coalification

Phase	Zone	Mineral facies	Degree of coalification	V^{daf}	R_{im}	Stability of primary and metamorphogenic minerals
biochemical phase	diagenesis (zeolite facies)	analcime	peat		<0.25	kaolinite, pyrophyllite, muscovite-illite, montmorillonite, vermiculite-biotite, albite, chlorite, glaucophane, actinolite, laumontite, prehnite, analcime, heulandite, pumpellyite, epidote
			brown-coal stage			
			brown coals	>42	<0.6	
		heulandite	high-volatile bituminous coals			
			bituminous hemitype	36–42	0.6–0.8	
geochemical phase	epigenesis	laumontite	bituminous gas + fat coals	36–29	0.8–1.5	
			bituminous orthotype			
		pumpellyite prehnite (metagreywackes)	bituminous coking coals + anthracitic coals	9–29	1.5–2.6	
			bituminous metatype			
		glaucophane (glaucophane schists)	anthracites	4–8	2.6–4	
regional metamorphism	epizone	albite chlorite actinolite (greenschists)	metaanthracites to semigraphites	<4	4–5 →	

109

Every coal seam is in a particular degree of coalification capable of attaching only a definite amount of gaseous and liquid components (methane, water). Polar molecules of water are easily attached, particularly in the superficial parts of the coal, covered with oxygen complexes. Škuta (1972) claimed that every bituminous coal type seems to be geochemically a natural equilibrium system including the following components:

(1) Coal matter (solid constituent) — natural colloidal sorbent with a large inner active surface.

(2) Bound water — either sorbed in the polymolecular layer, or condensed in the capillaries of coal matter.

The capillary water shows the properties of normal water, but the behaviour of the sorbed water is quite different. It freezes at a lower temperature (up to $-78\,°C$), and the substances that are readily soluble in normal water do not dissolve in it. The most different is its first monomolecular layer, which is most firmly attached to the active surface of the coal matter. The more distant are the molecular layers from the surface of coal matter, the less they are attached and the more their properties approach those of the normal water. The ratio of these two water types differs depending on the number of oxygen complexes of polar character. With the increase in their number increases also the proportion of the sorbed water. With the increasing coalification (from $V^{daf} = 40\,\%$ to $V^{daf} = 10\,\%$) the proportions of sorbed water decrease as a result of the decreasing number of active centres with oxygen complexes, and of capillary water as well (owing to the compression of capillaries, i.e. intermicellar spaces during the proceeding compaction of coal matter). According to Škuta (1972) the two components do not decrease uniformly, since during the maturation of the coal matter the contraction of capillaries occurs at a greater rate than the breaking off of oxygen complexes from the active surface of the coal matter.

(3) Bound CH_4 — either sorbed or gaseous, dissolved in capillary water.

In the equilibrium system of coal matter–H_2O–CH_4, water can be regarded as antagonistic to CH_4, which suggests that water to a certain extent controls the gas content of coal seams.

As a result of progressive homogenization in the course of coalification, and of the changes in the chemical composition of the coal matter, its active surface also underwent changes and decreased gradually. The total sorption capacity of coal was thus reduced. In the bituminous hemiphase ($V^{daf} = 43-36\,\%$) and orthophase ($V^{daf} = 36-29\,\%$) the polar surface of the coal matter gradually loses activity up to the coalification jump at $V^{daf} = 29\,\%$. In this phase oxygen separates from the polar oxygen complexes and the coal surface can adsorb a steadily decreasing amount of polar molecules of water. During this process the coal surface is relatively enriched in carbon centres of a polar character. Its methane content increases gradually to attain a maximum at the coalification jump.

After the coalification jump the remaining O-complexes liberate oxygen in a much lower degree. The coal matter can sorb a steadily decreasing number of H_2O and CH_4 molecules, which in the bituminous coal metaphase is reflected in relatively small proportions of these constituents. As all the three components mentioned above are genetically interconnected, H_2O and CH_4 may serve as an indicator of the degree of coalification and their ratio may characterize more precisely the coal of a particular coalification degree.

According to Škuta (1972), methane is released from coal seams at a smaller depth than H_2O and CO_2 because of the different affinity of these gases to the coal matter. Instead of desorbed CH_4, N_2 becomes bonded in this surface zone on account of its high partial pressure in the atmosphere. With increasing depth N_2 disappears, H_2O decreases and CO_2 increases. At greater depths H_2O continues to decrease, CO_2 begins to drop and CH_4 appears, which at a still greater depth is usually the only gaseous constituent. Moore (1940) writes that in the Ruhr Basin the proportion of gases increases with depth; the higher rank coals contain a larger amount of methane. The coals mined in this area yield 177 to 710 cubic feet of methane and 353 to 710 cubic feet of CO_2 per ton.

The amount of methane released from bituminous coal and anthracite during the coalification process is estimated by Colombo et al. (1970) at about 200 l/kg coal. Part of it is captured by coal and the remainder migrates towards the surface. If the overburden provides a suitable environment, methane may accumulate in economically profitable deposits.

The rank of coal depends, among others, on the mode of deposition of the overlying rocks and their petrographical character. The porous rocks above the coal seam permit the gases to escape, therefore the coal is then poorer in gases.

Rogoff (1959) believed the differences in the biochemical changes of plant remains to be important for the formation of the individual coal types. Slow microbial decomposition of plant material for long periods under anaerobic conditions would produce coals with small amounts of N and O, i.e. high-value coal kinds. On the other hand, aerobic oxidation processes involve a smaller loss of N from the system and the product may be richer in O; they may lead to the formation of brown coals. The role of various microbial groups in the decomposition of plant bodies has not yet been solved satisfactorily. Both N and O are important limiting factors in decomposition processes. The deficiency of N almost stops the biochemical processes and the final product of the decomposition of lignin depends strictly on the amount of oxygen.

Since the amino acids may play an important role in coalification, Enders (1943) thought that the products of the decomposition of nitrogenous polysaccharides are responsible for the formation of N-rich coal, whilst the coals poor in N are probably derived mainly from lignin. The chemical reversion of methylglyoxal to humic acids proceeds very likely through aldol condensation with gradual cyclization.

The decomposition products of glycides, the strongly reactive ketonic acids show a tendency to aromatization. The ring structure of aromatic oxiquinones can revert abruptly to reactive ketonic acids with a chain structure.

In some coals helium and small amounts of ethane have also been identified. Helium cannot come from the atmosphere because planet Earth lost its primordial atmosphere very early. Helium has been retained only in a zone at the very margin of the exosphere and could by no means be involved in the life processes of plants. Helium present in coal seams is a product of decay of the radium or thorium series.

Several parameters may be used as the criterion of coalification. The content of carbon in volatile matter (V.M.) is most reliable. Coalification leads to a gradual decrease in oxygen and hydrogen and to the increase in carbon. The components rich in fats, waxes and resins decompose during coalification so that they are impoverished almost exclusively in oxygen and, as the humic substances lose both O + H, they will be enriched eventually not only in carbon but relatively also in hydrogen. For example, the waxy coal from Pila near Karlovy Vary has higher hydrogen and carbon contents and a higher combustion heat than the humolith from the Sokolov Basin (Zelenka, 1973):

	A^d	C^{daf}	H^{daf}	Q_s^{daf}
Sokolov Basin Pila, Corona mine	8.96 %	74.10 %	8.35 %	34.0 MJ
Habartov, Gustav mine	4.63 %	72.01 %	6.21 %	30.5 MJ

Hard, bright, conchoidal humic coals are derived mainly from lignin and cellulose (derivatives of ligneous fibres), and dark dull sapropelitic coals originated from waxes, resins, fats and more resistant proteins. They contain similar substances to bitumens. Coals rich in hydrocarbons (some brown coals, cannel and boghead), are actually transitional to bitumen.

The coalification diagram based on the ratio of H/C to O/C is plotted in Figs. 36 and 37. Van Krevelen (1950) considers the changes as very marked (Fig. 38). The C/H ratio increases with the degree of coalification. Kessler (1974) recommended to use the atomic C/(H + O) ratio for the classification of the coalification degree in coals having a low atomic content of organic sulphur and nitrogen ($<1\%$).

With the increase in the carbon content the oxygen decreases abruptly, but the hydrogen content may even rise; it decreases only in anthracite. Gas coal, for example, may have higher H content than the less coalified subbituminous coal of the same genetic series. The N content increases slightly in bituminous coals and declines again in anthracite. The sulphur content in volatile matter is very low and its changes during coalification are minimal.

The carbon content which is the criterion of the coalification degree is influenced by the maceral composition of the volatile matter (Kessler, 1974), i.e. by the content of exinite and inertinite in bituminous coal and particularly by the content of humic substances in brown coal. The coalification degree is therefore estimated uniformly according to the composition of vitrinite which, however, need not be a criterion of the coalification of exinite and inertinite from the same coal. The chemical

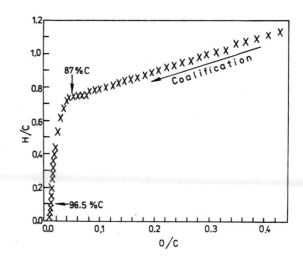

36. Changes in hydrogen and oxygen contents during coalification (according to Patteisky and Teichmüller, 1960).

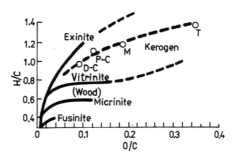

37. Coalification of microliths in relation to kerogen of different geological age. T — Tertiary, M — Mesozoic, P–C — Permo-Carboniferous, D–C — Devonian to Cambrian. According to Welt (1969), modified.

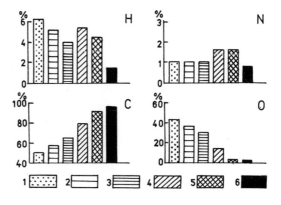

38. Coalification diagram (after van Krevelen, 1950). 1 — wood, 2 — peat, 3 — brown coal, 4 — bituminous coal of lower rank, 5 — high-rank bituminous coal, 6 — anthracite.

composition of macerals provides evidence that the degree of coalification in the same coal increases from exinite to vitrinite and inertinite.

The water content remains high in peat and low-rank coal and their hydrophilic character persists into the stage of bituminous coal. The change into the hydrophobic character of higher-rank coals implies the elimination of OH and COOH and other polar groups during coalification.

During the gradual alteration from the brown coal into the bituminous coal the number of functional groups of the molecule of coal matter decreases. The content of O and later also of H declines with the increase in carbon content. Oxygen cumulates in the marginal parts of coal molecules, i.e. in the hydroxyl, carboxyl and carbonyl functional groups. The study of functional groups deserves particular attention also because it is of assistance in developing the model of coal structure.

Table 16
Functional groups of coal in wt. percentage of oxygen (after Blom et al., 1957)

C	O_{COOH}	O_{OCH_3}	O_{OH}	$O_{C=O}$	Oxygen not included in functional groups	Total oxygen
65.5	8.0	1.1	7.2	1.9	9.6	27.8
70.5	5.1	0.4	7.8	1.1	8.2	22.6
75.5	0.6	0.3	7.5	1.4	6.4	16.2
81.5	0.3	0.0	6.1	0.5	4.2	11.1
85.5	0.05	0.0	5.6	0.5	1.75	7.9
87.0	0.0	0.0	3.2	0.6	1.3	5.1
88.6	0.0	0.0	1.9	0.25	0.85	3.0
90.3	0.0	0.0	0.5	0.2	2.2	2.9
90.9	0.0	0.0	0.6	0.25	1.15	2.0

Blom et al. (1957) reported that with the increase in carbon from 65.2 to 92.8 % the content of —COOH, —OCH$_3$,.—OH and =C=O functional groups decreased (Table 16). At the transition from the brown coal to bituminous coal (70—82 % C in volatile matter) the methoxyl groups are the first to disappear, being followed by carboxyl groups, whilst the content of carbonyl groups only drops considerably and the percentage of hydroxyl groups decreases to a less extent. Within the range of 81—89 % C in volatile matter the OH groups disappear rapidly and at 92 % C (in V.M.) practically all oxygen is present only in heterocyclic form.

The following conclusions can be drawn from the coalification degree and the presence of functional groups in the coal series brown coal — bituminous coal — — anthracite:

(1) The content of functional groups decreases considerably depending on the degree of coalification. Their amount is higher in younger coals.

(2) The oxygen of hydroxyl groups cover 1/4 to 1/8 of total oxygen in coal. Hydroxyl groups of phenol character greatly prevail over alcohol or slightly acidic hydroxyl groups (Blom et al., 1957; Friedman et al., 1961; van Krevelen, 1961). Brown coals may contain up to 9% hydroxyl oxygen. The content of hydroxyl groups decreases with progressive coalification. With the C content about 80%, the hydroxyl groups make up ca. 7%, and with some 90% C the content of functional groups is <1% or nil.

(3) The carboxyl groups are usually present in brown coal and absent from bituminous coal and anthracite. Brown coal contains up to 5.5% COOH (Blom et al., 1957; Manskaya and Drozdova, 1964). With 85% C in coal the carboxyl groups are usually lacking.

(4) Carbonyl groups occur in all coal types, but their content in bituminous coal is very small. Brown coal contains ca. 3.5% carbonyl groups, coal with 85% and more carbon only about 0.6%. Van Krevelen (1961) presumes the main portion of carbonyl groups to enter the structures of orthoquinone coal systems.

(5) The amount of methoxyl groups is usually small and decreases with the increase in polymerization. In case the brown coal contains 2.8% of methoxyl groups, their amount decreases to 0.4% with 72% C and to 0.2% with 80% C.

The data given above present the average values of oxygen in coals containing 70–90% carbon (in V.M.), but the question remains whether the heterocyclically bound oxygen also enters the structure of coal molecules. When the carbon content exceeds 92%, practically all oxygen is present in inactive and stable forms. The functional oxygen groups in various coal types are shown in Fig. 39.

Karweil's correlation diagram (Karweil, 1973) defines the relations between depth, age, percentage of volatile matter and the number of reactive functional

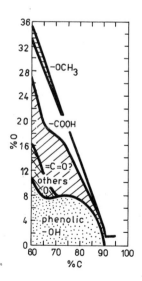

39. Functional groups with oxygen in different coal types (van Krevelen, 1961).

groups (Fig. 40). The number of the reactive groups labelled Z was defined by Huck and Karweil in their paper of 1955 (which virtually started the discussion on the physico-chemical problems of coalification) and is a criterion of coalification along with V^{daf}. However, Karweil's diagram does not solve the problem entirely; for example, the changes in geothermal gradient are not taken into consideration. Nevertheless, some of the examples he quoted are very illustrative.

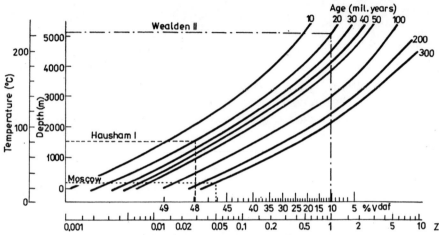

40. Karweil's coalification diagram (1973). Curves of coals from Hausham I and Wealden II boreholes and from the Moscow Basin. Z — number of reactive groups.

Table 17
Optical constants of vitrinites at 546 nm
(F. J. Huntjens and D. W. van Krevelen, Fuel, 33, 88, 1954)

Coal No.	C^{daf} %	Reflectance in cedar oil		Reflectance in air		Refractive index		Absorption index	
		max.	min.	max.	min.	max.	min.	max.	min.
0	58.0	0.26	0.26	6.40	6.40	1.680	1.680	0.02	0.02
1	70.5	0.35	0.35	6.80	6.80	1.705	1.705	0.03	0.03
2	75.5	0.51	0.51	7.25	7.25	1.730	1.730	0.04	0.04
3	81.5	0.67	0.67	7.85	7.85	1.775	1.775	0.05	0.05
4	85.0	(0.92)	(0.90)	8.50	8.45	1.815	1.815	0.06	0.06
5	89.0	(1.26)	(1.18)	9.50	9.30	1.88	1.87	0.07	0.07
6	91.2	1.78	1.55	10.60	10.00	1.95	1.90	0.08	0.08
7	92.5	2.37	1.84	11.70	10.55	2.00	1.93	0.12	0.11
8	93.4	3.25	2.06	12.90	10.80	2.02	1.93	0.20	0.14
9	94.2	4.17	2.22	14.05	11.05	2.02	1.93	0.26	0.16
10	95.0	5.2	2.64	15.35	11.55	2.02	1.93	0.33	0.20
11	96.0	6.6	3.45	17.10	12.55	2.02	1.93	0.39	0.26
Graphite	(100)	11.0	—	22.10	—	—	—	—	—

Table 18
Principal parameters in the alteration of organic matter during coalification
(Raynaud and Robert, 1976)

Rank of coal	% V^{daf}	Reflectance in oil (% R_v)	% C^{daf}	T.A.I.*	LOM**
peat	70 60	0.20 0.25		1	0
lignite and brown coal		0.30	71		2
	50	0.40		2	4
bituminous coals					6
	46	0.50		2.50	7
	44	0.60	77		8
	40				
	37	0.80		3	10
	33	1.0			11
	28	1.25	87	4	
	22	1.50			
	19	1.60			12
	15	1.90			
	12	2.1			
	10				14
	9	2.3	91	5	
	8	2.6			
anthracite	6	2.9 3.0			16
	5	3.5			
	4		93.5		18
metaanthracite	3	4–5			20

* T.A.I. = thermal alteration index according to Staplin, 1969.
** LOM = level of organic metamorphism in the sense of Hood et al., 1975.

The values of parameters of coalification become increasingly more precise thanks to the use of instrumental methods. The most important and most widely used parameter is the reflectance, termed by some authors "microreflectance" (R). It is measured on polished sections in immersion (R_{im}) but can be measured also in air. For correlation purposes it must of course be measured on the same maceral, which is generally vitrinite. The reflectance value is then labelled R_v and given in percentage. In Table 17 the reflectance is correlated with the refractive index and the index of absorption (absorption of light of a particular wave length is measured on vitrinite, less frequently on exinite, and is expressed by transparency or opaqueness; there are still insufficient data available). In Table 18 the reflectance is related to the thermal alteration index (T.A.I.) after Staplin (1969) and the level of organic metamorphism (LOM, according to Hood et al., 1975), which denotes the ratio of the depth of coal seam burial to the mean geothermal gradient.

Information about the coalification degree can also be derived from the values of index of refraction n, which are well comparable in coals of the same rank, or from ultraviolet and infrared spectra (Figs. 41 and 42). The reflectance is at present often measured on MOD (= matières organiques dispersées). In coal studies it is measured on colinite or sporinite, and on kerogen in exploration of oil fields. In this way the coalification stages can be correlated with the formation of hydrocarbons. Both these processes are due to the increase in temperature towards the depth. It appears that the upper limit of the accumulation of oil hydrocarbons from the mother rocks agrees with the transition from the bright brown coals to subbituminous coals (about 45% volatile matter and $R_{im} \sim 0.4\%$). The highest formation of oil hydrocarbons is currently connected with 30–40% volatile matter and R_{im} 0.7% in colinite in the category of subbituminous and gas coals. The coalification facies of bituminous fat coals with 26% volatile matter and 1.2% R_{im} would then represent the "dead line", i.e. the lower limit for the genesis of oil. At a higher alteration degree only methane forms instead of oil hydrocarbons. These values are obviously greatly distorted by differences in thermal gradient, and neither the lower limit of liquid hydrocarbons can be determined accurately. At present the thermal gradient varies within the range of 1–4 °C per 100 m, but the palaeothermal regime of deep profiles varied depending on the time of temperature influence, accumulation of sediments, their nature, thermal conductivity, the character of the original organic matter, etc. (Powell et al., 1976; Malán, 1976). However, it is likely that in oil prospecting by drilling to depths of about 8 km we really approach the absolute depth of hydrocarbons occurrence.

Recently Morishima and Matsubayashi (1978) claimed that electron spin resonance (ESR) parameters of coals changes systematically with increasing temperature. Samples consisting of a homogeneous vitrinite show a good ESR correlation of spin concentration with geothermal parameters and a graph of ESR line width vs. g-value can be useful in the determination of the coalification rank and also in the identification of the maceral (Fig. 43). ESR spectrum usually gives three

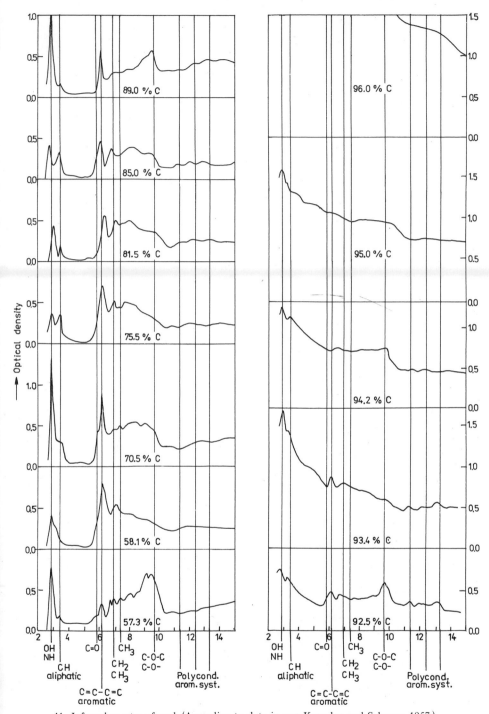

41. Infrared spectra of coal. (According to data in van Krevelen and Schuyer, 1957.)

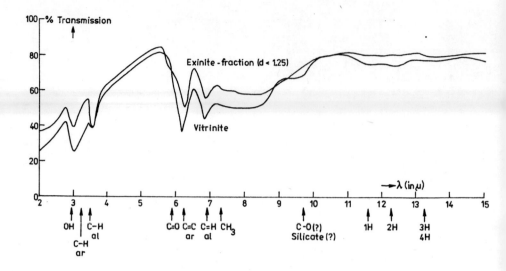

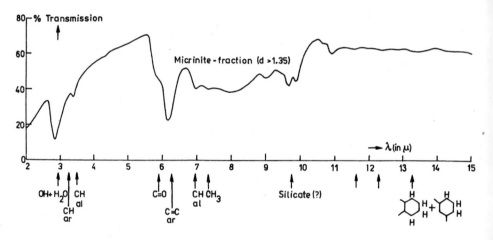

42. Infrared spectra of macerals from Yorkshire coal (van Krevelen and Schuyer, 1957).

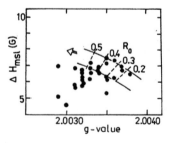

43. ESR diagram with line width plotted against g-value for coaly particles from the MITI-Niikappu well. Vitrinite-rich coals (●) occupy a special zone. Arrow indicates the direction of increasing coalification as shown by dotted lines of equal vitrinite reflectance value R_0 (%) (Morishima and Matsubayashi, 1978).

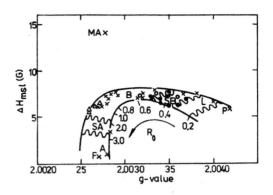

44. ESR diagram of several series of coals consisting mainly of vitrinite. ● — selected coal particles from the MITI-Niikappu well; ○ — coals from Japanese mines; ▵ — coals from other world mines; × — American coals after Retcofsky et al. (1968). P — peat, L — lignite, SB — subbituminous coal, B — bituminous coal, SA — semi-anthracite, A — anthracite, MA — meta-anthracite, F — fusinite (maceral). Reflectance value R_0 (%) is also indicated (Morishima and Matsubayashi, 1978).

parameters, g-value, line width, and spin concentration. Hence it is theoretically possible to identify the maceral species and the coal rank by the combination of ESR parameters themselves.

It is known from the ESR studies of coals, which were mainly on vitrain, that the signal intensity generally increases with increasing rank until it reaches a maximum, the g-value decreases nearly to the value of free spin, 2.0023, and the width behaves in a rather complicated way, depending on the individual macerals (Fig. 44). The scatter of ESR data along the burial depth trend is therefore very likely the result of the coexistence of different macerals of different ranks in the prepared samples (Fig. 45).

Vitrinite is known as an oxygen-rich maceral. Because the ESR g-value is related to oxygen content, the vitrinite group will have larger g-values at earlier coalification stages, as Retcofsky et al. (1968) reported. The ESR parameters depend on both

45. The depth-dependent change of spin number N_s of vitrinite-rich homogeneous coaly particles of the MITI-Niikappu well. ● — vitrinite-rich coal (Morishima and Matsubayashi, 1978).

macerals and coal rank. Vitrinite has also a larger aromatic hydrocarbon content than exinite at an equal rank of coalification.

In analysing the geothermal history with the use of ESR parameters it is essential to select homogeneous samples as concerns maceral and rank. Similar problems have also been studied by Toyoda et al. (1966), Yokokawa (1969) and Yen and Sprang (1977).

The results of individual coalification stages as established in Czechoslovak caustobioliths are listed in Table 19.

In summary, it can be stated at first that the discussion whether cellulose or lignin are the source materials of coal seems to be closed. Only in extreme conditions (formation of the durite constituent) is cellulose fully decomposed (it has been found in lignites of Early Tertiary age) and lignin with exines is the parent material of coal. Since coalification is a chemically highly complicated process, it is obvious that coal as its end product will have a complex heterogeneous structure, too. The evaluation of the sources of energy necessary for the processes of conversion of plant material into coal has shown that neither bacteria nor hydrostatic pressure and locally increased temperature were the controlling geological factors in the development of immense coal reserves on the Earth. The needed energy could have been supplied by tectonic stresses and the activity of low temperatures over long ages.

It has been recognized that aromatic structures and various oxygen functional groups play a relatively important role. The colour of coal can be explained in terms of free radicals it contains in large amounts. Dehydration, decarboxylation and demethanation represent the succession of the first coalification reactions. Humic acids are dominantly the products of oxidation. Resins and waxes resist to decomposition in the first stages of coalification but later on they undergo polymerization and aromatizing reactions (Timofeev – Bogolyubova, 1971).

The stage of hypergenesis occurs when the coal deposit is exposed to the effect of air oxygen. It may be intensive when part of the deposit is uncovered. Owing to the differences in temperature and pressure, in gas and water regimens, the Eh changes and the deposit is weathering; the coal is oxidized and loses its luster. As a result of oxidation humic acids are formed. In bituminous coal hydroxyl groups accumulate and subsequently carboxyl groups and still more humic acids are produced. When the denudation is slow, coal is converted into gas products and only inorganic components remain.

In brown coals the secondary humification process does not occur, because they themselves contain humic substances, the amount of which rises in the course of oxidation. The capability of oxidizing must be considered in open-cast mining. With this working procedure, some brown coals in northern Bohemia occasionally lost more than 0.8 MJ/kg of calorific value and the yield of tar decreased more than 2%.

The sulphuric acid resulting from the oxidation of pyrite is a strong oxidizing

Table 19
Elemental analyses of wood, peat and Czechoslovak coals of different coalification grades (Hubáček, 1948) in %

	Water (W_t^r)	C^{daf}	O^{daf}	H^{daf}	N^{daf}
Wood	20–60	49–52	43–45	5.9–6.1	0.2
Peat	85–91	56–62	30–36	5.3–6.3	1–3
Lignites (in S. Moravia and S. Bohemia)	40–52	62–69	25–30	5–6.2	1
Earthy brown coals (Sokolov-Cheb coals)	40–50	68–72	19–23	5.6–6.2	0.9–1.2
Brown coals (Sokolov Basin)	35–48	68–75.6	16–23	5.1–6.5	0.8–1.4
Brown coals (North Bohemian Basin)	25–38	69–76.5	14.5–22	5.3–6.4	1–1.5
Brown pitch + bright coals (North Bohemian Basin)	10–20	75–80	11–16	5.5–6.5	1–1.4
Brown coal (Sokolov Basin)	10–20	73–78	11–15	about 7	1
Waxy brown coals (Karlovy Vary and Cheb Basins)	48	73–75	12–15	8–9	0.3–0.5
Subbituminous coals (Rakovník-Slaný and Radnice-Břasy coals)	12–25	73–80	11–19	4.8–5.6	1.3–1.7
Subbituminous coals (Kladno and Plzeň coals)	8–15	80–83	10–14	4.9–5.2	1.3–1.4
Bituminous gas coals (Ostrava and Nýřany coals)	3–10	80–85	7–13	5.3–5.9	1.3–1.7
Bituminous coking coals (Ostrava and Rosice coals)	1–5	82–90	3–11	4.5–5.5	1.4–1.9
Anthracitic coals (lower Ostrava seams)	1–3	86–92	3–8	4–5	1.1–1.5
Anthracite (S. Bohemia and Brandov)	1–3	92–95	1–4	2–3.5	1.1–1.3

Coals of the Rosice-Oslavany and Žacléř-Svatoňovice Basins have more than 3% S (in volatile matter).

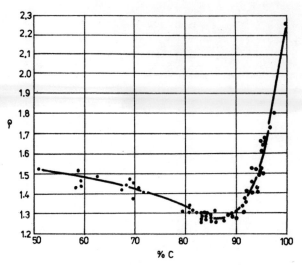

46. Density of coal. Summarized by van Krevelen and Schuyer (1957) from several papers.

agent and its presence facilitates oxidation of the coal. The fundamental cause of the heating is the oxidation of the coal itself. Heat generated during oxidation may effect spontaneous ignition if it is not lost to the surroundings quickly enough.

Coal as an organogenic rock is characterized by physical properties such as colour, streak, lustre, fracture, hardness and density, which in many cases safely distinguish coal from other rocks. Index of refraction and reflectance have been discussed on p. 118 (see also Fig. 47).

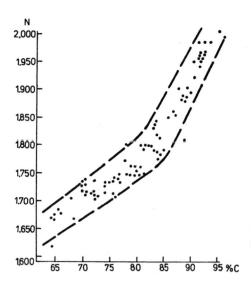

47. Refractive index (N) of vitrinite in relation to coalification grade (in white light). (According to Ivanov, 1975, modified.)

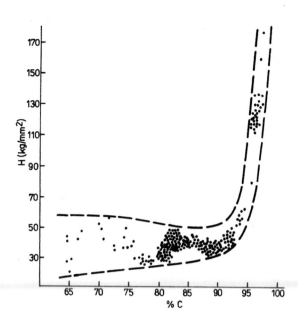

48. Microhardness of vitrinite from coals of various grades of coalification. (According to Ivanov, 1975, modified.)

Table 20

Chemical compositions and densities of some vitrinites of the coalification series
(D. W. van Krevelen and H. A. G. Chermin, Fuel, 33, 79, 1954)

Coal No.	C	H/C	O/C	N/C	d^* g/cm^3
1	70.5	0.862	0.247	0.015	1.425
2	75.5	0.789	0.181	0.015	1.385
3	81.5	0.753	0.108	0.017	1.320
4	85.0	0.757	0.071	0.016	1.283
5	89.0	0.683	0.034	0.018	1.296
6	91.2	0.594	0.021	0.015	1.352
7	92.5	0.509	0.016	0.015	1.400
8	93.4	0.440	0.013	0.015	1.452
9	94.2	0.379	0.011	0.013	1.511
10	95.0	0.307	0.009	0.013	1.587
11	96.0	0.223	0.007	0.012	1.698
Graphite	(100)	–	–	–	2.250

* All density values listed in the table have been related to dry, ash-free coal, using the correction formula

$$d = \frac{(100 - A) d' d_a}{100 d_a - d' d_a},$$

where d' = true density of the ash-containing, dry coal; d_a = average density of the ash constituents; A = weight percentage of ash in the coal.

Hardness of coal is generally determined microscopically. It is not determinable in softer coals, where it varies from point to point, but in anthracite it attains 2.75−3 and in semianthracite 2.5−2.75 on Mohs' scale. The microhardness values are more precise and increase slowly with progressing coalification, and at a greater rate from bituminous coal to anthracite (Ivanov, 1975; Fig. 48). The density of coal also depends on the degree of coalification (Fig. 46; Tables 20 and 21 for vitrinite). Generally, lignite and brown coals have a density of ca. 0.05−1.30; bituminous coals commonly 1.22−1.35 but the range may be wider, i.e. 1.15−1.5; cannel coal 1.2−1.3 and anthracite 1.40−1.70.

The parameters of coalification are also observable microscopically. The three main macerals − vitrinite, inertinite and exinite − change colour and lustre with increasing coalification. At first the colour and lustre of vitrinite blend with those of exinite, and in meta-anthracite the vitrinite and inertinite may have the same colour and lustre.

The colour of brown coals varies from light-brown to dark-brown; bituminous coals are greyish black to jet black in the higher ranks. Waxy brown coals are yellowish brown to yellow. The coals of lighter colours are richer in humic acids. Dark to pitch-black colours indicate a higher degree of coalification. Cannel coals and coals having an increased amount of mineral components are usually dull; bright and semi-bright coals are derived from ligneous plants. Anthracite is greyish black with a typical yellowish metal tint. The colour of coal depends on the original matter (humus or sapropel), the degree of coalification, petrographical composition and the content of mineral substances. The same parameters influence the streak. Both colour and streak provide only qualitative data and broad information on

Table 21
Densities of coal (R. E. Franklin, Trans. Faraday Soc., 45, 274, 1949)

	Densities (after 24 hrs) measured in:				"Helium density"
	methanol	water	n-hexane	benzene	
↓ Increasing rank of coal ↓	1.327	1.370	1.374	−	1.341
	1.387	1.328	1.272	1.321	1.317
	1.387	1.326	1.392	1.342	1.305
	1.402	1.333	1.297	1.292	1.302
	1.387	1.307	1.262	1.276	1.301
	1.334	1.297	1.272	1.286	1.293
	1.333	1.219	1.286	1.297	1.311
	1.352	1.305	1.297	1.299	1.337
	1.374	1.318	1.300	1.293	1.361
	1.549	1.475	1.425	−	1.497
	1.556	1.488	1.433	1.450	1.517
	1.700	1.630	1.497	1.518	1.645

the composition of coal. Streak is defined as the colour of mineral in its powdered form and is usually obtained by rubbing the specimen on a plate of unglazed porcelain. The streak of brown coals and lignites is yellow to brown, of bituminous coals brownish to black, and of cannels brown to black. Coke coals have a dark cinnamon to black coloured streak. Anthracite has a velvet-black streak. In general, the streak of higher ranks is dark-brown to black.

Another qualitative physical parameter is the lustre. Lignites and brown coals are of earthy to dull lustre, as are cannel, durite and fusite; wood components make bright lustred bands. Natural coke has a submetallic lustre and anthracite a bright to submetallic lustre; some anthracites are dull.

Jones et al. (1973) studied the optic character of vitrinite and found out that in case the vitrinite bands are parallel to stratification, vitrinite is uniaxial negative. Anisotropy was probably caused by the weight of overburden. Aromatic lamellae which make up the greater part of the molecular structure of vitrinite show form birefringence and specific birefringence. When the vitrinite bands trend at right angles to stratification, their optic character rather resembles that of biaxial materials. Optic anisotropy of coals is seen only in higher ranks.

As concerns the fracture, lignite tends to break irregularly or along roughly tabular surfaces. The fracture of brown coal is uneven or check-like to subconchoidal. Bituminous coals break with a blocky or cubic fracture, anthracite and cannel with conchoidal fracture.

ELEMENTARY COMPOSITION AND CHEMICAL COMPOUNDS OF COAL

The essential organic elements C, H, O, N and S are usually determined in all coal types. Numerous analyses of Czechoslovak and foreign coals are listed in the works of Pavlíček (1927), Včelák (1962), Havlena (1963, 1964, 1965), Hubáček (1948) and Hubáček et al. (1962). Several analyses providing a survey of the elementary composition of various coal types are presented in Table 22.

Elementary composition (C, H, O, N, S in volatile matter) of vitrinite, exinite and inertinite in the coal of the Ostrava-Karviná coalfield was recorded by Kessler (1972). Exinite is characterized by a high proportion of hydrogen.

Of particular interest is the water content, which in fossil fuel varies within a wide range, depending on the degree of coalification. The higher the coal rank, the lower is the water content. This relationship is defined by the structure of coal and its

Table 22
Chemical analyses of some Czechoslovak coals (in percentage) (Hubáček, 1948)

Type of coal, locality	Ash	Volatile matter				
		C	O	H	N	S
lignite, Ratíškovice, S. Moravia	16.57	68.55	23.37	5.71	1.05	1.32
brown coal, Bžany-Hradiště, N. Bohemia	13.23	71.06	20.01	5.31	1.22	2.40
brown coal, Jeníkov, N. Bohemia	8.91	73.41	17.97	5.95	1.20	1.47
brown coal, Handlová, Slovakia	10.30	72.39	19.35	5.44	1.39	1.43
Kounov subbituminous coal, Lhota p. Džbánem, Bohemia	19.46	76.84	11.57	5.37	1.67	4.55
subbituminous coal, Lužná, Bohemia	28.28	78.72	14.06	5.43	1.13	0.66
bituminous baking coal, Svatoňovice (Rtyně)	10.21	83.68	6.65	5.24	1.29	3.14
bituminous baking coal, Ostrava	9.18	82.02	5.76	5.46	1.84	0.92
bituminous coking coal, Zastávka near Rosice, Moravia	15.14	88.78	1.75	4.23	1.60	3.64
anthracite, Úsilné near Č. Budějovice	14.80	92.32	1.30	2.79	1.10	2.49

colloidal nature. Špetl (1956) gave the following values of water content: 80—95 % in peat, 38—45 % in brown coal from the Sokolov Basin, 28—37 % in brown coal of the North Bohemian Basin, 17—22 % in coal from Handlová, 9—15 % in subbituminous coal from the Kladno and Plzeň Basins, and 4—7 % in the bituminous coal from Ostrava. Anthracitic coal and anthracite contain only 1—3 % water. Water in coal is either free or physically and chemically bound to the volatile matter or to ash matter. Free water is not characteristic of coal; it trickles out and can be removed by simple mechanical procedures (filtration or centrifugation). Water bound to the volatile matter is either capillary or hygroscopic (constituent of the colloidal coal system) and constitutional water (bound chemically). The character of water bound to ash matter is determined by its mineral content.

It has been recognized that coal and oil have a higher $^{13}C/^{12}C$ ratio than limestone, dolomite and other carbonate minerals. Wickman (1953) found that the younger coals have only a little higher $^{13}C/^{12}C$ ratio than the older coals and in many cases the difference is not demonstrable at all. He did not manage to establish a relationship between the degree of coalification and the isotopic composition of carbon, the content of volatiles (2.5—40 %) or the depth of deposition. This is at variance with the finding of Geissler and Belau (1971), who claimed the increase of $\delta^{13}C$ with the coal rank on the basis of the study of peat—brown coal—bituminous coal—anthracite series. Jeffery et al. (1955) determined a wide range of values for coals of different ages and explained it in terms of the varying composition of the atmosphere and hydrosphere during the geological periods. Craig (1953, 1957a, b) stated a range of -21.1 to -26.7 ‰ and the average value -23.7 ‰ for $^{13}C/^{12}C$ ratio in coal (Fig. 49). The distribution of this ratio in coals is analogous to that of present-day terrestrial plants. $\delta^{13}C$ does not show any

Ash													
SiO_2	TiO_2	Al_2O_3	Fe_2O_3	FeO	MnO	CaO	MgO	Na_2O	K_2O	FeS	SO_3	P_2O_5	CO_2
17.63	0.13	10.19	16.20	0.14	0.49	22.46	2.95	0.44	0.22	0.06	28.57	0.09	—
12.39	0.57	9.61	30.25	—	—	20.39	1.94	1.49	1.49	—	23.22	—	0.14
33.40	0.37	17.11	31.61	—	—	9.81	2.50	0.63	0.63	—	3.69	0.57	0.31
40.58	0.55	20.93	13.63	—	—	13.09	2.80	2.29	2.29	—	5.56	0.19	0.38
48.79	0.43	21.02	23.89	—	—	1.84	1.00	1.57	1.57	—	1.21	0.12	0.13
53.99	0.49	40.66	3.06	—	—	0.52	0.49	0.23	0.41	—	0.02	0.12	—
37.63	0.31	22.37	13.16	—	0.86	10.76	4.20	0.58	0.58	0.74	7.58	1.43	0.38
41.68	0.63	26.68	12.94	—	—	5.91	2.38	0.76	0.77	—	5.08	1.50	1.02
44.22	0.29	9.23	16.03	—	—	14.82	1.69	2.38	2.38	traces	10.63	2.33	0.66
27.09	0.31	9.38	18.91	0.41	0.41	26.79	0.84	0.76	1.05	0.33	11.46	0.52	1.61

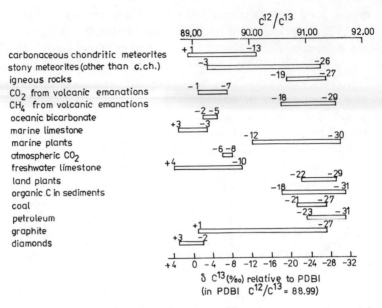

49. Variations in the isotopic carbon composition of some natural materials. Data mainly from Craig (1953) and Polański (1961).

dependence on the coal facies, or rank and type of coal. It is remarkable that coalification did not produce any measurable fractionation, as far as can be inferred from the available data. Isotopic fractionation of carbon decreases with increasing temperature (anthracite; contacts with igneous rocks), as was recorded by Colombo et al. (1970). The isotopic composition of carbon in coal and CH_4 and CO_2 obtained from it by desorption was studied by Alekseev and Lebedev (1977). They did not state any relationship between the isotopic composition of carbon in coal and the degree of coalification, but they thought that the appearance of lighter methane might be explained most plausibly by fractionation at the synchronous bacterial production of methane in the coal seam. They determined $\delta^{13}C_{CO_2} = -1.88$ (in ‰ to RDV) and $\delta^{13}C_{CH_4} = -3.38$ to -5.59 (in ‰ to RDV).

Chemically, the coalification process leads to the alteration of the primary material into humic substances that are present in coal in colloidal form (Cooper – Murchison, 1969).

The humic acids of brown coals differ from humic acids of peats in a higher condensation degree and a smaller number of additional ligands and functional groups. The formation of aromatic chains in humic acids (Manskaya and Drozdova, 1964) can be ascribed to natural phenolic substances, to the aromatization of hydrocarbons or to the melanoid reaction.

The following aromatic and heterocyclic chains may be presumed in humic acids: benzene, naphthalene, anthracene, furane, pyrrole, indole, pyridine and quinoline.

Most authors believe nitrogen to be an indispensable component of humic acids. It is present in the form of:

(a) end group NH_2,
(b) in acyclic bond —NH or —N—,
(c) in a closed chain of heterocyclic substances such as —NH— (pyrrole, indole) or =N— (pyridine, quinoline).

Varossieau (1950) and Varossieau and Breger (1952) after studying the buried and decayed wood both biologically and chemically, arrived at the conclusion that hydrological processes are active in decomposition and removal of cellulose, whilst demethoxylized lignin relics may be classified as humic substances, which formed and cumulated synchronously.

In the early stages of coalification cellulose and lignin undergo partial depolymerization and decomposition. Whereas a molecule of recent cellulose contains 2,000—3,000 glucose units, the Miocene brown coals contain only 40—800 glucose units.

Chemical characteristics of the least carbonized types — wood from peat and lignite — are listed in Tables 23 and 24 (Manskaya – Kodina, 1975).

On the basis of chemical analyses Gillet (1948, 1955, 1956) derived a general formula of coal, presuming the cyclization of lignin chain structures into large cyclic units. Lignin is resistant to microbial decomposition and has an aromatic structure as is the case of coal. Cellulose decomposes into CO_2 and various acids during coal formation; it has not an aromatic structure. Miller (1949) wrote that the structure of coal was usually regarded as a condensed complex either of polymer or polymer-like compounds with the exception that the units cannot repeat regularly, which is characteristic of true polymers. The position of nitrogen is not clear but it may be a regular component of the polymer units or may substitute for carbon in aromatic nuclei. Fuchs (1942) maintained that the presence of hydroxyl and carboxyl groups in brown coal indicates that the oxidation processes influenced the conversion of plant material into this fuel. The absence of these groups in bituminous coal and anthracite combined with the low oxygen content shows the importance of the reduction processes in the development of high-rank coals.

The optical properties of coal also depend on its aromatic type. According to Miller (1949) the index of refraction is highest in pure aromatic coal types. It is related to the C/H ratio. In the densified aromatic nuclei having the highest C/H ratio, the index of refraction is also the highest.

The study of the chemical compounds of peats has revealed a great difference in the primary and partly decomposed plant constituents. The material of peats in the Quebec area contained ca. 7—10% hemicellulose, 6—7% bitumen, 5—10% cellulose, 29—49% humic substances and 7—17% lignin. From these analyses it is evident that cellulose and lignin remains decrease with the increase in humic substances.

Table 23

Chemical composition of peatified wood and lignites (Manskaya and Kodina, 1975)

Sample	Ash %	C^{daf} %	H^{daf} %	N^{daf} %	OCH_3^{daf} %	Humic acids in dry matter %	Vanilline* in lignite	n-Oxybenz-aldehyde* in lignite
Peatified wood, *Pinus silvestris* L.	0.13	50.1	5.45	tr.	8.03	3.6	27.5	1.12
Lignites								
Pinaceae:								
Pinuxylon paxii Kräusel	4.37	47.2	6.15	—	6.16	45.1	8.55	<1
Pinuxylon paxii Kräusel	10.60	53.9	4.78	—	5.95	66.3	3.1	traces
Pseudolaricixylon firmoides Chudajb.	11.00	59.7	5.33	1.25	6.06	69.8	3.24	traces
Cupressaceae:								
Cupressinoxylon durum Kräusel	6.70	57.1	4.45	0.56	6.97	61.1	4.6	1.7
Cupressinoxylon hausruckianum Hofm.	6.55	50.8	4.53	0.81	5.96	93.9	4.57	2.1
Taxodiaceae:								
Taxodioxylon taxodii Goth.	20.00	53.5	4.94	2.36	6.92	49.7	10.05	0.86
Taxodioxylon anthratoxoides Chudajb.	25.80	58.8	5.72	—	2.82	76.7	2.1	traces
Taxodioxylon ishikuraense Takamotsu	5.16	58.2	6.20	—	6.15	70.7	4.66	0.97
Glyptostroboxylon tenerum Conw.	6.97	56.5	5.14	0.57	4.58	86.0	4.31	1.27
Glyptostroboxylon tenerum Conw.	22.00	51.5	5.80	—	4.47	48.3	3.78	<1

* Aromatic aldehydes given in mg/g dry matter of lignite or wood.

Table 24
Chemical composition of lignites (Manskaya and Kodina, 1975)

Petrographical characteristics	Ash %	C^{daf} %	H^{daf} %	OCH_3^{daf} %	bitumens	lignin	humic acids	OCH_3 in lignin %	sum of aromatic aldehydes	
					% in dry matter				mg/g lignite	mg/g lignin
lignite with *Taxodioxylon sequoianum*, Kräus., Miocene–Oligocene	2.68	60.19	6.60	8.09	5.0	65.0	5.0	10.05	2.5	4
lignite with *Podocarpoxylon severzovii* (Merkl.) Jarm., Palaeogene–Upper Cretaceous	3.60	60.14	5.97	8.28	0.53	69.1	12.5	7.05	1.8	11
Tertiary lignite (light brown)	1.66	65.73	6.47	4.45	14.0	35.0	3.6	8.27	1.0	12
Tertiary lignite (dark brown)	7.39	51.19	6.59	5.51	0.87	42.1	17.6	8.66	8.3	25
coalified wood (lignite), Jurassic	1.82	57.40	5.92	7.33	1.61	50.6	11.5	9.25	1.7	15
coalified wood (vitrain), Triassic	7.72	54.86	5.31	–	0.12	77.0	8.6	5.67	5.5	8

Francis (1961) recorded that the brown coal from Victoria (Australia) having the elementary composition of $C = 66.1-71.8\%$, $H = 4.4-5.8\%$, and O, N, S = $= 23.6-32.5\%$, contains the following amounts of components: resins, waxes and hydrocarbons $= 3.9-7.1\%$, cuticle remains $= 0.65-10.4\%$, opaque matter $=$ nil to traces, fusite = nil to traces, humic substances $= 85.5-95.45\%$. In 22 samples of brown coal from Czechoslovakia the amount of humic acids decreases with the increase in coalification (Hubáček – Lustigová, 1962). Diaconescu and Doma (1960) established that younger coals contain more humic acids than the older coals and also show a higher capacity for ion exchange.

Cellulose remains occur in peat and brown coal in various amounts, which decrease with depth. In the lignite of southern Bohemia and southern Moravia Hubáček (1962) determined $6-11\%$ cellulose. Mueller (1972) reported preserved leaf cuticles and cellulose from the ca. 20 cm thick seam of "paper coal" from Park Co., Indiana. Hemicellulose is occasionally also present. The amount of lignin in peat is generally $10-35\%$; it decreases with proceeding coalification to $3-10\%$ in brown coal.

Humic acids occur in both peat and brown coal, but are rare in bituminous coal

Table 25
Contents of amino acids in Upper Carboniferous coals in the Ruhr and Saar Basins
(Degens and Bajor, 1962)

		Bituminous coal Ruhr Basin	Bituminous coal Saar Basin
C %		85.47	80.40
H %		5.07	4.93
N %		0.78	0.95
H/C	$(\times 10^2)$	6	6
N/C	$(\times 10^3)$	9	12
Amino acids	$\times 10^{-5}\%$	128.29	120.59
Amino acids/C	$\times 10^{-5}\%$	1.50	1.50
Alanine	$\times 10^{-5}\%$	12.05	15.26
Tyrosine	$\times 10^{-5}\%$	3.25	3.25
Valine	$\times 10^{-5}\%$	8.76	12.25
Cystine	$\times 10^{-5}\%$	6.25	5.76
Asparagine	$\times 10^{-5}\%$	8.75	3.25
Glutamine	$\times 10^{-5}\%$	12.50	10.75
Serine	$\times 10^{-5}\%$	16.02	15.03
Threonine	$\times 10^{-5}\%$	5.00	6.25
Lysine	$\times 10^{-5}\%$	10.50	6.76
Arginine	$\times 10^{-5}\%$	4.25	7.76
Proline	$\times 10^{-5}\%$	7.26	12.27
Glycine	$\times 10^{-5}\%$	33.70	21.50
Lysine + arginine	$\times 10^{-5}\%$	14.75	14.52
Lysine + arginine/alanine		1.22	0.95

(in which rather humins are present). Humic acids are in peat either free or bound to metals such as Ca, Fe and Al. According to Kukharenko (1959) the humic acids of peats contain 53—61 % C in volatile matter, those of earthy brown coals 62 to 67 % C and of bituminous coals 67—71 % C. The hydrogen content in humic acids dropped from 5 to 4 % with the increase in C content. Hubáček et al. (1962) reported 63.84 % C and 3.77 % H with 2.5 % N in humic acids isolated from the brown coal of the Fučík mine in the North Bohemian Basin.

Humins are more complicated substances than humic acids. Some authors think them to be a more advanced product of carbonization of humic or other substances. Humins have a higher C content (75—90 %) and 4—6 % H at the most. The functional groups are represented by the carbonyl groups. The humins of humus character contain less hydrogen and are easily converted to humic acids. The humins of sapropelitic coal, cannel and boghead are richer in hydrogen and are not easy to convert into humic acids. They are monocyclic or polycyclic anhydrides of carbonic acids and neutral oxygen compounds of cyclic nature. Humins of sapropelitic coals developed obviously both from plant and animal materials. Humins of anthracite coals and anthracite have a relatively small hydrogen amount; they consist of condensation aromatic systems deficient in functional groups.

Although the humic acids are absent from higher-rank coals, they are produced by their weathering (Kukharenko, 1959, 1970). The humic acids in weathered coals are a product of oxidation, whereas those in peat and brown coal are for the most part the result of biochemical decomposition of plant remains. The humic acids formed by weathering lack the methoxyl groups and have a lower H, a higher content of carboxyl groups, a lower percentage of hydroxyl groups, a higher optic density and a lower coagulation capacity of alkaline solutions than the humic acids of biochemical origin (Kukharenko — Ekaterinina, 1960).

The modern methods of isolation and identification enable us to gain insight into the chemistry of coal and its components and to determine even substances that are present in minute to trace amounts.

In 1962 Degens and Bajor stated remarkable differences in the contents of amino acids in the Upper Carboniferous lacustrine and marine sediments of the Ruhr and Saar Basins. The total amino acids were higher in marine than in freshwater clayey and silty deposits, altough the freshwater sediments contained a higher amount of carbon. This fact can be applied in the determination of the salinity of environment (see Bouška et al., 1979, and the chapter "Palaeosalinity of the environment" in this book). Degens and Bajor (1962) also determined the contents of amino acids in coal (see Table 25). Detailed study of amino acids in coal was carried out by Bajor (1960). Diamino- and monoamino acids occur in peat and brown coal; bituminous coal contains only more stable monoamino acids.

Prashnowsky (1971a) studied the biochemical composition of the material from the Münsterland I, the deepest borehole in Europe (5,965 m). The Upper Carboniferous rocks were claystones and sandstones with coalified plant remains and coal lenses and contained saccharides, amino acids and fatty acids. The most stable saccharides were glucose, galactose, arabinose and xylose, the most abundant being hexoses. Pentoses increased in amount with depth and thus with the age of sediments. The most stable amino acids in the Carboniferous claystones were hydroxymonoamino carboxylic acids (serine, threonine), monoamino dicarboxylic acids (asparagine and glutamic acid) and amino acids containing sulphur (cystine, methionine). Their content decreased with depth. In studying the contents of saccharides and amino acids in the coal and "tonstein" of the Ruhr Basin, Prashnowsky (1971b) ascertained a varying representation of saccharides in the coal seams and in the sequence studied. In coal, hexoses were usually most abundant. In addition to glucose, galactose, mannose, arabinose, xylose, ribose and rhamnose, the stratigraphical sequence of the Essen Formation (Westphalian B_1) also contained fructose and maltose, and the coal of the Bochum Formation (Westphalian A_2) fructose, maltose, lactose and glucosamine (see also the paper of Prashnowsky and Burger, 1966).

Saturated aliphatic compounds (alkanes C_{24} to C_{32}), aromatic hydrocarbons, heterocyclic compounds, and fatty acids have been determined in fractions extracted from a bituminous coal in a formation with a fossil flora dominated by gymnospermous plants (Pedersen and Lam, 1975). Aliphatic compounds are known from recent gymnosperms and dibenzofuran derivatives from lichens.

Distribution maxima of the n-alkanes, as shown by gas chromatography, range from C_{27} and C_{29} in lower ranks to as low as C_{16} in higher ranks (Allan and Douglas, 1977). The distribution also shows a progressive decrease in the preference of odd-carbon-number homologues with increasing rank. Quantitative data indicate that the loss of the odd-carbon-number preference occurred, for the most part, while individual long-chain homologues increased in concentration. There is a progressive increase in the amounts of shorter-chain n-alkanes with increasing rank. See also the papers of Allan et al. (1975), Brooks and Smith (1967), Leythaenser and Welte (1969) and Schnitzer and Neyroud (1975).

Among other substances identified in coal there are deuteroethio-porphyrin, and in boghead coals desoxyphyllerythrin and vanadium-porphyrin complex (Swain, 1970). Creosol, xylenol, phenol and naphthol occur in coal tars. Dilcher et al. (1970) separated methyl pheophorbide from the Eocene coal from Geisel near Halle (G.D.R.). It is the oldest fossil pheophorbide reported in the literature.

Bituminous substances are represented in caustobioliths of the coal series in various amounts. Two kinds of waxes are distinguished: peat wax and montan wax (bitumen from brown coal). The montan wax was discussed in detail by Včelák (1959). Waxes and resins are either liptobiolithic (true natural waxes and resins) or sapropelitic (formed by polymerization of plant and animal remains rich in fats and proteins). The best known waxy coals are in Halle (G.D.R.) and in Devonshire; in Czechoslovakia they occur at Pila near Karlovy Vary and at Uhelná near Javorník.

Zelenka (1974) recorded the following proportion of waxes and resins in the bitumen extract from four samples of brown coal from the North Bohemian Basin: with 9–12 % bitumen in volatile matter (extracted by the benzene–alcohol 4 : 1 mixture, for 4 hrs.) waxes cover 2.64–3.66 % in volatile matter and resins 6.31 to 9.23 % in volatile matter. Typical liptobioliths have not been found in the North Bohemian Brown-coal Basin so far, unless we assign to them the fossil resin – duxite.

Ivanov (1975) analysed the Tertiary brown coals from the Aleksandrovskoe locality in the Dnieper Basin and obtained 8.71 % extract of dry coal (33.3 % resins and 66.7 % waxes); from the Babaevskoe locality in the South-Uralian Basin he obtained 14.31 % extract (63.2 % resins and 36.8 % waxes). Both these coals have a high content of extract. The differences in the resin/wax ratio are most probably due to the primary materials.

The content of bitumen is generally highest in peat and brown coal, being much smaller in bituminous coal; with the 90 % carbon content it disappears completely. As a result of dehydrogenation essential chemical changes take place, leading to the loss of compounds extractable by organic solvents. Moore (1940) quoted several papers which mention the finds of paraffins in the coals from England (Yorkshire, North Staffordshire etc.). In most cases the incomplete series $C_{13}H_{18}$—$C_{18}H_{38}$, $C_{10}H_{22}$, $C_{11}H_{24}$, $C_{13}H_{26}$ and $C_{32}H_{66}$ were concerned. The cited papers are of earlier date and complete series have not been differentiated. P. P. Bedson (1907; in Moore, 1940) writes: "It is probable that members of the paraffin series are much more common in coal than they were formerly believed to be, but they are likely to be overlooked and not separated in analyses."

Ouchi and Imuta (1963) identified a number of hydrocarbons in the extract from coal of Yubari. They determined paraffins C_9–C_{31} with dominant C_{20}. Benzene, naphthalene, phenanthrene, chrysene and picene were the main aromatic substances identified; their alkylated derivatives were also found. In addition, fluorene, anthracene, pyrene, benzofluorene, fluoranthene and 1,2-benzanthracene were determined. Inert oxygen was present chiefly in heterocyclic compounds such as diphenylene oxide and benzophenylene oxide.

The bitumen from Sangara coal (U.S.S.R.) contained more condensed aromatic hydrocarbons than non-aromatic (Danyushevskaya, 1959). Rushev and Dragostinov (1965) using infrared spectral analysis identified in the bitumen from brown coal alicyclic hydrocarbons, $COCH_2$ groups or cyclic hydrocarbons; Spence and Vahrman (1965) found in olefins from the bituminous coal extract a complete series C_{12}–C_{38}; C_{20}–C_{27} formed the largest proportion, in which C_{23} and C_{24} slightly predominated. Given and Peova (1960) determined 1–2.5 % carbonyl groups in the coal extract, which belonged mainly to ketones.

Extracts from brown coal rich in fusain contained 50 % substances of the phenolcarboxylic acid (Karavaev et al., 1965).

In the brown xylitic coal from Nováky (Slovakia) Streibl et al. (1972) identified the diterpene hydrocarbon iosene ($C_{20}H_{34}$; Fig. 50a), which is also known from other brown coal deposits (Simonsen and Barton, 1952; Streibl and Herout, 1969) and has been described under various synonyms such as hartite, hofmanite, branchite, rhetenite, krantzite and bombiceite. Fossil iosene is identical with α-phyllocladane which appears in the recent Australian coniferous genera *Phyllocladus*, *Dacrydium*, *Podocarpus*, *Cryptomeria* and others. Brown coal also contained tricyclic aromatic hydrocarbon simonellite, whose structure resembles that of diterpenes (Ghighi and Fabbri, 1965). The total yield of hydrocarbons from a coal was found to bear a linear relationship to its hydrogen content.

Streibl and Herout (1969) provided a very detailed information on the terpenoids contained in coal. In their summarizing study they also quoted numerous references related to this subject. The authors are of the opinion that most terpenoids and their derivatives are of plant origin. They mentioned mono- and sesquiterpenoids from young coal and from the buried pine trunks, several thousand years old at the most (p-cymene $C_{10}H_{14}$, p-menthane $C_{10}H_{20}$, α-pinene and 4-iso propylidenecyclohexanone $C_9H_{14}O$, and $C_{15}H_{20}$, $C_{15}H_{26}$, $C_{15}H_{28}$ etc.) and diterpenoids, which are more abundant in coal and particularly in fossil resins (they were given many names, e.g. branchite, fichtelite, phylloretine, hartite, hatchetite, scheererite, tekoretin). Nearly all these resins, fossil remains of conifers in peat or brown coal, are white or yellowish and of paraffin appearance, and represent a mixture of hydrocarbons with a predominance of fichtelite and retene. Triterpenoids are also relatively frequent constituents of coal (Streibl and Herout, 1969; see Table 26).

Besides the montan wax (extract obtained from coal by organic solvents, e.g. benzene) from Czechoslovakia also montan waxes from Germany have been studied in detail. Their composition is summarized in the paper of Wollrab and Streibl (1969). Friedelin (Fig. 50b) and ursolic acid (Fig. 50c) were also isolated from other brown coals of Central Europe (Ruhemann and Raud, 1932). Friedelin and friedelan-3β-ol (Fig. 50d) were identified in the brown coal of Scotland (Ikan – McLean, 1960).

50. a — Iosene, b — friedeline, c — ursolic acid, d — friedelan-3β-ol.

Table 26

Triterpenoids and similar compounds isolated from montan wax (Pila near Karlovy Vary); according to Streibl and Herout (1969)

Compound	Melting point (°C)	Formula	Functional groups
?	227	$C_{30}H_{52}$	hydrocarbon
Octahydro-2,2,4a,9-tetramethylpicene	233–235	$C_{26}H_{30}$	hydrocarbon
α-apoallobetulin	203–206	$C_{30}H_{48}O$	unsaturated ether
Allobetul-2-ene	249–250.5	$C_{30}H_{48}O$	unsaturated ether
Tetrahydro-1,2,9-trimethylpicene	230–231.5	$C_{25}H_{24}$	hydrocarbon
?	275–275.5	$C_{25}H_{24}$	hydrocarbon
?	204–205	$C_{25}H_{24}$	hydrocarbon
Tetrahydro-2,2,9-trimethylpicene	251–252	$C_{25}H_{24}$	hydrocarbon
Friedelin	253–255	$C_{30}H_{50}O$	ketone
1,2,9-trimethylpicene	277	$C_{25}H_{20}$	aromatic hydrocarbon
23,25-bisnormethyl-2-desoxyallobetul-1,3,5-triene	261	$C_{28}H_{40}O$	aromatic ether
α-apooxyallobetul-3-ene	289–291	$C_{30}H_{46}O_2$	unsaturated lactone
Oxyallobetul-2-ene	363	$C_{30}H_{46}O_2$	unsaturated lactone
23,24,25,26,27-pentanormethyl-2-desoxyallobetul-1,3,5,7,9,11,13-heptaene	250	$C_{26}H_{28}O$	aromatic ether
Friedelan-3β-ol	285–286	$C_{30}H_{52}O$	alcohol
Allobetulon	228–229	$C_{30}H_{48}O_2$	keto-ether
Friedelan-3α-ol	310	$C_{30}H_{52}O$	alcohol
3-dehydro-oxyallobetulin	338	$C_{30}H_{46}O_3$	keto-lactone
Allobetulin	266	$C_{30}H_{50}O_2$	hydroxy ether
Oxyallobetulin	347	$C_{30}H_{48}O_3$	hydroxy lactone
Betulin	254–255	$C_{30}H_{50}O_2$	unsaturated diol
?	267–269	$C_{30}H_{50}O_2$	polyalcohol
?	196–198	$C_{30}H_{52}O_4$	hydroxy ketone
Ursolic acid	285–287	$C_{30}H_{48}O_3$	hydroxy carboxylic acid
?	253–256	$C_{30}H_{50}O_3$	polyalcohol

Streibl and Herout (1969) further reported steroids and polyterpenoids from the brown coal. Free sterols, particularly β-sitosterol (R = C_2H_5; Fig. 29) and its saturated analog β-sitostanol were recorded from the brown coal by Belkevich et al. (1963).

Treibs (1936) determined porphyrins with V, Ni and Fe. They consist of four pyrrole rings linked by methine bridges (—CH=). Various functional groups may be attached to the porphyrin nucleus. Brooks et al. (1969) identified isoprenoids in bituminous coal and Kochloefl et al. (1963) in brown coal. Isoprenoid compounds in the geological materials are generally without functional groups. Most isoprenoids fall in the C_{15} (2,6,10-trimethyl dodecane = farnesane) to C_{20} (2,6,10,14-tetramethyl hexadecane = phytane) range, except for the C_{17} isoprenoid.

CHEMICAL STRUCTURE OF COAL

The present opinions on the chemical structure of coal are discussed in detail in the papers of Kasatochkin (1969), Lazarov and Angelova (1976), Speight (1971), Tingey and Morey (1973), Chakrabartty and Berkowitz (1976), Vavrečka et al. (1978) and some others.

The structural arrangement of coal increases with proceeding coalification up to the formation of organic macromolecules of flat shape with carbon atoms in the central part. The carbon atoms show a two-dimensional ordering, i.e. they lie in one plane, in the edges of more or less regular hexagons, which are not connected but float in the disordered gel matter of a disc-like molecule. Only in a part of the matter are the molecules arranged regularly, the remaining part is an unorganized confusion of molecules. The carbon hexagons are considered to be embryos of the

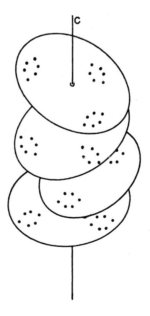

51. Turbostratic ordering of two-dimensional molecules with hexagonal carbon embrya.

graphite arrangement; their size is of the order of 0.01 μm. Every macromolecule has several carbon hexagons, which make up a planar pattern (Fig. 51). The disordered colloidal organic matter of a molecule is composed of various organic radicals and groups with linearly polymerized carbon atoms. At the higher stages of coalification the carbon embryos range themselves into parallel position and come together, which leads to the periodicity in the vertical direction. Flat, polynuclear aromatic molecules aggregate to form the turbostratic lamellar system (Blayden et al., 1944). The turbostratic arrangement is not typical of the crystalline matter but is found in colloidal gels (see Fig. 56). The two-dimensional turbostratic system is characteristic of the substances with two-dimensional periodicity. During coalification the size of molecules increases as well as the degree of their ordering; macromolecules range themselves laterally or vertically to form a lamellar system. On the basis of X-ray studies Hirsch (1954), Brown and Hirsch (1955) and Cartz et al. (1956) determined the dimensions of the "coal molecule", the packing of molecules and the dimensions of pores. Most coal types consist of 2–3 lamellae; the maximum number of lamellae rises from 4–5 with 80% C to 8–10 lamellae with 94% C. The interplanar spacing of carbon hexagons decreases from 3.65 Å (0.365 nm) to 3.43 Å (0.343 nm) with increasing coalification. The ordering is not perfect and the two-dimensional character still predominates, which is shown by diffused lines in X-ray pattern. Using the Debye-Scherrer method, we obtain first a strongly diffused diffraction line, which corresponds to d_{002} of graphite. With increasing coalification this line is less diffused and there appear additional diffraction lines, recalling a very diffused diffraction pattern of graphite (Fig. 52). Kessler and

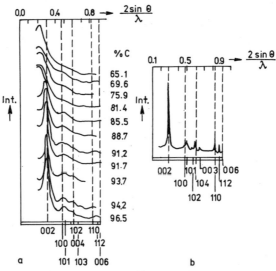

52. X-ray diffraction curves of (a) coal with increasing carbon content and (b) graphite (van Krevelen and Schuyer, 1957).

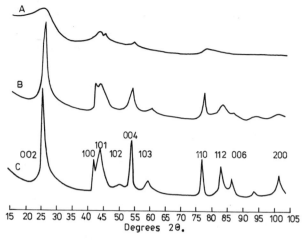

53. X-ray diagram. A — anthracite, Pennsylvania; B — meta-anthracite, Leoben, Austria; C — graphitic carbon (Quinn and Glass, 1958).

Večeříková (1954, 1960) determined $d_{002} = 3.58$ Å ($= 0.358$ nm) in brown coal, $3.58 - 3.53$ Å ($0.358 - 0.353$ nm) in bituminous hemitype, $3.53 - 3.50$ Å ($= 0.353$ to 0.350 nm) in bituminous orthotype, 3.45 Å ($= 0.345$ nm) in bituminous metatype and 3.43 Å ($= 0.343$ nm) in anthracite. The value d_{002} in graphite is 3.36 Å ($= 0.336$ nm). X-ray patterns of anthracite–meta-anthracite series are plotted in Figs. 53 and 54, after Quinn and Glass (1958).

The interpretation of X-ray patterns for brown coal is still questionable in many instances. Fusain has a higher d_{002} than would correspond to its chemistry or

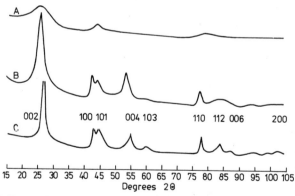

54. X-ray diagram of coal from the Narragansett Basin. A — anthracite from Pawtucket (least crystallized); B — meta-anthracite, Portsmouth; C — meta-anthracite, Cranston (best crystallized) (Quinn and Glass, 1958).

rank, which may be because fusitization occurred in an oxidizing environment and thus condensation and polymerization of hydrocarbon rings were limited.

Hirsch (1954) established the following structural stages of the coalification process (Fig. 55): (a) coal with less than 85% carbon has an "open structure"; lamellae are rarely orientated and are connected by cross-links; (b) coal with 85 to 91% C has a "liquid" structure, many cross-links of lamellae are interrupted and the lamellae show some orientation; (c) coal with more than 91% C has an "anthracitic" structure with both lamellae and pores orientated.

During anthracitization the number of crystallites increases but the three-dimensional hexagonal structure appears first in the course of the metamorphic process of graphitization. Optically, the anthracites behave like uniaxial minerals with the optical axis perpendicular to the stratification of lamellae. Coal and graphite differ essentially in their structure: coal has a laminar structure (Fig. 56) and graphite a three-dimensional structure.

Infrared spectroscopic analysis has shown that the C=O bonds are almost totally lacking in coal; it has isolated C=C bonds and C≡C bonds. Aliphatic groups CH, CH_2, CH_3 and aromatic ring groups are abundant. Free carbon is present only in coal that was intensely metamorphosed.

Theoretically, every coal molecule can contain some or all constituents in various combinations and spatial distribution, which implies that many diverse molecules may exist. However, considering that the coal molecules consist of the same and not too many components, all the coal types necessarily possess basic chemico-structural properties in common. A part of the hypothetical typical molecule of bituminous coal is shown in Fig. 57 (after Given, 1960, 1964). Of the four microlith types of the bituminous coal — vitrinite, exinite, micrinite and fusinite — it was

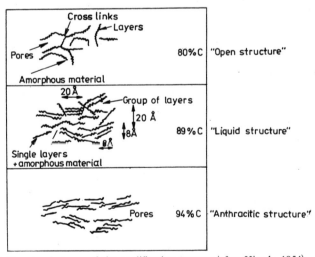

55. Structural stages of the coalification process (after Hirsch, 1954).

vitrinite whose chemical structure was studied in detail (Given, 1960, 1964; Fig. 58), with the purpose of gaining an insight into the chemical structure of coal. Figure 58 is a planar projection of the structure of a vitrinite molecule with 82% carbon. To the concept of the spatial arrangement of the proposed model it should be added that two rings of 9,10-dihydroanthracene make an angle of about 135°, whilst three rings of tripticene are mutually inclined at 120°. Dashed lines indicate the bonds of tripticene, in which the orthopositions of the third benzene ring are

56. Turbostratic lamellar model of the coal structure (a — less ordered type, b — more ordered type), called Riley's model. (After van Krevelen and Schuyer, 1957, modified.)

57. Part of a vitrinite molecule (according to Given, 1960).

58. Chemical structure of vitrinite (according to Given, 1960).

substituted by CH_2 groups of dihydroanthracene. Given's model is based on the ratio of aromatically and aliphatically bound hydrogen, determined by infrared spectroscopy and paramagnetic resonance method. The formula of the model is $C_{102}H_{78}O_{10}N_2$ at the molecular weight 1,490. Elementary composition is 82.1 % C, 5.2 % H, 1.9 % N and 10.7 % O. The hydroxyl content equals 6.25 % O_{OH} and 4.15 % O_{C+O}. The $(C-H_{arom})/(C-H_{aliph})$ ratio is 0.26 and the ratio of $CH_3 : CH_2 : CH : C = 2 : 23 : 3 : 1$. Given (1960) based his model of an average vitrinite molecule on the 9,10-dihydroanthracene, but after detecting the absence of the methylene bridge carbons by the nuclear magnetic resonance method (^{1}H NMR) he chose dihydrophenanthrene as the basic aromatic unit of coal (Given, 1964).

The properties of the soluble extract (obtained by solvents such as benzene, acetone, acetonphenone, mixture of acetonphenone and dimethyl formaldehyde, dimethyl formaldehyde and ethylene diamine up to 100 °C; yield 50—60 %) were compared using infrared spectroscopy, X-ray spectrographic analysis, nuclear magnetic resonance analysis, mass spectroscopy and other methods. The infrared spectra of extracts proved to be very similar, which implies that the basic structural pattern of coal is identical.

The infrared spectra of coal have confirmed its aromatic and most probably also its polycyclic character. The presence of OH groups, aromatic CH groups, aliphatic CH_2 and CH_3 groups, polycondensed aromatic groups and benzene rings have been substantiated. As the spectra are relatively poor, the explanation is not yet unequivocal. The intensity of the absorption zones changes depending on the content of carbon in coal, i.e. on the degree of coalification. The proportions of aromatic and aliphatic groups in a coal molecule can also be assessed relatively. Aliphatic groups become eliminated with 89—93 % carbon in volatile matter. With 94 % C in volatile matter aromatic rings with isolated CH bonds predominate so that condensed aromatic systems may be presupposed.

The relation between the intensities of infrared absorption bands of the aromatic and the aliphatic CH groups shows clearly that the number of aliphatic groups greatly decreases with the increase in coalification (van Krevelen, 1963). With 93 % C coal is completely aromatic.

Detailed chemical study, IR and NMR have proved that the content of methyl groups in coal is greater than was presumed (Vavrečka et al., 1978). In the aliphatic groups as much as about 10 % C of volatile matter has been determined. The highest content is ascribed to the methyl groups; the maximum length of the aliphatic chain is estimated at C_4. Experimental methods, however, have so far not distinguished satisfactorily between the aliphatic and alicyclic groups, and the solution of this problem will require to combine several methods, for example, ^{1}H NMR with ^{13}C NMR (Retcofsky and Friedel, 1976), or with ^{14}N NMR and some other procedures (e.g. selective oxidation). Using the ^{1}H NMR analysis of pyridine extracts from some Czechoslovak coals, Vavrečka et al. (1978) have ascertained that some brown coal extracts contain six condensed aromatic rings in an average molecule and the bituminous coal extracts ten to twelve such rings.

The X-ray study of vitrinites has shown that their aromatic systems are small, consisting prevalently of one to three connected rings. At the beginning of coalification, the aromatic systems are probably linked through the non-aromatic layers (van Krevelen, 1963). This would account for the polymeric character and a more or less open structure of coals of lower ranks. The concept of polymer has been considerably extended and generalized for the purposes of the chemical structure of coal until it nearly lost its original meaning. t is, however, difficult to speak about a polymer arrangement when the structure of a monomer has not yet been fully recognized. The structure of individual macerals is another still unravelled problem which has to be solved before the concept of a macromolecule is justifiably restricted to a polymer (Vavrečka et al. 1978). With increasing condensation and aromaticity of the primary material, the bridge structure becomes unstable as a reaction to increasing forces in the aromatic nucleus. Molecules may differ in molecular weight and may contain several types of geometric and positional isomers.

Although resins, hydrocarbons, acridine and other derivatives of naphthalene are important constituents of coal, they cover only about 1 % of pure vitrinite. It appears that some 60—85 % carbon in vitrinite occurs in the form of aromatic three-ring nuclei (Francis, 1961) formed by benzol, naphthalene, diphenyl and phenantrene, or of major nuclei, which are their multiples. In individual vitrinites any of the four nuclei may predominate.

The remaining carbon in the vitrinite occupying 10—35 % of the total carbon exists in the non-aromatic form and is involved in the hydroaromatic rings (Given, 1962, 1964).

Van Krevelen (1963) established three important structural parameters for coal:

(1) aromaticity (f_a = number of atoms in aromatic form or the ratio of atomic portion of aromatic carbon to total carbon; C_a/C);

(2) condensation index (i_R = number of carbon atoms involved in the ring structure);

(3) average dimensions of aromatic groups (C_a = number of aromatically bound carbon atoms that belong to one condensed ring system) (Figs. 59, 60, 61).

The aromaticity value of all microliths increases abruptly with the degree of coalification (ordering). At the beginning of coalification exinites are slightly condensed, vitrinites intermediately and micrinites quite strongly. With 91 % C the values i_R and C_a rise rapidly, depending again on the degree of coalification.

The major part of hydrogen in vitrinite is plausibly attached to carbon in the aromatic rings and only a small part is present in the form of methylene groups. Oxygen in vitrinite and obviously in coal altogether occurs in phenol hydroxyl groups. Part of its amount may be contained in carboxyl or carbonyl groups, which are bound to hydroxyl groups. The remainder of oxygen is presumed to connect

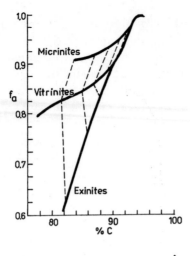

59. Aromaticity f_a of coal macerals (van Krevelen, 1963).

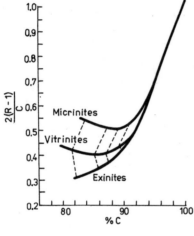

60. Condensation index of coal macerals (van Krevelen, 1963).

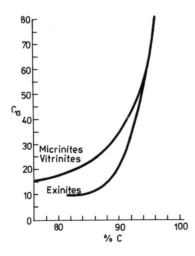

61. Number of aromatic carbons in a structural unit (van Krevelen, 1963).

aromatic nuclei in ethers or is present in heterocyclic furane or pyran chains (Francis, 1961). About 80 % vitrinite is organized in molecules, chemically in aromatic and hydroaromatic rings, whose molecular weight ranges from 300—400 to several thousands.

In the opinion of many authors the chemical structure of coal consists of condensed aromatic rings with a small number of hydroaromatic rings. For the time being van Krevelen's model (1954) seems most reasonable (Fig. 62).

Van Krevelen proposed a structural model for an average vitrinite macromolecule, which contains around the groups of condensed aromatic nuclei (6, 9 or 11 rings) side radicals and connecting bridges — the naphthene, heterocyclic and aliphatic groups. In Fig. 62 the dashed lines indicate the thermally unstable bonds of a macromolecule, where splitting to simpler basic units may occur. The model issues from the concept of a highly condensed polyaromatic structure of the volatile matter, which is today favoured by the majority of workers engaged in the study of the coal structure. A different theory was put forward by Friedel and Queiser (1959), who inferred from the UV spectra that the fundamental elements of the coal structures are of aliphatic, alicyclic and benzoid characters and form large systems of tetrahedral C—C bonds. Chakrabartty and Berkowitz (1976) consider the bridge tricyclo-alkanes (e.g. polyamantanes) attached to the individual benzene as the dominant motif of the coal structure. Their hypothesis is based on the finding that about 60 % C is present in coal in non-aromatic sp^3 valence stage.

According to Kröger and Bürger (1959) exinites contain more H and less O and are less aromatic than vitrinites (Table 27). Given et al. (1960) disclosed that spore-rich exinites have less hydroxyl groups than the associated vitrinites but

$C_{135}H_{96}O_9NS$ $H/C = 0,72$

62. Model of the chemical structure of bituminous coal (macromolecule of vitrinite) (according to van Krevelen, 1954).

Table 27

Typical data of maceral compositions (van Krevelen and Schuyer, 1957)

% C of the vitrinite	Maceral*	C^{daf} %	H^{daf} %	O^{daf} %	N^{daf} %	S^{daf} %	H/C	O/C	$d_{corr.}$	f_a	$2(R-1)/C$	n	C_a	R_{im} (max.)	V^{daf} %
81.5	V	31.5	5.15	11.7	1.25	0.4	0.753	0.108	1.259	0.83	0.42	1.788	18	0.67	39
	E	32.2	7.4	8.5	1.3	0.6	1.073	0.078	1.120	0.62	0.31	1.628	9	0.13	79
	M	33.6	3.95	10.5	1.35	0.6	0.563	0.094	1.380	0.90	0.54	1.910	21	1.27	30
85.0	V	85.0	5.4	8.0	1.2	0.4	0.757	0.071	1.240	0.84	0.40	1.836	23	0.92	34
	E	85.7	6.5	5.8	1.4	0.6	0.905	0.051	1.168	0.75	0.35	1.671	10	0.24	55
	M	87.2	4.15	6.7	1.35	0.6	0.566	0.058	1.357	0.92	0.51	1.952	27	1.50	24
87.0	V	87.0	5.35	5.9	1.25	0.5	0.732	0.051	1.243	0.86	0.41	1.864	26	1.07	30
	E	87.7	5.85	4.4	1.45	0.6	0.793	0.038	1.207	0.84	0.37	1.729	13	0.44	42
	M	89.1	4.2	4.7	1.4	0.6	0.561	0.040	1.352	0.93	0.51	1.982	31	1.66	20
89.0	V	89.0	5.1	4.0	1.3	0.6	0.683	0.034	1.262	0.89	0.43	1.899	31	1.26	26
	E	89.6	5.2	3.3	1.3	0.6	0.691	0.028	1.255	0.89	0.42	1.817	18	0.82	29
	M	90.8	4.1	3.2	1.3	0.6	0.537	0.026	1.363	0.94	0.52	2.008	37	1.90	16
90.0	V	90.0	4.94	3.2	1.35	0.5	0.655	0.027	1.277	0.91	0.43	1.921	34	1.39	23
	E	90.4	4.9	2.8	1.3	0.6	0.646	0.023	1.281	0.91	0.44	1.873	22	1.12	23
	M	91.5	3.95	2.6	1.35	0.6	0.514	0.021	1.380	0.95	0.54	2.017	40	2.08	14
91.2	V	91.2	4.55	2.6	1.15	0.5	0.594	0.021	1.314	0.93	0.48	1.960	39	1.64	18
	E	91.5	4.5	2.3	1.2	0.5	0.586	0.019	1.320	0.93	0.48	1.960	30	1.64	18
	M	92.2	3.65	2.2	1.35	0.6	0.471	0.018	1.416	0.96	0.57	2.020	43	2.44	11
85–91	F	94.2	3.0	1.4	0.9	0.5	0.379	0.011	1.499	1.0	0.62	–	–	–	10

* V – vitrinites, E – exinites, M – micrinites, F – fusinite.

probably more hydroaromatic groups. Although the molecular weights of exinites are very likely higher than those of vitrinites (less soluble in organic solvents), the chemical structure of these two coal constituents is not supposed to differ greatly.

Micrinites have high carbon (Brown, 1959) but low hydrogen and oxygen (Kröger and Bürger, 1959). Their aromatic nuclei are greater and the number of non-aromatic groups seems to be small (van Krevelen and Schuyer, 1957). In the content of hydroxyl groups they stand roughly midway between vitrinites and exinites.

Fusinites are richest in carbon, which is rather amorphous than having a true organic chemical structures (Brown, 1959). From Table 27 it follows that fusinite contains 85—91% C and this composition seems to persist from the first coalification stage, which demonstrates its difference from other microliths, particularly in genesis.

Tschamler and de Ruiter (1966) inferred from the relative frequence of aliphatic and aromatic structures that the coal microliths (exinite, micrinite and vitrinite) formed by various combinations of analogous chemical complexes (Table 28).

Table 28
Percentage of hydrogen in aliphatic and aromatic structures
(Tschamler and de Ruiter, 1966)

Microliths	% H	% H_{aliph}	% H_{arom}	C_{arom}/C	R_{arom}
Exinite	7.0	5.4	1.4	0.62	2.2 — 3.5
Vitrinite	5.5	3.3	1.9	0.77	2.75—4.1
Micrinite	3.9	1.7	2.0	0.89	3.6 — 5.2

C_{arom}/C — aromaticity per unit.
R_{arom} — number of condensed aromatic rings per average unit.

The knowledge of the positions of nitrogen and sulphur in the coal structure is rather poor. Nitrogen is assumed to appear in bituminous coal almost exclusively in heterocyclic structures. Among the heterocyclic substances isolated from coal there is, for example, the nicotinic acid. Organic sulphur is probably present in thioether and thiophenol groups.

Coal is a mixture of molecules with a medium molecular weight. With 93—94% carbon, the coal molecule is altogether free of aliphatic carbon and the molecule weight varies about 1,000.

Most structural models of coal are based on the data obtained on the extractable constituents of coal or on the products of thermal destruction of the volatile matter, and on the presumption that the coals of the same coalification degree have the same chemical structure. It is thus evident that a generally valid simple basic unit for the condensed aromatic system has not yet been found. Some analogies have been revealed between the chemical structure of the coal volatile matter and that of heavy oil asphaltenes, which are the subject of recent studies.

MINERALOGY OF COAL SEAMS

Most bituminous and brown coal seams are rather poor in minerals. In Czechoslovakia, the Kladno coal Basin is relatively richest in minerals, which occur on joints in pelosiderite and on fractures. North and Howarth (1929) found a similar paragenesis in the coal of South Wales. Mineral paragenesis from the Illinois coal was described by Rao and Gluskoter (1973). The coal in the Ostrava and Rosice Basins contains pelosiderite concretions, but they are for the most part compact; the mineral filling as far as it occurs on joint surfaces is poor, usually comprising the most common carbonates and quartz.

More than 95% of inorganic components of coal consists of five minerals: kaolinite, illite, calcite, pyrite (marcasite, melnikovite) and quartz. More than 95% of coal ash is formed by Al_2O_3, SiO_2, Fe oxides and CaO. The remaining 5% are Mg, Na, K, and Ti oxides and chlorides, sulphates and phosphates of common elements.

Beautiful crystals are not so important in assessing the properties of coal as is the fine admixture, often difficult to discern. In defining the whole mineral paragenesis of coal seams it is necessary to consider all the minerals present, including those that are generally not visible with naked eye and got into coal by mechanical transport.

The first genetic classification of the mineral admixture of coal was proposed by Lessing (1922, 1926). He distinguished:

(1) residues of mineral substances contained in the plant material;

(2) detritus washed or blown into the basin during the accumulation of sediments;

(3) salts precipitated from water that was in contact with the organic material before or during coal formation;

(4) crystalline substances precipitated from water solutions migrating along fractures and joints of coal during its formation;

(5) products of decomposition or of interactions between the substances mentioned above; organic coal matter may have also taken part in these reactions;

(6) substances that usually originate in other environments and that were introduced into coal or associated rocks after its development had been completed.

A modern genetic classification (e.g. Svoboda and Beneš, 1955) divides the inorganic substances into (A) internal ash matter (primary, proper), (B) external ash matter (secondary, free).

(A) The first group comprises particles present in original plant bodies, trace elements of plants and, according to Mackowsky (1945), also phosphorus in tricalcium-phosphate bond.

(B) The external ash matter is divided into syngenetic and postgenetic (Beneš, 1954). Clayey and siliceous substances, pyrite, marcasite, melnikovite, siderite, ankerite and other carbonates are syngenetic. They were formed during the cumulation of inorganic substances from dissolved salts, products of the decomposition of minerals transported into the peat bog from its environs, and products of the decay of organic substances such as carbonic acid. Organic minerals are usually represented in the peat bogs by mellite (aluminous salt of benzene hexacarbonic acid), whose presence points to a considerable amount of dissolved Al salts in peat waters. The postgenetic minerals originated mainly from the later mineral solutions and form incrustations of fissures and joints in the adjacent rocks. They are as follows: pyrite, marcasite, galena, sphalerite, arsenopyrite, halite, melnikovite (described from the durites of the Kounov coal seam by Beneš and Dopita, 1954), various sulphates, phosphates, chlorides and others.

Minerals can also be divided into allogenic (quartz, feldspars, apatite, zircon, rutile, garnets, epidote, amphiboles, pyroxenes, magnetite) and authigenic. Authigenic minerals are important for the recognition of facies conditions; they are subdivided into diagenetic (pyrite, quartz, siderite, ankerite, dolomite, calcite) and epigenetic (calcite, dolomite, quartz, feldspars, pyrite, chlorite, micas, gypsum). Barite, apatite and others may also be authigenic components.

As far as FeS_2 is concerned, it usually forms during diagenesis by the reduction of sulphates by organic substances; marcasite and melnikovite occur rather in brown coal, and pyrite in bituminous coal and anthracite. Organic substances function as reducing factors at the contact with substances capable of reduction. Reduction of sulphate solutions may produce elementary sulphur in addition to sulphides. The distribution of pyrite, marcasite and sphalerite in the Pennsylvanian coal from the Chinook mine, Indiana, has been studied by Boctor et al. (1976). Chalcopyrite was described from bituminous coal of Leicester, South Yorkshire, Tyne and North Staffordshire by Reynolds (1948). The same author identified galena in the coal of North Wales. Francis (1954) reported pyrite nodules from Yorkshire, and pyrite stalactites from the Salt Range, Pakistan, even more than several inches long.

Some minerals (pyrite and other sulphides) may have formed both synchronously with the formation of a coal seam and in a later stage. Still more debatable is the classification of those minerals which may have formed syngenetically with coal or may have been transported into the coal-forming environment together with clastic and detrital material.

All coals, when burnt, leave back some mineral residue — ash. The ashes from coals differing in type and origin, and those from identical coals differing in the ash content, often show a diverse distribution of the essential components. Generally, the higher the ash content in coal, the larger is the amount of SiO_2 and Al_2O_3. In other words, the composition of ash approaches the composition of the adjacent rocks with the increasing ash content in the coal. The relationships between sediments of the claystone-coal series were defined, for example, by Havlena (1959). The composition of ash may vary in one coal seam both vertically and laterally.

The ash of bituminous coals consists prevalently of SiO_2, Al_2O_3 and Fe oxides; it is usually poorer in CaO than the brown coal which contained CaO in humic substances. Of the four principal petrographical coal components, fusain has the highest content of ash and vitrain the lowest. From the results of the analyses of the Hamstead bituminous coal Lessing (1926) inferred that the ash of clarain and vitrain is essentially the original ash of coal-forming plant material; ash from durain consists of particles of the overlying claystone, and the ash from fusain is composed of infiltrated material from the overlying beds (Table 29).

The composition of the primary ash matter is indicated by the chemical composition of the ash from coal having the minimum ash content. It is rich in the Ca, Mg, Fe, Na and K oxides, which are bound in coal matter most probably

Table 29
Chemical composition of ash from the main petrographic constituents of coal in (wt.%)
(after Lessing, 1926)

	Fusain	Durain	Clarain	Vitrain
SiO_2	8.84	50.54	9.44	6.08
Al_2O_3	8.66	42.34	16.58	15.49
Fe_2O_3	3.37	1.36	3.31	3.09
MnO	0.51	–	0.23	0.13
TiO_2	0.04	0.44	0.50	0.24
CaO	57.00	3.69	12.98	15.22
MgO	1.30	–	10.52	1.87
Na_2O	3.24	–	15.71	17.67
K_2O	0.67	–	–	0.20
SO_3	14.65	3.23	32.18	30.89
P_2O_5	–	–	0.01	traces
CO_2	2.98	–	–	6.69
Sum	101.26	101.60	101.46	97.57
Soluble in H_2O	16.57	3.48	65.24	69.52
Soluble in HCl	71.38	23.81	17.68	20.46
Insoluble in HCl	12.05	72.71	16.90	10.02

in the form of humates. The proportion of these oxides decreases with the increase in ash content (A^d), depending mainly on the increase of clay mineralization (mainly kaolinite, illite, less montmorillonite). It will be manifested by a higher proportion of SiO_2 and Al_2O_3. Sulphide and carbonate mineralizations are intensive in some cases, lacking altogether in others. The relationship between $SiO_2 + Al_2O_3$, $CaO + MgO + Fe_2O_3 + Na_2O + K_2O$ and A^d was examined by Zelenka (1972). He computed the relations from 65 boring data obtained in the North Bohemian Brown-coal Basin. In the coal ash with A^d above 30 %, the sum of $SiO_2 + Al_2O_3$ was nearly constant and was near to the sum of these oxides in the overlying and underlying clayey sediments. The increase in the $SiO_2 + Al_2O_3$ sum was greatest between 0 and 10 % A^d, which suggests a considerable proportion of primary ash matter in the ash of coal with $A^d < 10 \%$.

Minerals forming the admixture in bituminous coal of Czechoslovakia were studied by Šindelář (1957), Malkovský (1959), who determined clay minerals in some coals of the Ostrava Formation, Králík (1959), Dopita and Králík (1977, tonsteins from the Ostrava coalfield) and others. The mineral composition of some American coals has been described by O'Gorman and Walker (1972).

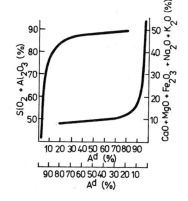

63. Relations between $SiO_2 + Al_2O_3$, $CaO + MgO + Fe_2O_3 + Na_2O + K_2O$ and the ash content in coal dry matter; borehole Ht 4 in the North Bohemian Brown-coal Basin (Zelenka, 1972).

Survey of minerals in bituminous coal (Šindelář, 1957)

Very frequent	Frequent	Less frequent	Rare	Very rare
gypsum	ankerite	dolomite	halloysite	montmorillonite
kaolinite	limonite	sphalerite	orthoclase	hydrargyllite
illite			chlorite	boehmite
plagioclase			muscovite	biotite
calcite			magnesite	amphibole
quartz			apatite	pyroxene (augite)
siderite			sphene	tourmaline
marcasite			garnets	zircon
pyrite			staurolite	
			goethite	
			galena	

Among rather scarce minerals there are diaspore, magnetite, rutile, hematite, disthene, epidote, topaz, protochlorite, penninite, witherite, barite, halite, arsenopyrite, chalcopyrite and antimonite.

Most of these minerals occur in coal as extremely fine grains, which in many cases are identifiable microscopically only after separation and concentration.

In association with coal sometimes occur minerals that do not belong in the mineral paragenesis of coal seams and are not genetically related with coal. Near Bytom in Upper Silesia, for example, Pb and Zn ores lie in the proximity of a coal seam; in the Rhine valley Permian salt deposits occur above the Carboniferous coal seams; in the Permian at the foot of the Krkonoše Mts. (Bohemia) Cu ores are associated with coal and bitumens; near Stříbro (western Bohemia) galena veins pierce a coal seam (Sokol, 1919).

The origin of some iron ores is thought to be connected with the formation of coal, as, for example, Fe carbonates with an admixture of clayey, bituminous and humic substances in Westphalia and England. In Czechoslovakia, pelosiderite concretions form even fairly thick layers in the Karviná Formation (Ostrava-Karviná coalfield) and in the Kladno Basin. Concretions in coal seams are generally rich in iron; calcareous concretions do not occur in coal. Silicification of coal seams is relatively frequent. Quartz is usually an authigenic mineral; it forms either from material liberated by the action of humic acids on feldspars of the underlying rock or crystallizes from the colloidal SiO_2 derived from the adjacent weathering rocks. Therefore, quartz often mineralizes the cells of tissues, fusite and parts of coal seams (e.g. in the Ruhr Basin).

Anthracite from Úsilné near České Budějovice (Bohemia), yielded calcite, muscovite, anorthite, albite, kaolinite and quartz (A. Orlov in Hubáček, 1939). Epigenetic calcite has also been found in anthracite from the Brandov Basin (Bohemia).

The Kladno Basin in central Bohemia afforded rich mineral finds, in Westphalian (Radnice and Nýřany) and Stephanian (Kounov) coal seams. The whole cycle of Carboniferous beds rests on the Algonkian. Kašpar (1939) published a monograph on the minerals of the Kladno coal seams, in which he also discussed the numerous earlier papers. The richest finds of minerals were obtained from the lower Radnice seam; between the upper seam and the overlying whetstone horizon there are abundant pelosiderites. The typical mineral paragenesis (a similar one is also known e.g. from Westphalia, Rhine area, Belgium, England) is developed mainly as a filling of cracks of the pelosiderite septaria, on fissures and fractures in coal and the neighbouring rocks. Minerals are found particularly in fissures which are wide enough for minerals to crystallize, but sporadically also in tiny microtectonic fractures in coal. They contain nacrite, occasionally pyrite and galena. Branches and trunks with quartz and chalcedony incrustations occurred in the Kounov seam.

Kašpar (1939) reported the following minerals from the Kladno coal seams:

sulphides — galena, sphalerite, pyrrhotite, millerite, pyrite, marcasite, chalcopyrite, arsenopyrite, linnaeite;

hydroxides — opal (remarkably, quartz is not present);
carbonates — siderite, ankerite, dolomite and calcite;
silicates — nacrite;
sulphates — barite;
organic compounds — whewellite and resins.

This list does not include the secondary minerals, which were formed by oxidation of the above mineral association as a result of the disturbance of the seam by mining (limonite, calcite, zaratite, gypsum, epsomite, melanterite, and others). The minerals of burning pit heaps have not been listed either.

Minerals that are not genetically associated with coal but belong to the Algonkian or Cretaceous complexes, comprise antimonite (Kašpar, 1939) cited from the Max (today Gottwald 1), Prago (today Zápotocký) and Vaněk (today Nejedlý) mines (Rost, 1942b); galena and sphalerite (Kratochvíl, 1932) from the Ronna (today Gottwald 3), Max and Vaněk mines; tetrahedrite (Rost, 1942b) from the crosscut between Max and Mayrau (today Gottwald 2) mines, and galena, sphalerite, chalcopyrite and pyrite found in a small ore lens in quartz porphyry. In this group are also placed minerals found in the cavities in spilite (Jan mine), and zeolites, aragonite, calcite, siderite and gmelinite associated with the basalt effusion of Vinařická hora (mainly in Mayrau mine, Rost, 1942b). Nováček (1932) described almandine from the underlying conglomerate. Some pyrrhotites are mentioned by Kašpar (1939).

Slavík (1925) recorded the following paragenetic succession of minerals from the Kladno mines:

(1) earlier carbonates: siderite, ankerite, dolomite, rarely calcite;

(2) barite;

(3) sulphides: galena, sphalerite, pyrite, millerite, beyrichite, marcasite, chalcopyrite, arsenopyrite, linnaeite;

(4) whewellite;

(5) nacrite, kaolinite;

(6) secondary minerals: later carbonates and sulphates — calcite, zaratite, malachite, gypsum, epsomite, melanterite, and others.

Kašpar (1939) noted that this paragenesis cannot be regarded as valid for the whole Kladno Basin, since the succession of minerals depends mainly on the place of their origin and the distance from the coal seam. Paragenesis is more regular in the close neighbourhood of the seam. Kašpar mentioned his finds of millerite older than barite, chalcopyrite grown on barite crystals and other examples, and plotted the succession of the most important minerals in a graph. Bernard and Paděra (1954) supplemented the sulphides from Kladno by bravoite (17.12 % Ni) found in the Gottwald mine. On the basis of Nováček's analysis (1931), these authors and Kutina (1956) identified the linnaeite as siegenite $(Co, Ni)_3S_4$.

An interesting mineral paragenesis has also been described from burning pit heaps of the Kladno mines, where the waste material slowly burns and is oxidized.

The minerals thus formed cannot be regarded as characteristic of coal seams; they formed by oxidation of primary minerals, when material from underground was transferred on the pit heaps. I think it convenient to mention these minerals in order to complement the overall picture of chemical composition.

The minerals of burning pit heaps in the Kladno district have been described in detail by Rost (1937, 1940, 1942a, b). He found out that they formed mainly from sulphides, particularly pyrite and marcasite. Prerequisite for the origin of a typical paragenesis are the fires of dumps, as the buried sulphides cannot weather and calcine to Fe_2O_3; during this process sulphur escapes as sulphur vapours and less as SO_2. At cooler places sulphur vapours precipitate on the surface of pit heaps as native sulphur (similarly as native selenium also derived from sulphides) and partly oxidize to form sulphuric acid, which attacks adjacent rocks and minerals from which various sulphates originate. Distillation of coal gives rise to salmiac and various organic compounds. According to Rost (1937), sulphur and sulphates may also form from the organic sulphur contained in coal. The minerals crystallize on pit heaps chiefly on cool walls of fractures and cavities, which resemble minute fume-exhalating volcanic craters. Hot air, smoke, water vapours, SO_2, salmiac and sulphur fumes escape in gaseous form. Rost (1937, 1940, 1942a, b) recorded the following minerals from the pit heaps in the Kladno coal district:

products of pyrite calcination: sulphur, selenium (gamma modification with traces of tellurium), realgar;

sulphates: tschermigite, epsomite, pickeringite, halotrichite, alunogen, rostite (a new mineral — orthorhombic $Al(SO_4)(OH).5H_2O$; it was originally described by Rost, 1937, as lapparentite from Libušín near Kladno, Czechoslovakia. — Commission on New Minerals and Mineral Names, Memorandum: August 25, 1975), copiapite, mascagnite, letovicite, cryptohalite, hexahydrite, rhomboclase;

products of dry distillation of coal: salmiac;

organic compounds: kratochvílite $C_{13}H_{10}$, kladnoite $C_6H_4(CO)_2NH$.

Of interest is the find of Thénard's blue from the fired volcanogenic argillitic rock (Rost, 1942). Rost interpreted the origin of Thénard's blue in terms of a partial decomposition of siegenite grains in the clayey sediment by infiltrating water, during which Ni and Co pass into solution most probably as sulphates. The concentration of solutions is fairly high close to the siegenite crystals. The solution of cobalt sulphate with the aluminous component of fired argillite produces the bluish tint of the Thénard's blue.

As concerns the derivation of heavy metals involved in the structure of sulphides formed on fracture surfaces and in pelosiderites, Slavík's opinion (1925) is essentially still accepted, i.e. that the metal elements come from the living matter of Carboniferous plants. They were supplied in very diluted solutions into swamps, peat bogs and growths together with elements from weathered rocks, very likely in the form of soluble compounds, such as sulphates, hydrocarbonates and hydrosilicates. Carbonitization and sulphidization are easy to explain, since during coalification a sufficient amount of water and gases, particularly CO_2 and H_2S is released from the decaying plant material. Sulphidization is characteristic of the coal seams with regard to a high reduction potential of the environment. According to Balme (1956) pyrite forms in the coal-forming environment at pH > 6.5 and marcasite within the range pH 5.8 — 6.5. Mineralization invariably occurred under strongly reducing conditions both in peat bogs and in developing coal. The presence of sulphides outside coal seams is less constant. Sulphides oxidize producing sulphate anions. Owing to their solubility some sulphates can migrate a considerable distance from the coal seams. Sulphates generally formed later than sulphides but they can also be older in case that sulphate solutions were brought from other parts of the seam before sulphidization. The variation of the succession of mineral precipitation may be explained in this way. Whewellite generally began to precipitate after carbonates, barite and sulphides. One of the youngest minerals is nacrite.

The largest Czechoslovak coalfield of Ostrava-Karviná is poorer in the occurrence and the amount of finds of minerals. Neither the very abundant pelosiderites have yielded remarkable finds. A monograph on the minerals from the Ostrava-Karviná coalfield was written by Kruťa (1951, supplements 1952).

The paragenetic series, succession and recurrence of minerals (valid particularly for pelosiderites) have been established by Kruťa (1951) as follows:

(1) quartz (rock crystal, smoky quartz)
(2) ankerite
(3) calcite
(4) pyrite
(5) sphalerite
(6) pyrite II
(7) sphalerite II
(8) galena
(9) quartz II (rock crystal, smoky quartz)
(10) ankerite II
(11) calcite II
(12) quartz III (rock crystal, smoky quartz)
(13) calcite III
(14) millerite
(15) pyrite III
(16) chalcopyrite
(17) sphalerite III
(18) galena II
(19) nacrite
(20) hatchetite

In the neighbouring sediments and in coal the sulphides — usually pyrite, less often marcasite and occasionally sphalerite and galena — set on the crystallized carbonate filling (ankerite, calcite). Minute pyrite crystals are also found on the rock crystal, and marcasite druses on sandstone. In Čs. Pionýr II mine, arsenopyrite was recovered from sandstone together with galena and sphalerite. Barite and siderite pseudomorphs after pyrite were found in pyrite-bearing coal. Halite in the form of stalactites derives from the marine bands of the Ostrava Formation. Their rocks were leached and water containing NaCl was transported along faults and fissures even to considerable depths, where halite precipitated from the oversaturated solutions. The salt has been analysed by Vysloužil (1927), who reported the following composition: $H_2O = 0.06\%$, insoluble residue $= 0.32\%$, $CaO = 0.28\%$, Na = $= 39.50\%$, Cl $= 60.72\%$, sum $= 100.88\%$ (average of two analyses). Iodine and bromine have not been identified. (The so-called "Salzkohle" with a high salt content are known from the brown-coal deposits of Germany.) Dopita (1955) described sideritized wood from seams nos. 11 and 13b of the Ostrava Formation.

The most frequent minerals of the Ostrava-Karviná coalfield are ankerite, calcite, quartz (rock crystal, smoky quartz). Most of the remaining minerals are rare. Secondary minerals and minerals formed by oxidation or calcination on burning pit heaps comprise tschermigite, sulphur, salmiac, realgar, epsomite, hematite, limonite, mirabilite, gypsum and possibly halotrichite.

In the text below a survey of minerals is given only from the remaining coal districts of Czechoslovakia that contain a richer and interesting paragenesses.

Radnice near Plzeň (Westphalian C) — hematite, partly oolitic, chamosite, white kaolinite coatings on coal, keramohalite, quartz, limonite, marcasite, melanterite, pyrite, retinite, gypsum, and siderite in the form of pelosiderite. Ankerite, galena, sphalerite and pyrite are known from coaly sandstone.

Nýřany near Plzeň (Westphalian D) — hematite and limonite ores, ankerite, barite, dolomite, chalcedony, calcite, quartz, millerite, pelosiderites, pyrite, sphalerite, galena and siderite.

Mirošov SSE of Rokycany (Westphalian D) — barite, galena, pyrite and sphalerite.

Radvanice (Upper Stephanian, Permian) — copper ore at the outcrop of arkoses of the Radvanice group of seams, azurite, galena in coal, chalcopyrite, chalcocite dispersed in coal matter and in pods jointly with pyrite, quartz, malachite in coal and adjacent rocks, pyrite and tetrahedrite.

Svatoňovice (Westphalian D — Lower Stephanian) — ankerite, azurite on coal, copiapite, galena in coal, halotrichite, hematite, quartz, malachite, millerite brush aggregates, pyrite, gypsum and sulphur, abundant spherosiderite. Bornite, galena, chalcopyrite, pyrrhotite and sphalerite have been identified in the Svatoňovice seams near Hronov.

Žacléř (Westphalian) — pyrite in aleuropelites and in cracks of spherosiderite, pyrite coatings of Sigillaria, calcite, malachite (powdery films on coal and in sandstones), galena; quartz and chalcedony incrustations of coalified wood.

Rosice-Oslavany (Stephanian-Autunian), according to Burkart (1953) and Kruťa (1951):

The mines Julius and Ferdiand (Zastávka u Brna) and Antonín mine (Babice, Zbýšov and Padochov): anhydrite, ankerite, barite, tschermigite, dolomite, epsomite, fluorite, galena, goethite, garnet in graphitic micaceous shale, graphite, hematite, halotrichite in pit heaps, hatchetite in pelosiderite septaria, chalcopyrite, chrysocolla, calcite, keramohalite, quartz in septaria, rock crystal, limonite, malachite, marcasite, mascagnite, melanterite, mirabilite, pelosiderite, psilomelane, pyrite, pyrrhotite (?), gypsum, salmiac, sulphur α, β, γ, sericite, siderite, brown sphalerite, thenardite and smoky quartz in pelosiderites. Additionally, elaterite (matter similar to ozocerite) and valaite (asphalt matter) have also been recorded.

Nosek mine in Oslavany near Ivančice: azurite, barite, dolomite, epsomite,

I/1
Pelosiderite concretion with ankerite filling the concentric and radiating cracks (septarium). Zápotocký mine, Kladno.
Photo by R. Rost.

I/2
Whewellite — colourless columnar crystals, 25 mm long, with white ankerite and barite on pelosiderite. Gottwald 3 mine, Kladno. Photo by R. Rost.

II/1
Ankerite filling a cavity in pelosiderite (natural size). Gottwald 3 mine, Kladno. Photo by V. Bouška.

II/2
Acicular millerite crystals on ankerite filling a cavity in pelosiderite (natural size). Gottwald 3 mine, Kladno. Photo by V. Bouška.

III/1
Marcasite twins (magn. 2×). Vintířov. Photo by F. Tvrz.

III/2
Tschermigite (10 × 2 cm). Želénky near Duchcov. Photo by F. Tvrz.

IV/1
Yellow sulphate efflorescences on desiccation cracks in peat. Soos near Františkovy Lázně. From the archive of V. Bouška.

IV/2
Ferruginous concretions from Cretaceous sediments (reduced to 2/3). Sand pit at Žíšov near Veselí nad Lužnicí. Photo by V. Bouška.

V/1
Active mofette. Soos near Františkovy Lázně. From the archive of V. Bouška.

V/2
Mofette at a low water level. Soos near Františkovy Lázně. Photo by F. Tvrz.

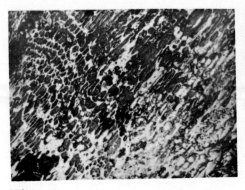

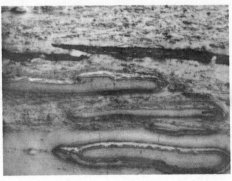

VI/1
Fusinite. Borehole CO 22 near Chotíkov (Plzeň Basin — Carboniferous), depth 489.3 m; photomicrograph, magn. 60×. Photo by J. Spudil and M. René.

VI/2
Megaspores in collinite. Borehole CO 21 near Chotíkov (Plzeň Basin — Carboniferous), depth 419.3 m; photomicrograph, magn. 120×. Photo by J. Spudil and M. René.

VI/3
Roots filled with fine-grained sandstone. Dobré štěstí mine, Plzeň Basin. Photo by S. Hatláková.

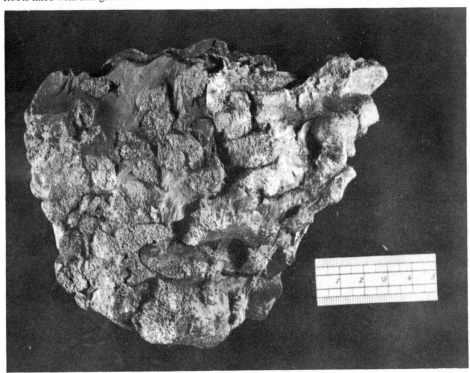

VII/1
Coalified log from the claystone of F seam, Nová Jáma mine, Březina. Photo by S. Hatláková.
VII/2
Lignite with preserved wood structure (size of the sample 10 × 15 cm). Mydlovary, southern Bohemia. Photo by F. Tvrz.

VIII/1
Cleats in coal filled with clay minerals (reduced to 1/2). Nejedlý mine, the main Kladno seam, Kladno-Rakovník Basin.
Photo by S. Hatláková.

VIII/2
The main Kladno seam contaminated by numerous partings of volcanogenic material. Kačice mine field, Nejedlý mine.
Photo by J. Pešek.

galena, garnet in micaceous shale, hematite, halotrichite, chlorite, calcite, keramohalite, quartz, rock crystal, limonite, malachite, marcasite, melanterite, pelosiderite, pyrite, gypsum, sphalerite and sulphur. Calcified tree trunks have been found.

Czechoslovak coal seams of Cretaceous age (freshwater Cenomanian) are developed mainly in Moravia, where they were mined impersistently from the fifties of the past century until 1921. At present coaly argillites are worked near Moravská Třebová. The Cretaceous coal-bearing area extends around Boskovice, Letovice and Moravská Třebová. The Cretaceous coal has the appearance of bituminous coal but the properties of brown coal. According to data cited in Burkart (1953) and Kruťa (1951), the mineral association consists of pyrite, marcasite, kaolin, calcite, quartz, glauconite (in overlying sandstones), gypsum, recent sulphates (gypsum and lublinite near coal seams), sulphur, hematite, limonite, psilomelane, cacoxene and vivianite. In the clay near Mladějov, unusual resinous substances form up to 30 cm large nodules. Fossilization of resins is associated with the decrease of the constituents soluble in organic solvents, as a result of polymerization and other reactions. In the claystones and transitional coaly sediments of the continental Cenomanian of Moravia the fossil resins are represented by walchowite, neudorfite and muckite.

Five samples of these fossil resins were subject to extraction with pyridine and to pyrolysis (walchowite from Hřebeč, Březina, Valchov; neudorfite from Nová Ves and muckite from Nová Ves). Infrared spectrometry, gas chromatography and mass spectra of the corresponding products have shown that all samples are probably of a closely related chemistry and of a common plant origin (Streibl et al., 1976). The terms retinite, schraufite (Skutíčko), alingite (Nová Ves) and succinite, which appear in the literature, are only synonyms for amber (Savkevich, 1970) and similar substances. Fossil resins, which are the most stable biogenic product, are numerous. They have appeared since the Mesozoic, although scleretinite was recorded by Mallet (Hintze, 1933, 1935) from the Coal Measures of Wigan, England. Fossil resins similar to amber have been found on lignite and brown coal deposits in North America, South America, Europe and Asia (e.g. retinite and ambrite in New Zealand, a large amount of retinite nodules as large as human head in western part of the U.S.A.). The fossil resins dissimilar to amber are more friable and do not liberate succinic acid on distillation. Piauzite is a resin found in lignite in the proximity of a porphyry intrusion near Trifail in Austria. Scheererite was described from lignite near a basalt dyke at Wilhelmszeche near Bach and from the Tertiary brown coal at Uznach near St. Gallen, Switzerland. Middletonite has been reported from England, wheelerite from New Mexico, ionite from lignite in the Ione Valley, California, tasmanite, trinkerite from Albona, Idria; rosthornite and rochlederite were also recognized. Most of the terms are synonyms but there is still confusion in the nomenclature. The best known waxes from brown coal are pyroretinite (Czechoslovakia), pyropissite (G.D.R., Halle a. d. Saale), geomicrite (Gesterwitz) and denhardite (East Africa).

A complete list of resins and waxes will be found in mineralogical compendia or in the survey compiled by Mueller (1972).

In the Cretaceous coal from the Letovice area two minerals were found and described by Sekanina (1932a, b): rosickýite (natural gamma sulphur) from Havírna and letovicite $(NH_4)_3H(SO_4)_2$ from Vísky. Both these minerals come from pit heaps.

The North Bohemian and Sokolov Brown-coal Basins (Miocene) are rather poor in minerals. The mineral associations comprise only common minerals and the finds are not frequent. The following overview is based on the works of Kratochvíl (1937—1943) and Brus (1965, occurrence of orpiment).

Most — ankerite in septaria, barite in the overlying clay, tschermigite (crusts and crystals on coal), nests of duxite in brown coal, epsomite, mirabilite, halotrichite, keramohalite, melanterite, pickeringite, calcite, marcasite, kaolin, pyrite, gypsum, spherosiderites, sulphur, whewellite (beautiful crystals at Lom), bilinite at Světec near Bílina.

Duchcov — barite, tschermigite, duxite, limonite, marcasite, pyrite, gypsum, whewellite.

Chomutov — oxalite, alum, halotrichite, marcasite, opal, pyrite, sulphur, gypsum, limonite, epsomite; silicified woods are known from the base of the main coal seam.

Sokolov — dolomite, alum, quartz, limonite, marcasite, melanterite on pit heaps, pyrite, realgar, gypsum, spherosiderite, sulphur, vivianite and amber. Marcasite twins occur mainly at Lipnice near Sokolov.

Mellite is recorded from Bílina, Korozluky, Lužice and Mezboltice; pyroretinite from Byňov and Zálezly; humboldtine from Lužice near Most and from Kolozuby.

Of mineralogical interest is the South Moravian Basin of Neogene age. The coal is typical hemixylite showing preserved wood structure. At the main localities Ratíškovice, Dubňany and Čejč, marcasite, calcite, retinite and some clay minerals not strictly defined habe been found. Kruťa (1951) reported a remarkable paragenesis of recent sulphates, tschermigite, epsomite, halotrichite, keramohalite, mascagnite, melanterite, gypsum, and from burning pit heaps alpha, beta and gama sulphur. The scarce hexahydrite found at Kelčany was identified and described by Kokta (1932). Sekanina (1948) identified and described the mineral koktaite

$$(NH_4)_2Ca(SO_4)_2 \cdot H_2O$$

from Žeravice.

A small Neogene basin at Uhelná near Javorník was opened by a surficial mine. Kruťa (1949) determined spherical concretions of marcasite and secondary limonite, gypsum and melanterite.

The remaining coal basins of Czechoslovakia of whatever geographic situation are very poor in minerals, or the lack of finds is due to insufficient mineralogical investigation. Calcified tree stems and silicified woods are reported from the main coal seam at Handlová. Quartz also forms minute clear crystals on bedding planes. Quartz and chalcedony are present on the Nováky coal deposit; Čech and Petrík (1972) think that silicification postdated the formation of coal; these authors described from the same deposit realgar and orpiment, which form fine films on coal and neighbouring rocks and contribute to a high As content in coal. The main source of As are supposed to be volcanic exhalations. Pyrite, marcasite, sporadic melnikovite and secondary gypsum have been found in both Handlová and Nováky coal basins. Bouška and Novák (1963) described opal and cristobalite from xylitic coal at Sejkov in eastern Slovakia.

The mineral contents of world coal deposits are practically identical. In some coal seams in the U.S.A. uranium minerals such as schroeckingerite and autunite have been identified (Breger et al., 1955). Vine (1959) reported dopplerite as a recent product of weathering of the Cretaceous coal from Fremont County, Wyoming, U.S.A. Coals of some deposits also contain rare hydrocarbons, resins, and organic minerals as, for example, whewellite (bituminous coal in Brux, Switzerland), or mellite (lignite in the Paris Basin). Brown coals of the Eastern Alps contain realgar and orpiment. Witherite and barite have been described from the Brancepeth Colliery, Durham (H. Briggs, Coll. Eng., 11, 304, 1934).

In the overall survey of the minerals of coal basins we cannot omit the rock-forming minerals that are constituents of the coal itself (they were mentioned in the introduction to this chapter), or of the rocks overlying and underlying the coal seams, of the partings and of any rocks that build up the filling of coal basins. These minerals include, among others, quartz, muscovite, feldspars (mostly kaolinized), calcite, dolomite, minerals from the chlorite group, accessory zircon, rutile, magnesite and garnet. Clay minerals represent a separate group; kaolinite is most common, and when newly formed it functions as a mineralizer of fusain cells. Carbonates and clay minerals occur on cleats. Of the relatively scarce clay minerals, metahalloysite was recorded by Kuhl (1959) from brown coal at Konin in Poland.

The comparison of the mineral associations of the coal basins with those of peat bogs shows that coal contains a greater variety of minerals.

CONTENTS OF CHEMICAL ELEMENTS IN COALS AND COAL ASHES

Within the scope of this manual not even the briefest descriptions of the investigations of the quantitative and qualitative distribution of chemical elements in coal can be given. Of the wealth of data available (e.g. Yudovich et al., 1972; Gulyaeva and Itkina, 1974; Ruch et al., 1974; Sheibley, 1975; Yudovich, 1978) only some most relevant facts will be mentioned. In the further text maximum or average values of the individual elements are presented. Those interested will find more references in the papers quoted.

Ag — silver

The Ag contents in coal are very low, according to Rankama and Sahama (1950) maximum 10 ppm. Goldschmidt and Peters (1933) gave an average value as 0.0005 to 0.001 % Ag in coal ash. In some coals of Germany Leutwein and Rösler (1956) established that Ag/Pb ratios varied from 1 : 90 to 1 : 2,000. Rajský found 0.0113 % and 0.0078 % Ag in anthracite at Lhotice near České Budějovice in Bohemia. His results are cited in Koblic (1950). Several data on Ag contents are available from the Ostrava-Karviná coalfield. Kessler et al. (1965) determined 3–6 ppm Ag in coal ash. Minchev and Eskenazi (1966) analysed some Bulgarian coals for silver and determined 0.25 ppm, 0.6 ppm, and 1.2 ppm Ag in ashes. Eskenazi (1974) studied the Ag concentration in a light fraction with a low ash content (0.3 to 1 ppm Ag) in the Eastern Maritsa and Pkhelarovo Coal Basins. Experiments with Ag adsorption on peat, xylenite and vitrinite have shown that 90 % of the initial silver is adsorbed at pH 3 to 6, when dissociation of the sorbents, carboxyl and phenol groups, takes place.

The presence of cations like K^{1+}, Ca^{2+} and Al^{3+} in the solution has a negative effect on the degree of silver adsorption, the cations with a higher charge being more active in this respect. In the coal of the Yakutsk region 0.001 % Ag has been determined by Pavlov (1967).

Al — aluminium

Aluminium is an essential component of ash of most coal kinds. Deul and Annell (1956) found 6 to 12 % Al in the ash of brown coal from Harding County in the north-western part of South Dakota. The presence of mellite, the aluminous salt of organic acids in some brown coals provides partial explanation of the rich Al contents in coal ash. Kreulen (1928) reported 31.5 % Al_2O_3 in the ash of the Ruhr coal. Kessler and Valeška (1957) developed a direct spectrographic method for determination of Al in coal without previous burning. They obtained 0.78—12.76% Al_2O_3 in 13 samples. Unless the organic matter is purely separated, the Al content should be ascribed mainly to the mineral content, particularly the clay matter.

As — arsenic

Arsenic is usually concentrated in sulphidic minerals of coal. Goldschmidt and Peters (1933) referred 0.05—0.1% As from bituminous coal. Many authors interpret the As accumulation in terms of As concentration during mouldering of plant matter in the humic layers. Leutwein and Rösler (1956) determined 1,900 ppm As in ash from the anthracite of Brandov, and 2,400 ppm from bituminous coal (Ebersdorf), with ash content of 16.6%. Duck and Rimus (1951) reported that the largest As amount in coal is usually concentrated in arsenopyrite, which may be very finely dispersed. In the Kladno coal district the average As content is 0.1% in the ash of the main coal seam and 0.01% of the basal coal seam (Schejbal et al., 1972). According to Hokr (1971a, 1975), the arsenic in brown coal of the North Bohemian Basin is associated mainly with pyrite. Maximum arsenic contents correspond to maximum sulphur amounts. Arsenic accumulated in the North Bohemian brown coal both syngenetically during coalification and epigenetically after the termination of the Miocene basinal sedimentation. Syngenetic As is dispersed uniformly in the coal matter (5—10 ppm). Epigenetic As comes from several sources in the Krušné hory Mts. and is associated only with pyrite accumulations; it amounts to 500 ppm, occasionally to 3,000 ppm. In the most enriched samples the As:S ratio = 1:500. Mecháček and Petrik (1967) studied the presence of arsenic and other trace elements in coals from Handlová and Nováky in Slovakia. In the ash of Nováky coal they found As amounts of the order of hundredths of one per cent, and ascribed it to arsenopyrite, realgar, orpiment and some organic compounds, which they did not define precisely. Kessler et al. (1965) determined 83—300 ppm As in ash from the Ostrava-Karviná coalfield (for other examples see Table 30).

Table 30
Contents of arsenic in coal

	As (ppm)	Authors
Coal, U.S.S.R.	1.3—60	Chinovnikov et al. (1964)
Coal, South Africa	0.26—10.3	Kunstmann and Bodenstein (1961)
Bituminous coal, N.S. Wales, Australia	<1—17	Clark and Swain (1962)
Brown coal, Victoria, Australia	1—3	Brown and Swain (1964)
Coal ash, Virginia, U.S.A.	<80—450	Headlee and Hunter (1953)
Ash of brown coal, North Dakota and South Dakota	max. $1 \times 10^3 — 1 \times 10^4$	Deul and Annell (1956)

Au — gold

There are not many data available on its accumulation in coal. According to Goldschmidt and Peters (1933) the concentration of Au in coal ash ranges from 0.02 to 0.05 ppm. Gold occurred in association with small amounts of platinum and palladium. In Czechoslovakia, Koblic (1950) reported 35 ppm Au from anthracite (Lhotice near České Budějovice) and Bouška (1961) trace amounts of Au in the ash from the Anežka seam in the Sokolov Basin. Sheibley (1975) mentioned 0.146 ± 0.048 ppm Au in coal ashes as an average.

B — boron

Boron concentration in coal derives either from plants, or is due to the effects of thermal springs (New Zealand), or is present in inorganic compounds (presumably borosilicates) of coal. Leutwein and Rösler (1956) thus correctly placed boron among the elements that are bound to coal matter only to a small extent and by far predominantly to extraneous ash substances. Goldschmidt and Peters (1932c) gave the maximum B content as 0.3 %. Fortescue (1954) determined a substantial amount of boron in ash of some American coals; Inagaki (1957) reported B contents in the Japanese coal of the order of tenths of ppm. Vasilev et al. (1977) identified boron in ash from the Jurassic coal in the southern and south-western parts of the Gissarskogo khrebet. The content in coal ash varied between 18 to 130 ppm, which recalculated to coal equals 1—28 ppm. The B content was 3 to 33times smaller in coal than in the neighbouring rocks. It is bound to the silicate component of coal ($B_{silicates} = 33-95\%$), to the organic matter (5—40 %), and sorbed boron is also present (15—59 %). The authors have proved that the amount of boron bound to organic substances and of sorbed boron decreases from bituminous coal to anthracite. Davies and Bloxam (1974) also stated that the B content decreases with the increase in coalification. From the Ostrava-Karviná coalfield Kessler et al. (1965) reported 240—1,683 ppm as average contents in coal ash; several coal seams in the Žofie mine gave up to 5,000 ppm B in ash. Somasekar (1971) studied the distribution of boron in the main seam of the Kladno coal district and found increased contents 194—476 ppm in ash of bright banded coal. He believes the source of boron to be the original plant material. Schejbal et al. (1972) determined 0.051 % B in ash (coal of the main seam) and 0.032 % B in ash (coal of basal seam) of the Kladno coal district. Beneš et al. (1964) recorded relatively high B concentration in Handlová coal (0.32—1.40 % in ash), the highest content being in the xylitic coal. Geochemistry of boron in coal of the West Carpathians is studied by Mecháček (1975). In general, boron is concentrated in coal in another way than in the neighbouring claystones. The boron contents in coal thus cannot be assessed unequivocally in terms of salinity of the basin.

Ba — barium

The barium content in coal ash is positively correlated with the content of strontium, but barium strongly prevails over strontium. The presence of barium in coal is accounted for by extraneous ash substances. Bouška and Havlena (1959) reported a higher Ba content in some coal ash samples from Lampertice near Žacléř, which they ascribed to a small amount of minute barite fragments. Part of barium may be bound to the organic matter of coal by absorption. Katchenkov and Flegontova (1955) determined 0.04 % Ba in coaly clays of Jurassic age; in another paper (1955) they referred that the barium contents in Jurassic coals are very near to the content of barium in the Earth's crust. The extremely high Ba contents determined in the ash of coals (up to 4.76 %) of Great Britain by Reynolds (1939) were caused by secondary barite mineralization of coal seams. Leutwein (1966) found out that there exists a positive relation between the Ba contents and clay proportion in brown coal. From the Ostrava-Karviná coalfield Kessler et al. (1965) gave 860—2,800 ppm Ba as average values in coal ash. The average Ba content in ash of the Kladno coal is 0.014 % in the main seam and 0.056 % in the basal seam (Schejbal et al., 1972). Kyshtymova (1976) studied the distribution of Ba and Sr in the stratigraphic succession of coal-bearing basins. (For additional data see Table 31.)

Be — beryllium

The concentration of beryllium in coal ash and thus also in coal is thought to belong to the coal matter itself. Jedwab (1964, Table 32) referred that the higher the ash content in coal, the lower is the content of Be. The degree of coalification does not influence the Be amount. Silberminz and Rusanov

Table 31
Contents of barium in coal

	Stratigraphy	Ba in ppm (in ash)	Author
Brown coal			
Australia	Permian—Tertiary	up to 800 (concentration in dry coal)	Swaine (1967)
Czechoslovakia	Tertiary	100—1 000	Honěk and Jiřele (1965)*
	Tertiary	245—2 200	Mecháček (1975)
German Dem. Rep.	Tertiary	200—2 800	Rösler and Lange (1965)
U.S.A.	Cretaceous	100—1 000	Deul and Annell (1956)
U.S.S.R.	Mesozoic	140—2 730	Tkachev et al. (1965)
Bituminous coal			
Australia	Permian	1 000—10 000	Clark and Swain (1962)
German Dem. Rep.	Carboniferous	100—27 000	Leutwein and Rösler (1956)
	Permian	100—500	Leutwein and Rösler (1956)
	Triassic, Jurassic	100	Leutwein and Rösler (1956)
United Kingdom	Carboniferous	90—710	Puchelt (1967)
U.S.A.	Carboniferous—Permian	270—22 000	Puchelt (1967)

* In Puchelt (1972): Barium (Ed. Wedepohl 1969).

Table 32
Average Be contents in coal and coal ashes from the mines of Belgium, the Netherlands and France
(Jedwab, 1964)

Locality	Country	Number of samples	Average Be content in ppm	
			in ash	in coal
Messeix, St. Eloy	France	8	15	1.1
Helchteren-Zolder, seam 19	Belgium	15	17	1.0
Helchteren-Zolder, seam 20	Belgium	11	33	1.6
Merlebach, Cécile seam	France	46	—	1.0
Wendel, Cécile seam	France	10	—	0.8
Bore LXXVII, Mijngebied at Peel	Netherlands	78	22	1.0
Helchteren-Zolder, seam 34/19	Belgium	3	54	1.5
Zwartberg	Belgium	9	26	1.7
Waterschei	Belgium	6	24	1.2
Eisden	Belgium	8	26	2.8

(1936) recorded 0.1 % Be as a maximum value in ash from the Donets coal (U.S.S.R.). Beus (1966) determined the following average Be contents from the Donets Basin (in coal ash): Lisikhanskii district 15 ppm Be (19 analyses); Almazno-Marevskii district 11 ppm Be (118 analyses); Centralnyi 44 ppm Be (94 analyses) and Stalinsko-Makeevskii district 9 ppm Be (54 analyses). Stadnichenko et al. (1956) carried out spectrochemical determinations of Be in coal ashes of the U.S.A. Of 1,123 samples examined, 95 % contained Be in concentrations about 45 ppm. The greatest concentration of Be was found in two low-ash coal pieces from the eastern part of Kentucky (140 ppm BeO in coal). Stadnichenko et al. (1961) reported Be contents in coal from different parts of the U.S.A. (Table 33); they inferred from their studies that beryllium concentrations are higher in vitrain than in fusain (Table 34). The Be contents in coal ash are also given in the papers of Goldschmidt (1935 and 1944), Goldschmidt and Peters (1932b), Fortescue (1954), and many others. In the coals of Czechoslovakia beryllium was found in most samples but only in trace amounts. Kessler et al. (1965) mentioned it from the Ostrava coalfield (average content = = 50 ppm in coal ash), Bouška and Havlena (1959) from Žacléř, Zahradník et al. (1959), Schejbal et al. (1972) from the Kladno district (on the average 0.003 % Be in ash from the main seam and 0.001 % in ash from the basal seam), Macháček et al. (1962) from the Josef seam in Chodov near Karlovy Vary (max. 0.28 % Be in ash), and Hak and Babčan (1967) from the Sokolov Basin (tens to hundreds ppm Be in coal). Macháček et al. (1966) determined 0.5 – 57 ppm Be in ground water of the Sokolov Basin.

Bi — bismuth

Goldschmidt (1944) gave the maximum bismuth concentration in coal ash as 0.02 %. Weninger (1965) measured 1 and 10 ppm Bi in two samples of the anthracite from the Eastern Alps. Coal and bituminous sediments in eastern Austria contained 0.02 – 4.1 ppm Bi, on average 1.1 ppm Bi, in 25 samples (Brandenstein et al., 1960).

Ca — calcium

This element occurs in coal basins in minerals whewellite $Ca(C_2O_4) \cdot H_2O$ and erlandite $Ca_2(C_6H_5O_7)_2 \times 4H_2O$. There are numerous data available on calcium in coal, which is often found in considerable amounts; Kreulen (1928) gave 1.8 % CaO in the ash from the Ruhr coal. The amount of calcium in coal is generally increased by the content of the inorganic mineral component, but part of it can be present in organic matter (Ca humates). In the North Bohemian Brown-coal Basin, for example, about 0.5 % Ca is bound to organic matter, as is evidenced by the regressive relation to ash $Ca^d = 0.5 + k \cdot A^d$ (Hokr, 1971a).

Cd — cadmium

The presence of cadmium in coal was recorded by Goldschmidt (1937) and Slutzer (1940). Gluskoter and Lindahl (1973) studied the occurrence of this element in the coal of Illinois. Gordon (1952) found 3 – 30 ppm Cd in the peat ashes (50 samples).

Cl — chlorine

Leutwein and Rösler (1956) reported on the Cl contents in the coal from the Zwickau-Oelsnitz district; six samples contained 0.43 to 0.77 % Cl in coal. The mine water of this district contained 0.5 % NaCl.

Co — cobalt

Cobalt is generally thought to be inherent in inorganic matter, although part of it may concentrate in plants during their life and growth. It is classed with the inorganic components of coal also because

Table 33
Contents of Be in coal in various regions of the U.S.A.
(Stadnichenko et al., 1961)

Locality	Be in coal (ppm)	
	maximal content	minimal content
Eastern province, Appalachian region		
Northern part:		
Kittanning	4.2	1.5
Southern part:		
Alabama	4.6	0.5
Eastern Kentucky	31	0.4
Tennessee	11	0.1
Virginia	3.6	0.2
Interior province		
Eastern part:		
Illinois	6.3	0.7
Indiana	12	1.5
Western Kentucky	9.5	0.5
Western part:		
McAlester Basin	2.9	<0.1
Province of the Rocky Mountains		
Sweetwater County, Wyoming	13	0.1
Other areas	31	<0.1

Table 34
Contents of Be in vitrain and fusain of some coal samples from the U.S.A.
(Stadnichenko et al., 1961)

Sample	Ash (%)			Be in ash (%)			Be in coal (ppm)		
	coal	vitrain	fusain	coal	vitrain	fusain	coal	vitrain	fusain
0-Mal-Mk-4	6.56	3.50	13.04	0.001	0.01	0.0004	0.66	3.5	0.52
0-368	10.80		19.60	0.002		0.0005	2.2		1.0
Va-310	3.42	3.00	5.80	0.002	0.01	0.0006	0.7	3.0	0.4
WVa-215		0.90	9.60		0.004	<0.0001		0.4	<0.1
Ill-P-1-9	5.26	2.02		0.005	0.03		2.6	6.1	
Mont-K_1-6		3.98	7.08		0.003	0.0001		1.2	<0.07

it may be absorbed in clay minerals or may derive from pyrite, which is common in coal. In the coal from the Šverma mine (Czechoslovakia) microconcretional pyrite was a constituent of almost every sample (Bouška and Havlena, 1959), as it was in many other cases (Čadková, 1971, in coal of the Radvanice seams). Leutwein and Rösler (1956) established a positive correlation between the Co amount and the content of ash matter. Otte (1953) determined maximum Co content in coal ash (Dickebank seam in the Ruhr Basin) as 0.2 %. Goldschmidt and Peters (1933) also reported cobalt from coal ash. In Czechoslovakia, Kudělásek (1959) measured the Co contents and the Co : Ni ratio in coal from the Žacléř Formation in the Lower Silesian Basin. He established that Ni almost invariably predominated over Co in a ratio of 10 : 1. The maximum content of cobalt was 92 ppm (seam No. 28). In the Kladno coal seams the average Co content was 0.001 % in ash (Schejbal et al., 1972). Average Co contents in the Ostrava-Karviná coalfield were 53—134 ppm in coal ash (Kessler et al., 1965). Tkachenko and Tretenko (1958) studied the contents of some elements in the ash of brown coal from the Dnieper Basin; cobalt was present at a maximum concentration of 0.005 %.

Cr — chromium

Chromium is fairly common in coal and is recorded in almost all qualitative spectral analyses. According to Goldschmidt and Peters (1933), the chromium can amount to 1,000 ppm. Gibson and Selvig (1944) determined 0.04 % Cr_2O_3 in ash of one American coal. Reynolds (1948) cited the following Cr_2O_3 values in ash from the Peacock seam coal: 0.5 % — ash content 2.3 %; 1.0 % — ash content 1.8 %; 0.7 % — ash content 1.6 %. Otte (1953) reported far higher Cr contents from the vitrain matrix, in the Katharina seam, Ruhr Basin, i.e. 0.5 to 1.3 % in ash. The presence of Cr in coal and coal ash is almost universally associated with organic matter. Tkachenko and Tretenko (1958) determined spectroscopically 0.01—0.02 % Cr in ash of brown coal. Stadnichenko et al. (1950) described brown coal containing germanium with 0.1—0.8 % Cr in ash. Katchenkov and Flegontova (1955) determined 0.014 % Cr in coaly clays of Jurassic age. The ash of coal from the Ostrava-Karviná coalfield contained 213 to 449 ppm Cr (Kessler et al., 1965). Kudělásek (1959) determined a relationship between V, Cr and Mo; Mo and Cr decrease synchronously with V. Vanadium often prevails strongly over molybdenum and locally also over chromium.

Cs — cesium

See potassium.

Cu — copper

Copper often concentrates in plant bodies, which suggests that the greater part of it found in the coal ash will be a component of organic matter. Otte (1953) recorded 0.4 % Cu in ash from the Helena coal seam. Leutwein and Rösler (1956) determined 1,500 ppm Cu in the ash of Brandov anthracite. Čadková (1971a) reported 150 ppm Cu as an average from the Radvanice coal seams and Schejbal et al. (1972) 0.021 % Cu (in ash) from the Kladno main seam and 0.023 % Cu (in coal ash) from the basal seam. Kudělásek (1959) studied Cu concentrations in the coal of the Lower Silesian Basin. The highest Cu concentration was found in the Radvanice Member (up to 3,320 ppm in coal ash); in the Svatoňovice Member the maximum concentration was 3,000 ppm and in the Žacléř Member 1,310 ppm (in coal ash). He has also established that the oldest complex of the Lower Silesian Basin, i.e. the Žacléř Member, is poorest in copper; the maximum content was in the uppermost seam (No. 10), and upwards of it the Cu content increases appreciably. The average contents of Cu in the Ostrava-Karviná coalfield range from 150 to 310 ppm in coal ash (Kessler et al., 1965). Katchenkov and Flegontova (1955) measured 0.005 % Cu in the coaly clays of Jurassic age and 0.003 % Cu in the samples of Upper Cretaceous clays. Copper may be partly associated with pyrite.

F — fluorine

As the identification of fluorine requires special analytical methods, only a few data are available. Crossley (1946) inferred from the correlation of fluorine and phosphorus contents in coal that they come from apatite. The most detailed study on the fluorine content in coal was published by Bradford (1957). On examining the coals from the Western United States he recognized that the samples with a high fluorine were also rich in phosphorus (fluorapatite). This relationship has been stated already before in coals from England. (Francis, 1954, referred that the proportions of fluorine in British coals vary between 20 and 150 ppm and are usually proportional to the phosphorus content.) For information several Bradford's analytical data are tabulated below:

F content (ppm) in dry coal	Samples					
	1	2	3	4	5	6
(a) coal burnt in O bottle	53	99	45	132	40	57
(b) coal burnt at 475 °C on addition of CaO	43	24	38	116	34	54
(c) coal burnt at 475 °C without adding CaO	31	80	15	85	23	39
(d) coal burnt at 800 °C	11	43	7	27	4	15
(e) P content (%) in dry coal	0.010	0.007	0.009	0.018	0.008	0.005
P(e) : F(a) ratio in dry coal	1.9	0.7	2.0	1.4	2.0	0.9

Correns (1956), who analysed 120 samples of British coals and determined 0–175 ppm F in coal (most samples contained below 80 ppm), thought that apatite is a carrier of fluorine. Hokr (1975b) thinks that fluorine in the coal of the North Bohemian Brown-coal Basin is associated with the inorganic component.

Fe — iron

The coal ash generally contains an appreciable amount of iron; it varies depending on the amount of inorganic component to which it is bound. The mineral humboldtine, hydrous iron oxalate $Fe(C_2O_4) \cdot H_2O$ is also known to occur in brown coal. Iron is an important biophile element. In the North Bohemian Brown-coal Basin Hokr (1971a) differentiated iron bound to disulphides (chiefly postgenetic), to carbonates (predominantly siderite), both syngenetic and postgenetic, to oxides and hydroxides and partly also to organic compounds, most likely humates (0.5 % Fe).

Ga — gallium

Goldschmidt (1944) gave 0.4 % as the maximum Ga content in coal ash. Otte (1953) determined 0.1 % Ga in ash from the coal of the Katharina seam (Ruhr Basin); the increased amount was in the vitritic component. Leutwein and Rösler (1956) observed a relationship between the contents of gallium and vanadium. Gallium appears or increases in samples that contain vanadium. Ramage (1927) determined 0.04 to 0.05 % Ga in the British coal and coaly claystones. Tkachenko and Tretenko (1958) reported trace amounts of gallium from the brown coal of the Dnieper Basin. The coal from the Zyryanski Basin contains 1–7.5 ppm Ga (12–100 ppm in ash), the coal from the Donets Basin 0.6–12.6 ppm (7.5 to 270.5 ppm Ga in coal ash) (Ratynskii and Zharov, 1976). The maximum concentration was established in the Ekibastuzskoe coal — 17.6 ppm and in the ash from the Garbagataiskoe coal — 398 ppm. The enrichment is of a regional scale. The relationship between gallium and ash contents has been evidenced statistically. In Czechoslovakia the presence of gallium in coal was studied by Zahradník et al. (1959) on the coal seams near Vejprnice (Plzeň Basin). They have established that the Ga content rises with the increase in Al (see also Bouška and Honěk, 1962), and that it is associated rather with the clayey rocks than with the coal itself. The Ga content also rises with the increase in ash amount. The dependence on the amount of Fe sulphides, on the position of the coal seam and on the site of sampling was not clear-cut. The study of the relationship between Ga and Ge in the cross-sections of the deposit has shown that Ga is strikingly increased in the claystone beds, whilst Ge is bound to pure coal matter. Ga has not been detected in sandstones and arkoses or only in very small amounts. These findings indicate that gallium belongs to the inorganic coal component. The Ga concentration decreases markedly with the increase in silicates and, on the other hand, it rises with the content of alumosilicates. The investigations of Bouška et al. (1963) of the upper Doubrava Member in the Ostrava-Karviná coalfield led to quite opposite results. The Ga content in a coal seam was invariably higher than in the adjacent claystone. The maximum Ga amount was 130 ppm in the ash from seam No. 12. Gallium may be obviously bound either to organic or mineral component. A direct proportionality between ash and gallium contents exists in case Ga is bound to the mineral portion. Volodarskii et al. (1976) arrived at similar conclusions. Several analytical results are presented in Table 35.

Table 35
Gallium contents in coal

Occurrence	Ga (ppm) in ash	Author
West Virginia, U.S.A.	8 – 1 000	Headlee and Hunter (1953)
U.S.A., brown coal	20 – 36	O'Neil and Suhr (1960)
Japan	35 – 75	Inagaki and Yamaguchi (1958)
Norway	<10 – 135	Butler (1953)
Italy	30 – 150	Bertetti (1955)
England	9 – 282	Dalton and Pringle (1962)

Ge – germanium

After major amounts of germanium had been discovered in coal ash by Goldschmidt (1930) and Goldschmidt and Peters (1932a), much attention was centred to its occurrence in coal, particularly when some coal ashes proved to be feasible economic sources of this element. In a number of countries such as the U.S.A., U.S.S.R., Great Britain, Japan, the F.R.G. and the G.D.R., the investigations have advanced sufficiently to ensure the production of germanium. Goldschmidt (1930) determined the maximum concentration of Ge in coal ash as 1.1 % (bituminous coal, Hartley Seam, England), and gave 0.05 % Ge in coal ash as the median value. On comparing these values with the average Ge content in the lithosphere (0.0004 – 0.0007 %), Rankama and Sahama (1950) stated that in the first case (1.1 % Ge in coal ash) the coal is enriched in germanium 1,600 – 2,800 times, and in the second case (0.05 % Ge in coal ash) 70 – 120 times. Rotter (1952) and Švasta et al. (1955) summarized in their papers numerous records on Ge contents in Czechoslovak coals. The team of workers under the guidance of Zahradník continued in investigations. In Zahradník et al. (1959a, b, 1960a, b) only a minor part of the results has been published. According to these authors, the largest Ge amounts are in bituminous coals of the lowest rank in central Bohemia. They believe that germanium is bound to organic components and its concentration is related to the absorption capacity of the coal matter. On the basis of the studies of Stadnichenko et al. (1953), who found the highest Ge concentration in the samples of Cretaceous coal and coalified wood, Zahradník (1955) carried out investigations of the Cretaceous coals in Bohemia.

He obtained interesting results from the Mladějov power plant, where Cretaceous coal from Moravská Třebová was burnt; the ash from beneath the boilers contained 0.9 % Ge; ash from the mine locomotive heated with the same coal contained 0.08 % Ge. The content of germanium has also been studied preliminarily in the coal of the West Bohemian mines. The highest concentrations have been determined at the contacts with the over- and underlying beds and in lenses or pieces of coal enclosed in sandstone or claystone. Germanium was more abundant in vitrain than in fusain. Coal from a drill hole near Touškov (Plzeň Basin) contained 0.18 % Ge; ash of this coal 0.8 – 0.9 % Ge. The upper seam of the Eliška mine gave 0.001 and 0.005 % Ge in coal. In one sample from the Obránců míru mine 0.00847 % Ge was determined in coal. Coal of a low specific gravity proved to be richer in Ge than coal having a high specific gravity. The opinion prevails that vitrain having the lowest specific gravity contains the largest amount of germanium. Of interest is the content of 0.0178 % Ge in a rotted wood (allegedly birch trunk), thousands or tens of thousand years old, which was found in the sandy alluvium of the abandoned Radbuza river bed between Mantov and Chotěšov (Zahradník et al., 1959).

Votavová (1958) studied the germanium content in the coal of the main seam in the Kladno district. The largest amount of 120.7 ppm Ge (in coal) was in one sample of banded coal from the Nejedlý mine (Votavová and Král, 1959). The average content of 0.022 % Ge in coal ash (215 analyses) was determined colorimetrically in the main seam of the Nosek, Gottwald and Zápotocký mines (Somasekar, 1969). In the Kounov seam 30 – 50 ppm Ge in coal was measured by Kadaňková-Votavová (1960). The geochemistry of germanium in the Radnice Basin was examined by Tyrolerová (1959), who established the largest amounts in the lower Radnice seam in the central part of the Vejvanov Basin (240 ppm in coal; the ash content 13 %). The Ge content increased towards the wedging-out of the seam and the over- and underlying beds, and along the N–S trending fault bounding the mine field. The Ge distribution in the coal seams at Týnec near Plzeň was studied by Tyroler (1958).

The earliest reports on the germanium contents in Czechoslovak coals were published by Šimek (1940) and Šimek et al. (1948). They reported 1 – 10 ppm Ge from the Ostrava coal. The coal of the Ostrava-Karviná coalfield was investigated in this respect also by Macháček (1957), Králík and Polický (1960), Králík et al. (1960), in great detail by Kessler et al. (1965), Bouška et al. (1964) and Bouška and Krejčí (1958, 1965). Macháček et al. (1955) carried out reconnaissance study of trace elements and heavy minerals in the hanging wall, footwall and partings in the Karviná Formation. Leutwein and Rösler (1956) recorded 57 ppm Ge in the ash of the Brandov anthracite. Vlasák (1949, in Rotter 1952) analysed the coal from Mirošov and found 0.18 % GeO_2 in ash. Kudělásek (1959a, b, 1960) examined the coals from the Lower Silesian Basin and determined the highest Ge concentrations in the youngest coal seams worked – in the Radvanice group of seams. In the Svatoňovice group of seams there was only 100 ppm Ge in ash; in the Žacléř group of seams the germanium decreases downwards from the uppermost seams. The highest seam studied contained 280 ppm Ge in ash, and the amount decreased through 103 ppm to 28 ppm Ge in ash. It has been recognized that there is a relationship between the stratigraphic succession and the Ge content: the oldest Žacléř Formation contains least germanium and its amount rises gradually upwards. This relationship is apparently disturbed in the Svatoňovice group of seams, but the samples from them contain many times greater amounts of ash than those from the Žacléř group of seams (see also Bouška and Havlena, 1959).

The average content (3.2 ppm in ash) of Ge in the coal of the Rosice-Oslavany Basin was determined by Kadaňková-Votavová (1960). Analytical results from the Cretaceous coals in Bohemia and Moravia were given by Šulcek (1956). The brown coal from Most contained only 0.6 ppm Ge (Pavlů, 1956), the coal of the Sokolov Basin 10 – 100 ppm Ge (Hak and Babčan, 1967), and the coal from the Cheb Basin 0.01 – 0.1 % Ge in coal ash at the most (Šantrůček, 1958, 1961). The maximum content of germanium in the coal from the Hrádek part of the Žitava Basin was 0.65 % in ash (Václ and Čadek, 1959; Čadek et al., 1961). Mrázek and Vlášek (1958) found anomalous Ge contents in isolated coalified woods of Cretaceous age from Branišov in southern Bohemia (0.25 % and 0.27 % in coal). In his report of 1957 Vlášek recorded 1 % Ge in ash from the coalified wood in the Zliv Formation. He did not identify germanium in the anthracite samples from Lhotice near České Budějovice and in the ash of lignite

Table 36

Germanium in coal and coal ashes from various world regions

Country	Locality	Content of ash in %	Ge in ash	Ge in coal	Author
Australia	New South Wales (279 samples)	3.4–34	6–2 000 ppm (aver. 50 ppm)	0.4–150 ppm	Clark and Swaine (1962)
Belgium	several localities (485 samples)			1–10 ppm	Rouir (1954)
Bulgaria	Plevno (L. Cretaceous) (10 samples)		aver. 0.043 %		Eskenazi (1965)
	W. Rhodope (Palaeogene, Oligocene) (21 samples)		aver. 1.40 %		Minchev and Eskenazi (1969)
	Volchepolskoe (Palaeogene, Oligocene)		3.79 %		Minchev and Eskenazi (1969)
Canada	Sydney coalfield, Nova Scotia (181 samples)	5.6–22.3	max. 0.2 %		Hawley (1955)
Fed. Rep. Germany	Ruhr Basin, seam Katharina (3 samples)	2–14.8	0.1 %		Otte (1953)
German Dem. Rep.	Meisdorf-Oppenrode	20	0.0023 %		Leutwein and Rösler (1956)
	Zwickau	8.6	0.034 %		Leutwein and Rösler (1956)
	Oelsnitz	7.7	0.042 %		Leutwein and Rösler (1956)
India	Assam	2.1	0.16 %		Mukherjee and Dutta (1949, 1950)
	Assam (70 samples)		500–600 ppm		Banerjee et al. (1974)
	Hyderabad	7.1	0.1 %		Mukherjee and Dutta (1949, 1950)
Japan	several localities		max. 0.3 % GeO$_2$	max. 3.57×10^{-2} %	Inagaki (1958)
					Inagaki et al. (1956)
				23–102.5 ppm	Oka Kanno–Ayusawa Nada (1955)
	Cretaceous coals				Eguti (1955)

Country	Location/Description	Value	Value	Reference
	brown coal		max. 0.4 %	Oka et al. (1956)
	Tertiary lignite		0.197 %	Sakanone (1960)
Spain	Asturia		0.01 % GeO$_2$	Lopez de Ascona and Ping (1948)
United Kingdom	Northumberland, Yard seam	0.9	11 100 ppm	Goldschmidt and Peters (1933)
		1.1	700 ppm	
		1.2	700 ppm	
	Northumberland, Yard seam	1.8	1 500 ppm	Morgan and Davis (1937)
		2.4	1 000 ppm	
	Staffordshire	11.15	230 ppm	Reynolds (1948)
		20.20	700 ppm	Reynolds (1948)
	Dorset lignites (Lower Lias)		>1: some samples up to 8.4 %	Hallam and Payne (1958)
	all main coalfields of England (207 samples)	up to 20	10–1 000 ppm	Aubrey (1952)
U.S.A.	Columbia (Cretaceous)		up to 6 %	
	Pennsylvania (6 samples)	1.8–37.5	20–120 ppm	Stadnichenko et al. (1950)
	Kentucky (14 samples)	0.8–14.4	4–1 430 ppm	Stadnichenko et al. (1953)
	Ohio (35 samples)		100–1 000 ppm	Stadnichenko et al. (1953)
	Colorado (3 samples)		100–500 ppm	Stadnichenko et al. (1953)
	Montana (19 samples)	2.4–23.4	50–100 ppm	Stadnichenko et al. (1953)
	Prince George County, Columbia (Lower Cretaceous)	2.85	7.5 %	Stadnichenko et al. (1953)
		3.4	2.3 %	Stadnichenko et al. (1953)
		2.8	5.0 %	
		2.85	6.7 %	
		2.04	2.4 %	
		4.12	0.8 %	
	Tennessee and Ohio (Upper Devonian)		0.6 %	Breger and Schopf (1955)
			2.0 %	
			1.5 %	
			3.0 %	
	Illinois (3 samples) (Lower Carboniferous)		aver. 0.03 %	Zubovic et al. (1964)
	Tennessee (Lower Carboniferous)		0.18 %	Zubovic et al. (1966)

							100 ppm	
							77 ppm	
							84 ppm	
							27 ppm	
							21 ppm	
							46 ppm	
							77 ppm	
							1.7–46 ppm aver. 7 ppm	
							2–4 ppm	
							6–12 ppm	
							1–15 ppm	

Table 36 – continued

Country	Locality	Content of ash in %	Ge in ash	Ge in coal	Author
U.S.S.R.	Donets Basin (53 samples)	1.4–8.1	0.1–1.0 %		Silberminz (1936)
	Donets Basin (14 samples)		10–100 ppm		Silberminz et al. (1936)
	Kuznetsk Basin (31 samples)	1.8–5.9	100–1 000 ppm		Silberminz et al. (1936)
	Ural and Pechora valley (19 samples)		1 000–10 000 ppm		Silberminz et al. (1936)
	Ural Mts.		0.001 %		Ratynskii (1943)
	Caucasus Mts.		up to 1 %		Ratynskii (1943)
	Caucasus, vitrain (Lower Jurassic)	1.4	0.985 %		Ratynskii (1945)
	Urals, vitrain	2.3	0.815 %		Ratynskii (1945)
	Byloriskoe (ASSR)		up to 1 %		Korolev (1957)
	Siberia (450 samples)		max. 0.3 %		Travin (1960) and Lomashov (1961)
	C. and N. Timan (Mid. + Upper Devonian) (23 samples)		0.01 %		Kochetkov (1966)
	Yakutsk Basin (Upper Jurassic-Lower Cret.) (532 samples)		0.002–0.031 %		Pavlov (1967)
	Sangara, Lenskii Basin (Lower Jurassic-Lower Cret.) (9 samples)		0.005 %		Yudovich (1972)
	Lenskii Basin, S. part (Lower Cretaceous) (40 samples)		0.14 %		Yudovich (1972)
	Pechora (Upper Permian) coal beds in claystone (40 samples)		0.36 %		Yudovich (1972)
	coal beds in sandstone (13 samples)		0.10 %		Yudovich (1972)

from the Svatopluk and Václav mines in Mydlovary. In a drill hole near Petrovice in the Třeboň Basin a thin coal bed gave about 1 % Ge in ash (Vlášek admits sorption in situ in this case). This author observed that the highest germanium amounts are in coal samples that are encompassed by permeable clastic rocks. Malecha (1959) found high Ge in coal from drill hole M9, close to the Měcholupy railway station. Xylitic coal encountered at a depth of 75.6 – 75.65 m at the base of a 10-m layer of blackish grey strongly sandy claystones and clayey sandstones, contained 5.25 % Ge in ash (ash content = 6.29 %), and in coal itself 0.33 % Ge. The highest Ge content in the Bohemian coal — 5.35 % (ash content 3.71 %) was determined in the ash from a coalified tree trunk in the freshwater Cenomanian near Polička (Bouška et al., 1963). Peats from Borkovice, Horusice and Třeboň in southern Bohemia and some cannels were spectrographically analysed for germanium and other trace elements by Bouška (1959, 1961). The Ge contents established were very low.

The germanium contents from a number of world localities are listed in Table 36.

Of geochemical interest is the study of Žák (1951) referring to a find of germanium-bearing opal near the brown coal seam at Březno near Chomutov.

He — helium and other rare gases

Patteisky (1954) analysed mine gases in the Ruhr Basin and identified in addition to $10-20\%$ CH_4, $14-20\%$ CO_2, $65-75\%$ N also 4 % He, $0.1-1.2\%$ Ne and small amounts of Ar and Kr. A correlation has been established between helium and nitrogen. This He-rich mine gas was intended to be a source of helium.

Hg — mercury

The results of analyses for mercury in coal are cited by Stock and Cucuel (1934). They reported 8 ppb Hg in the bituminous coal from the Saar Basin, and 10 ppb and 22 ppb Hg in the bituminous coals from the Ruhr Basin. The British coal contained 12 ppb Hg and brown coals from various German localities 1.2 ppb to 25 ppb Hg. The coals from Illinois (U.S.A.) contain $0.10-12$ ppm Hg, and much of it is bound to pyrite (Ruch et al., 1971). These authors give a 0.2 ppm Hg mean for 101 U.S. coals. According to O'Gorman et al. (1972) Hg in the U.S. coals ranges from 0.07 to 33 ppm with a 0.18 ppm average. Dvornikov (1967, 1971) established a range of 0.7 to 1.33 ppm Hg in the Donbas coal with a mean of 0.86 ppm.

I — iodine and Br — bromine

Lissner (in Leutwein and Rösler, 1956) found 4.868 ppm I in a sample of bituminous coal from Karl-Marx-Werk in Zwickau, 3.9 ppm from Mühlsengrund and 2.67 ppm in the boghead from Oelsnitz. Itkina (1955) studied the distribution of iodine and bromine in the coal-bearing horizon in the Saratov area (U.S.S.R.). The I contents were 1 to 10 ppm and Br contents were $4.1-41.6$ ppm. In the clayey rocks the iodine amount slightly increased; no relation has been observed between bromine content and lithology. In Czechoslovakia, iodine is contained in waters at Darkov in the Ostrava area and in the Carboniferous waters around Slaný in the Kladno-Rakovník Basin.

Furher data on the content of iodine in coals from the Ruhr Basin, Saxony and Bohemia are given in the paper of Gulyaeva and Itkina (1962). The values range from tenths of ppm to 11.17 ppm. All the coals examined also contained F, Cl, and Br, which suggests a relationship between the halogens and the organic components of coal.

In — indium

Goldschmidt (1937) quoted up to 2 ppm indium in coal ash.

K, Na, Li, Rb, Cs — alkali metals

The alkali metals, particularly potassium and sodium, are common in coal and their contents are given in every analysis of coal ash. For example, Kreulen (1938) found 4.8% $Na_2O + K_2O$ in the ash of the Ruhr coal. Kessler and Dočekalová (1955) developed and modified a quantitative method for determination of Na and K in coal ash; in their paper they recorded many data on Na_2O and K_2O contents in brown coals, bituminous coals and anthracites from Czechoslovakia. The average contents were $0.41-2.19\%$ Na_2O and $0.445-3.165\%$ K_2O in coal ash. The content of alkalies in the plant matter is well known but the greater part of alkali metals is assumed to have derived from the inorganic admixtures, particularly muscovite and clay minerals. Lithium is recorded from Nowa Ruda in Silesia (1,100 ppm LiO_2 in coal, Goldschmidt, 1954). Smales and Salmon (1955) reported 30–250 ppm Rb from vitrain. Medek (1956) determined cesium contents within a range of 27–111 ppm in eight ash samples from various Czechoslovak coals; in his paper he gave a detailed description of the method applied. The presence of alkali metals in the coals of the Ostrava-Karviná coalfield was studied by Kessler et al. (1965, 1967). In the later paper they informed on the application of alkali metals in the stratigraphy and identification of coal seams. According to Kessler et al. (1965), Li is contained both in external and internal coal matter. The Li content decreases markedly towards the seams of higher numbers. In the Ostrava Formation it cumulates predominantly in the lower part, where a maximal average of 200 ppm Li in ash has been determined in the Petřkovice Member. The lowest content of 90 ppm Li in coal ash has been found in the upper layers of the Hrušov Member and in the Jaklovec Member. The lithium contents in coal ashes from various Czechoslovak coal basins were found to vary between 0.003 and 0.05% Li_2O by Kessler and Dočekalová (1960).

Mg — magnesium

Magnesium is usually present in coal ashes in an appreciable amount. Kreulen (1928) quoted 1.7% MgO from the ash of the Ruhr coal. The Mg amounts are related to the amounts of inorganic matter.

Mn — manganese

The increased Mn contents in coal should again be accounted for by the presence of inorganic components. Otte (1953) determined a large amount (2.2% Mn in ash) in the coal from the Dickebank seam in the Ruhr Basin; like many others she thought the manganese to be almost exclusively a constituent of external ash matter. Manganese is present in coal in the form of carbonates, silicates or oxides. In the Ostrava-Karviná coalfield the Mn contents in coal ash range from 105 to 2,508 ppm (Kessler et al., 1965).

Mo — molybdenum

Goldschmidt and Peters (1933) referred on its concentration in coal ashes. Otte (1953) determined the highest concentration of 0.6% Mo in ash of the vitrain from Meissen near Münden. Kessler et al. (1965) found an average amount 11–33 ppm Mo (in ash) in the Ostrava-Karviná coal. In the ash of Brandov anthracite 30 ppm Mo has been determined (Leutwein and Rösler, 1956). Molybdenum is thought to be bound mainly to the coal matter itself. Korolev (1957) and Razumnaya (1957) are of the same opinion; the latter author presumed that Mo and V are cumulated by sorption. Korolev reported average Mo contents in the Jurassic coals from the Bylym-Kabardinskoï district in the A.S.S.R. to be 9 ppm Mo. This average amount is 4.5 times the clarke of this element in sedimentary rocks. The same author determined 3 to 10 ppm Mo in strongly pyritized coal (syngenetic pyrite). Molybdenum may be adsorbed on the clay admixture of coal. Golovko (1960) found 10 ppm Mo in bituminous coal from the central part of the U.S.S.R., which also contained 2—2.5 % S (in coal). This author presumes that molybdenum accumulated by sorption on the organic matter in the initial phase of the coal-forming process and subsequently concentrated in the sulphides.

N — nitrogen

The coal contains 0.2—3 % N in dry matter; it is derived from plant and animal proteins, from the plant alkaloids, chlorophyll and others. The conversion of these compounds and fixation of nitrogen during coalification was studied by Flaig (1968). During distillation of coal the nitrogen is liberated as ammonia and is used extensively for the production of synthetic fertilizers. The amount of nitrogen in the world coal deposits is estimated at 2×10^{11} tons (Gorham, 1949). N^{daf} in brown coal is usually 0.4—2.5 %, in bituminous coal 0.6—2.8 % and in anthracite 0.2—1.5 %, very rarely attaining 3.0 %. Brooks and Kaplan (1972) cite the $^{15}N/^{14}N$ ratio to be -3.0 to $+2.0$ ‰ in coal, -8.0 to $+8.0$ ‰ in living organisms, $+1.0$ to $+16.0$ ‰ in crude oil and $+1.0$ to $+17.0$ ‰ in sedimentary rocks. The variation of $\delta^{15}N$ in coals of various ranks does not fluctuate uniformly with the degree of coalification (Drechsler and Stiehl, 1977).

Na — sodium

See potassium.

Nb — niobium

Pavlov (1967) recorded 10 ppm Nb in some ash samples from the coal of the Yakutsk Basin. Trace amounts of niobium (<0.01 %) were found in the ash from the Cretaceous coal at Březina (Bouška and Honěk, 1962).

Ni — nickel

Jones and Miller (1939) demonstrated 0.9—10.3 % NiO in the ash of vitrain. In the Katharina seam (Ruhr Basin) the maximum amount of 1.6 % Ni was found in the vitrain ash (Otte, 1953). Fortescue (1954) referred on a local enrichment in Ni, and Reynolds (1948) recorded 1.0 and 1.5 % NiO from the Peacock seam (ash content 1.6 or 1.8 %). Leutwein and Rösler (1956) stated a relationship between Ni and Co contents in coal. Tkachenko and Tretenko (1958) quoted 10 to 60 ppm Ni in the ash of brown coal from the Dnieper Basin; Katchenkov and Flegontova (1955) determined 0.05 % Ni in the Jurassic coaly clays. The nickel content in the coal of the Lower Silesian Basin in Czechoslovakia was studied by Kudělásek (1959). The highest content of 1,240 ppm Ni (in ash) was determined in the Žacléř seams; the coal seams of Svatoňovice and Radvanice had 237 ppm Ni (in ash) at the most. The contents vary greatly and according to Čadková (1971a) the average Ni content in the Radvanice seams was only 20 ppm in ash. Some nickel-bearing minerals such as millerite, bravoite or siegenite provide evidence of the increased Ni concentration. The coal ash from the Ostrava-Karviná coalfield contained 116 to 250 ppm Ni (Kessler et al., 1965). The average Ni content in the coal ash of the Kladno main seam is 0.01 % and 0.019 % in the basal Kladno seam (Schejbal et al., 1972). Kovács et al. (1960) are of the opinion that nickel is rather related to the organic matter than to external ash substances, although it also accumulates in coal with a high ash content. This is very likely due to its presence in pyrite.

P — phosphorus

Leutwein and Rösler (1956) presented many analytical data on its occurrence in coal. The coal from Zwickau contains 0.06—0.16 % P_2O_5, from the Martin Hoop mine 0.11 % P_2O_5 (in ash). Feigeľman

(1949) and Feigeľman and Voinalovich (1955) discovered a relationship between the petrographic composition of rocks, the coal and the P content in coals of Central Asia. Bright coal contained 8–10 times less phosphorus than dull coal. In Czechoslovakia, Hlavica (1924) determined phosphorus in the Nýřany coal, and Kessler et al. (1965) in the coal of the Ostrava-Karviná Basin. The maximum P value was there 7,300 ppm (in ash) and a negative correlation existed between P and ash contents. Šplíchal (1967) reported 0.01–0.11 % P_2O_5 from the coal mined at present. Other data are given in the book of Hubáček (1948).

Pb — lead

Otte (1953) determined the highest content of about 0.3 % Pb in the coal ash of the Katharina seam. Leutwein and Rösler (1956) quoted 90 ppm Pb from the Brandov anthracite ash, and 3,000 ppm Pb from the Mesozoic coal from Altenbeichlingen (ash content = 68.1 %). Kudělásek (1959) reported maximum Pb values of 0.1–1 % (in ash) from coal samples from Horní Verněřovice and Chvaleč. According to Čadková (1971a), the average Pb amount in the coal ash of the Radvanice seams is 520 ppm. From the Ostrava-Karviná coalfield Kessler et al. (1965) gave the Pb values 95–202 ppm (in ash). Lead in the coal ashes from the Kladno district was determined by Schejbal et al. (1972). The average contents are 0.02 % in the main seam and 0.025 % in the basal seam. Part of the lead amount may be associated with pyrite or bound to galena.

Pt — platinum and platinum metals Rh, Pd

Goldschmidt and Peters (1933) reported from coal ashes 0.5 ppm Pt, 0.2 ppm Rh and 0.2 ppm Pd.

Ra — radium

In 1933 Lloyd and Cunningham determined the highest Ra content in the ash of coal from the Pratt No. 5 mine in the Warrior coalfield. The sample richest in ash contained 7.05×10^{-12} g Ra/g ash, and the sample with the least ash content had 0.51×10^{-12} g Ra/g of ash. Muchemblé (1943) reported up to 10.5×10^{-12} g Ra/g rock from the coaly claystone rich in organic matter. Jeczalik (1970) gave the maximum content 56×10^{-10} g Ra/g bituminous coal from the Meszko deposit in Poland.

Rb — rubidium

See potassium.

Re — rhenium

Kuznetsova and Saukov (1961) determined 0.084 to 0.328 ppm Re in coal. Gorokhova and Pokrovskaya (1962) reported 0.095–0.327 ppm Re in brown coal from Central Asia, which contains large amounts of Mo (66–5,210 ppm in coal ash) and other trace elements. Re and Mo appear in coal in the form of organometallic compounds and in the sulphide microinclusions dispersed in coal. Martin and Garcia-Rossell (1970) reported on rhenium in the Spanish coals.

S — sulphur

Sulphur occurs in coal in four types:

(1) as **sulphides** (pyrite, marcasite and others),
(2) as **sulphates** (calcium sulphate, barite and others),
(3) **organic S** (component of organic compounds forming part of coal matter),

(4) elementary S — only in some brown coals, probably of secondary origin

$$2H_2S + SO_2 \rightarrow 3S + 2H_2O.$$

Sulphidic sulphur is sometimes called "pyritic" as pyrite is the most common mineral of all sulphides.

Pyrite may have originated in various stages of coalification. It forms wherever decaying organic substances come into contact with dissolved Fe sulphates. Primarily, amorphous FeS (melnikovite) is probably precipitated, which by aging alters into pyrite. The greater part of pyrite and marcasite developed in the early stages of coalification. During the decay of dead plants in a peat bog bacterial reduction produces H_2S, which precipitates FeS from the soluble iron compounds. Under the continuing activity of hydrogen sulphide, the crystallization of FeS_2 occurs. Marcasite forms from acid solutions under low temperatures and pyrite rather from alkaline solutions at various temperatures.

The sulphate sulphur is represented mainly by calcium sulphate. Pyrite-bearing coals upon oxidation contain iron sulphate. The reactions

$$FeS_2 + O_2 \rightarrow FeS + SO_2, \text{ later } FeS + 2O_2 \rightarrow FeSO_4, \text{ or}$$
$$2SO_2 + O_2 + 2H_2O \rightarrow 2H_2SO_4 \text{ and later } FeS + H_2SO_4 \rightarrow FeSO_4 + H_2S$$

are accompanied by the development of heat, which may lead to spontaneous ignition of coal. The soluble iron sulphate is usually leached from the coal seams by percolating water. Hubáček et al. (1962) identified Na_2SO_4 in the German coal (so-called Salzkohle). According to these authors bituminous coals contain only hundredths of one per cent of sulphate sulphur and brown coals about 0.2 %. An increased content of sulphate sulphur suggests that oxidizing processes occurred during the formation of the coal seam. Of the Czechoslovak coals only coal sample from the Roland seam (Jeremenko mine, Ostrava-Karviná coalfield) has increased sulphate sulphur — 0.76 % $S_{SO_4}^d$ (Hubáček et al., 1962).

Organic sulphur of coal is derived partly from the coal-forming plants (plant proteins have up to 1.3 % S) and partly is the product of reactions occurring during coalification. The content of organic sulphur in plants is too small to account for the amounts of up to several per cent, established in some coals. Sulphur is present in resinous substances, N-containing groups and even in thioether and dithioether bonds.

Elementary sulphur seems to be a secondary product of the decomposition of pyrite.

The total sulphur contents in coals may be fairly high. Kreulen (1952) and Hadzi and Novak (1954) reported 8.5—11 % organic S from the coal of Istria. According to Wandless (1954) the British coals contain 0.4—1.5 % organic sulphur on the average. The content of inorganic sulphur varies greatly. The average of 474 analyses of the Illinois coal is 3.57 % S (Gluskoter and Hopkins, 1970); sulphate sulphur ranges from 0 to 0.88 % (av. 0.08 %), organic sulphur from 0.27 to 2.98 % (av. 1.46 %) and sulphide sulphur from 0 to 4—9 % (av. 2.06 %). Savchuk (1971) found that in the Donets Basin the coal of a lower rank contains more sulphur (2 % total sulphur) than high-rank coal (only 0.7 %). According to Morozov (1971) total sulphur in the Kiselovsk Basin amounts to 2.45—14.91 %. The influence of the proximity of marine beds on the S content in coal has been observed. High sulphur coals are known from Pakistan; coal of the Salt Range contains up to 12 % of total sulphur with organic sulphur contents ranging from 1—7 %. In the Sor Range area there is a pronounced smell of hydrogen sulphide in the mine workings indicating that the decomposition of organic or inorganic sulphides is still proceeding (Francis, 1954).

The brown coals of Czechoslovakia usually contain 0.05 to 2.50 % sulphidic sulphur, the bituminous coals even above 10 % (Dalibor mine in the Slaný area; Hubáček et al., 1962). In some seams of the Žacléř-Svatoňovice district even above 5 % sulphidic sulphur has been determined. The coals of the Ostrava-Karviná coalfield and subbituminous coal from Kladno have a rather small amount of sulphidic sulphur. The least average amounts of sulphate sulphur (usually <0.1 %, rarely 0.2 %) have been determined in brown coal of the North Bohemian Basin, and somewhat higher in the Sokolov Coal Basin. In the lignite of southern Moravia the sulphate sulphur ranges from 0.2 to 0.3 %.

Hubáček et al. (1962) gave the average organic sulphur content in dry bituminous coal as 0.53 % (Ostrava-Karviná coalfield) to 0.69 % (Žacléř-Svatoňovice district). Only bituminous coal of the Rosice-Oslavany Basin contains more than 2 % organic sulphur with a relatively high sulphide sulphur. The average content of organic sulphur in the brown coal of the North Bohemian and Sokolov Basins is 0.56 %, exceptionally above 1 %. Higher organic sulphur was found only in the coal of Hrabák (2.95 %) and Michal (3.87 %) mines. The relatively largest proportion of organic sulphur has been ascertained in the lignite of southern Moravia (about 2 % on the average) and in the brown coal of the Nováky Basin in Slovakia (1.2 – 4.25 % in volatile matter).

The Rosice-Oslavany coal has high contents of total sulphur (up to 5 %). Hokr (1971a, b, 1975b) studied the genesis of sulphur in the North Bohemian Brown-coal Basin. Sulphur accumulated there most probably in two stages. The first-generation sulphur was predominantly of organic origin, which forms at present the more or less constant background of about 0.5 % in volatile matter. In the second stage, after the termination of Miocene basinal sedimentation, the overwhelming part of sulphidic sulphur formed due to the activity of anaerobic bacteria, which reduced the sulphates supplied by the ground water flow, very likely from the area of the Krušné hory Mts. At places of extreme concentrations of sulphidic sulphur (more than 3 %), the content of organic sulphur also increased. The sulphate sulphur is the product of oxidation of sulphides, which formed in the second stage. According to Špetl (1977, personal communication) the content of sulphidic sulphur at some localities of the North Bohemian Brown-coal Basin reaches 10 – 15 %. The distribution and isotopic composition of sulphur in coal have been studied by Smith and Batts (1974).

Sb – antimony

Otte (1953) reported 0.3 % Sb as a maximum content in ash from the Dickebank seam (Ruhr Basin). The antimony in coal has also been recorded by other writers. The coal enrichment in antimony is rather small. Headlee and Hunter (1953) found <40 ppm Sb in coal ash from Virginia (U.S.A.); Swaine (1962) gave <200 ppm Sb from the ash of bituminous coal from New South Wales (Australia).

Sc – scandium

Goldschmidt and Peters (1933) reported 0.04 % Sc in coal ash. The Sc contents in coals of the Soviet Union was studied by Dobrolyubskii (1962), who recorded 5 ppm as an average amount in coal and 24 ppm in ash. Vlasov (1968) found the maximum Sc content – 140 ppm in coal ash. The origin of Sc in coal is uncertain; it seems to have been absorbed on organic components from the circulating water. Scandium may form organometallic complexes.

Se – selenium

Selenium occurs in coal in trace amounts (10 – 30 ppm). The highest Se contents have been found in coals having large amounts of ash and sulphur, i.e. in coals rich in pyrite (Goldschmidt and Strock, 1935). Savelev (1964) determined 5.2 – 10.0 ppm Se in small isolated coal seams enclosed in the Upper Cretaceous sandstones in Uzbekistan.

Si – silicon

Almost all silicon in coal should be attributed to external ash matter, as its content is strongly influenced by the presence of inorganic substance. The Si content is prevalently equal to 30 – 40 % SiO_2 or more in ash.

Sn — tin

Borovskiĭ and Ratynskiĭ (1944) determined 0.001–0.03 % Sn in ash of coal from the Kuznetsk Basin (U.S.S.R.), using spectrographic analyses. Otte (1953) reported maximum 0.6 % Sn in the ash of pure organic coal substance from the Katharina seam (Ruhr Basin). Leutwein and Rösler (1956) found 20 ppm Sn in the Brandov anthracite; the authors think the tin in coal to be bound to organic matter. After Terebinin and Angelov (1962), tin enriches mainly the benzene and ethanol extracts. On the contrary, it has been reported that the Sn content in coal tends to increase with the ash content. In the Ostrava-Karviná coalfield 3–12 ppm Sn has been determined in ash (Kessler et al., 1965). In the Kladno coal district, the coal ash contained 10 ppm Sn (Schejbal et al., 1972).

Sr — strontium

The strontium contents in coal are low and are usually attributed to inorganic constituents. Leutwein and Rösler (1956) determined a relationship between strontium and barium contents. Bouška and Havlena (1959) confirmed this observation on the ash of the Žacléř coal. Mecháček (1975) determined 290 to 2,380 ppm Sr from the ash of the Tertiary coal in the West Carpathians.

Ti — titanium

Jones and Miller (1939) described the vitrain from Northumberland that contains an extraordinary amount of 7.0–24.3 % TiO_2 in ash, with a small content of ash matter. Reynolds (1948) gave 9.2 and 15 % TiO_2 (in ash) with 1.6 to 1.9 % ash content from the Peacock seam. Otte (1953) determined 3.0 % Ti in vitrain ash from the Osterwald coal. Minor to trace amounts of Ti in coal ash are common. Titanium is generally thought to belong to external ash substances. According to Kudělásek (1959), the coal from the Lower Silesian Basin contains up to 2.17 % Ti. The contents are higher in clayey shales and argillites. The Žacléř seam No. 9 is the only one that contains 7,000–9,000 ppm Ti. The highest average Ti content in the coal of the Ostrava-Karviná coalfield is 1.147 % (in ash) in the Jaklovec Member. Tkachenko and Tretenko (1958) obtained 0.2–1.0 % Ti in the brown coal ash of the Dnieper Basin.

Th — thorium

There is little known about the thorium in coal. Aramu and Uras (1957) measured the contents of uranium and thorium in bituminous coal and found 5.66×10^{-4} % to be the maximum amount. Bibliography on this subject is given in the paper of Cooper (1955).

Tl — thallium

Thallium has been discovered in coal only recently. Voskresenskaya et al. (1962) determined 0.3 to 1.5 ppm Tl (Table 37) in coal from Central Asia and traces to 2.3 ppm Tl in coal from the North Caucasus. Thallium is not firmly bound in coal. It forms soluble compounds of humic and fulvene types (humates and fulvenes) with organic coal components. Pyrite from coal was richer in Tl than pyrite from the neighbouring sedimentary rocks. Thallium came into the coal-forming basins during sedimentation or in the early diagenesis; its contents are not controlled by the petrography of coal and its contents in pyrite are constant throughout the vertical profile of the seam. In 1968 Voskresenskaya published another paper on the Tl contents in pyrites from the following coals:

Central Asia (30 samples)	average 34 ppm
Moscow Basin (14 samples)	3.3 ppm
Transcarpathian Basin (5 samples)	11 ppm
Dnieper Basin (23 samples)	<0.5 ppm

Table 37
Thallium contents in coal from the U.S.S.R. (Voskresenskaya, 1968)

	Tl (ppm)	
	coal	coal rich in ash
Central Asian Basin (Jurassic):		
Soguty	0.5	–
Kavak	0	0.4
Tashkumyr	0.1	1.0
Dzhergalan	0.5	–
Kara-Kiche	0.8	1.8
Aksai	0.6	–

U – uranium

Uranium is known to occur in coal in many world regions. It is universally believed that it was absorbed into the coal (Moore, 1954). Moore pointed out that coal possesses a high absorption capacity for uranium. It is also assumed that radioactivity produced by Th and U may contribute to the polymerization of organic substances (Breger and Deul, 1956).

The economic importance of uranium provoked investigations in many countries. Hoffmann (1943) recorded the U content in peat as 1×10^{-6} g/g, as 3.3×10^{-6} g/g in brown coal, and 3.72×10^{-5} g/g in bituminous coal, but these values have been in many cases surpassed.

Many data on U contents in coal are available. Eklund (1946) reported 1 % U in coal ash from Sweden as its maximum content. The Upper Cambrian "kolm" formation in the Vastergotland province contains layers rich in organic matter (black shales) and in uranium. Cobb and Kulp (1957) determined 0.5 % U, and in the paper of 1958 they reported 4,000 – 6,000 ppm U, with only 100 ppm U in the neighbouring shales. In Hungary the U contents in coal were studied by Foldvari (1952) and Szalay (1954, 1974). Analytical results from the German coals were recorded by Davidson and Ponsford (1954) and Leutwein and Rösler (1956). Radioactivity of brown coal in the G.D.R. was determined by Rösler and Zscherpe (1971). Jeczalik (1970) gave 0.99 % U (in ash) as the maximum value for Polish bituminous coals. Maximum contents of U were found in the fraction of small density ($1.7 - 2.0$ g/cm^3). The same observations are mentioned from the Pechora basin (U.S.S.R.) by Gipsh et al. (1971). The uranium contents in the brown coal from Granada and the Ebro Basin in Spain were reported by Martin and Garcia-Rossell (1971); the results from France have been published by Sarrot-Reynaud de Gresseneuil (1950) and Jurain (1968). Jurain described the uranium deposits in coaly claystones and coal near St. Hippolyte, where U is predominantly dispersed in organic matter of coal, which suggests the presence of uranium-organic complexes. In the richest part the U content ranges from 0.7 to 1.0 %. Mineralized layers with uraniferous ores have been found in several coal basins at the margin of the French Massif Central (Ronchamps, Fins) (Kervella, 1958).

An increased uranium concentration in brown coal is known from the F.R.G. (Pluskal, 1972). Near Wakkesdorf in eastern Bavaria, the Tertiary coal-bearing series lies on the granite bedrock. Increased uranium has been determined mainly in the clay partings inside the seam and in the claystones along the seam margins. The average U content is 0.03 – 0.04 %. Pluskal also described an increased U concentration in the pitch coal from some molassoid basins in Switzerland. Davidson and Ponsford (1954) measured a high radioactivity in a coal layer with pyrite in the Warwickshire Basin. They obtained 0.005 % U_3O_8, or 0.08 % U_3O_8 in ash (9.95 %). In Czechoslovakia uranium was found in coal of the Radvanice seams. Tomaňa (1957) studied the geochemistry of uranium in the Kladno coal and established the average concentration as 2 ppm to 90 ppm U. He stated that the maximum contents of U correspond to the maxima of other heavy metals. Uranium is concentrated mainly in the hanging and foot walls, decreasing towards the interior of the seam. Pure clarain contained the least amount of U. The U amount increased in the samples with the increasing ash content. The distribution of uranium in the sedimentary rocks and ground waters of the Kladno-Rakovník Basin was examined by Lepka (1967). The uranium mineralization of the coal in Permocarboniferous, Cretaceous and Tertiary basins of the Bohemian Massif is dealt with in the paper of Pluskal (1972). Several coal samples from the Ostrava-Karviná Basin, from Cretaceous basins and an anthracite sample from Lhotice were also examined (Bouška, 1961).

The only detailed data on the uranium contents in coal and coaly claystones were published in the U.S.A. Kehn (1957) published in his paper a map showing the occurrences of uranium-bearing coals and coaly claystones with at least 0.005 % U. Breger and Schopf (1955) studied small lenses and seams of bituminous coal from the Upper Devonian complexes in Tennessee and Ohio. Microscopically, the tissue of coalified woods (*Callixylon*) was identical with vitrinite. The ash from these woods contained up to 0.5 % U, up to 5 % germanium, up to 5 % vanadium and up to 1 % nickel in ash. Here are some of their results:

ash in %	uranium in ash in %
4.23	0.42
3.80	0.55
2.45	0.57
1.69	1.32

Breger, Deul and Rubinstein (1955) determined 0.039 % U (0.12 % U in ash 32.43 %) in the brown coal from the Mendenhall strip mine, Harding County, North Dakota. Another sample of brown coal had 0.31 % U in ash (13.8 %). The neighbouring rocks contained only tenths of these amounts. Uranium occurs in coal in the form of an uranium-organic complex or an ion uranium-organic compound, which is soluble below pH 2.18. The analyses of Mesozoic coals and coaly claystones in Arizona and New Mexico gave the maximum contents 0.62 % and 1.34 % U in coal ash (Bachman and Read, 1952). According to Bachman et al. (1957) all important uranium concentrations are intimately associated with coaly sediments and porous sandstones; the latter served as filters through which the uranium solutions migrated to underlying sediments containing coal or at least coal pigment. The results of investigations of two coal layers (about 1.2 m thick) from the neighbourhood of Sage Lincoln County are presented in the study of Beroni and McKeown (1952). The coal is of Cretaceous age and contains 0.004 to 0.013 % U in ash. In the north-western part of South Dakota and contiguous states Denson et al. (1957) studied uranium contents in brown coals in the Late Cretaceous Hellcreek Formation. Many samples contained 0.005 to 0.02 % U and some even 0.05 to 0.1 % U in ash. The authors think uranium to have been transported by ground water flow from the overlying leached Oligocene and Miocene volcanogenic rocks of the White River and Arikaree Formations. Denson and Gill (1955) also found 0.01 to 1.0 % Mo in ash of this coal, as well as autunite and zeunerite, which contributed to the increased uranium concentration (often 2–5 %). It was particularly the deposit in the Cave Hills area that was of economic interest (seams 45–75 cm thick). Ferm (1955) recorded 0.014 % U from the coal in the Darlington area. The uranium presence in coal and adjacent rocks in Colorado, Wyoming and Montana was studied by Gill (1953, 1954a, b, c, 1957) and Gill and Zeller (1957). In the Cambrian coalfield (Weston County) 0.0085 % U was determined in the ash from a coal layer enclosed in sandstone, and 0.05 % U (in ash) in the brown coal from Carter County. Samples from a 45 cm thick layer contained 0.057 % U in ash (34 %). Uranium is thought to be of epigenetic origin and to have been leached from the overlying radioactive tuffaceous rocks. In some brown coals of very small thickness, e.g. in the North Cave Hills (Harding County) 0.2 % U (in ash) was found. The mineral metaautunite was identified concurrently. A similar U content is reported from the Long Pine Hills.

Gott, Wyant and Berone (1952) presumed that uranium (up to 0.01 %) in brown coal has been transported into coal by water leaching the overlying volcanic ash. The Tertiary high-ash coal in Churchill County (Nevada) contains 0.05 % uranium. The coal seam of Late Cretaceous age in Old Leyden (Jefferson County, Colorado) gave maximum 0.1 % U in ash (Gude and McKeown, 1953). According to Hail and Gill (1953) the Cretaceous coals from western Montana contain 0.013 % U in ash and the adjacent claystones only 0.0006 % U. Mapel and Hail (1957) mentioned the richest deposit from the Goose Creek District (Cassia County, Idaho), which contained 0.12 % U at the roof of a 2.4 m thick layer of coaly claystone. They believe that uranium was leached from the overlying volcanic ash. Masursky (1956) found local higher concentrations of uranium in a Miocene coal seam near Creston Ridge (Red Desert, Wyoming); the U content was highest near the roof (0.051 %) decreasing to 0.001 % U some 12 m lower. The higher U is accompanied by higher Ga, Ge, Fe, Mo, Pb, V and REE. Weathered granite or tuffogenous rocks are thought to be the source of all these elements. The average U content of this brown coal is very low (0.003 %), increasing to 0.01 % (in coal) in the richest portions, which gives 0.03 % in ash (Masursky, 1962).

In the ash of the Lower Eocene coal in Sweetwater County (Wyoming, U.S.A.) Masursky and Pipiringos (1957) determined 0.005 to 0.19 % uranium. The largest U enrichment is at the north-eastern margin of the basin, where the coal is interlaid with coarse sediments. Moore (1954) found 0.024 to 0.12 % uranium in the ash of brown coal from Slide Butte in North Dakota. Coal layers and associated claystones in the northern parts of Illinois contained less than 0.001 % U; the largest amount of 0.125 % U (in ash) was obtained in coal from Harrisburg (Patterson, 1954).

Snider (1953) studied some bituminous and anthracitic coals of the Mississippi Formation in Montgomery County (Virginia). All samples of bituminous coal and semianthracite had less than 0.001 % U. Black shales of Late Devonian age contained 0.003 − 0.004 % U (Lee County). Staatz and Bauer (1954) determined the maximum content of 0.059 % U (with ash content of about 70 %) in the brown coal of Churchill County in Nevada. Many data are also in the papers of Vine (1953, 1955, 1957) and Vine and Moore (1952). The last named authors recorded 0.3 % (in ash) as the maximum amount of uranium in the Cretaceous coal from Fallcreek in Idaho. Welch (1953) determined 0.001 % U in the anthracitic coal of the Pennsylvanian Formation in eastern Pennsylvania. The neighbouring shales contained 0.001 to 0.003 % uranium. Zeller (1955) found 0.006 and 0.007 % U in the Cretaceous coaly clays of South Utah, and 0.002 % U or less in other sediments. In his paper of 1957 Zeller claimed a relationship between uranium and molybdenum. The uranium occurrence in the Japanese lignite of Misasa group was studied by Sakanone (1960); he determined 5.2 % U (maximum) in the ash of the Tertiary lignite.

V − vanadium

Vanadium occurs in coal and in all sediments usually in the pentavalent form. In Czechoslovakia, vanadium was found in the anthracite from Lhotice near České Budějovice: Koblic (1950) reported 3.48 % V_2O_5 (in ash) and Mazáček (1960) 0.03 % V_2O_5 (anthracite) and up to 3.0 % V_2O_5 in ash. Leutwein and Rösler (1956) determined 400 ppm in the Brandov anthracite. A detailed study of vanadium in the coaly argillites, shales, ashes and clinkers was made by Pokorný (1954). He found the highest content in bituminous argillites from Horní Vernéřov (up to 1,000 ppm V). The Kounov seam in the Rakovník area is very poor in vanadium; Rost (1950) determined amounts of only hundredths and thousandths of one per cent in ash. The content of vanadium in the Kladno coal district amounts to 0.002 % (in ash) in the main seam and 0.04 % (in ash) in the basal seam (Schejbal et al., 1972). The average vanadium contents in the Ostrava-Karviná coalfield are 192 − 362 ppm in ash (Kessler et al., 1965). In the Lower Silesian Basin the greatest V concentration is in seam no. 10 of the Žacléř seams (1,680 ppm V in ash); the Žacléř seams nos. 29, 8 and 9 have up to 1,000 ppm V, but the concentration drops below 400 ppm in the Svatoňovice and Radvanice seams (Kudělásek, 1959). Vanadium also accumulates intensely in the neighbouring clayey rocks; it is highest in the clayey siltstones, medium in sandy claystones and the lowest in sandstones, which suggests a direct relation to the pelitic material.

In some samples the vanadium concentration exceeds 1,000 ppm V (Kudělásek, 1959). Čadková (1971a) gave 100 ppm V as an average in the coal of the Radvanice seams.

In the U.S.S.R. the vanadium contents in coal was studied by Miropolski (1939), Katchenkov and Flegontova (1955). Tkachenko and Tretenko (1958) determined 0.005 % V in the ash of the brown coal from the Dnieper Basin. In the ash of bituminous shales (kukersite) from the Estonian S.S.R. up to 60 ppm V is reported by Chaldna (1958). Silberminz (1935) determined $1.22-8.64$ % V in ash (clarainvitrain, often with preserved wood structure). Uzunov (1965, 1967) studied the V contents in some coals of Bulgaria and on several examples demonstrated a relationship between V content and low-ash coals (the concentration of vanadium in coal was 10 times higher than in adjacent claystones).

Bøgvad and Nielsen (1945) referred on the V in brown coals in Denmark. Reynolds (1948) determined V contents in the Peacock seam (coal ashes) at 10.1 % V_2O_5, 14.1 % V_2O_5 and 12.4 % V_2O_5 with ash contents 1.6 %, 1.8 % and 2.3 %, respectively. According to Rankama and Sahama (1950), the ash of some Argentine coals contains up to 21.4 % V_2O_5. Stadnichenko et al. (1950) determined $0.7-5$ % vanadium in the ash of Lower Cretaceous brown coal containing also germanium. The highest concentration was found in the coalified trunks of *Cupressinoxylon wardi*. Breger and Schopf (1955) determined $1-5$ % V in the ash of the Devonian coalified wood, whilst the adjacent shales contained only $0.005-0.01$ % V.

W — tungsten

In the Cheb Basin (Bohemia) anomalous W contents were found by Vyjídák (1967). The highest amount 5 % W (in ash) corresponds to 0.19 % W in dry coal sample. Since tungsten occurs only in coal, it is assumed to be bound to the organic coal matter, which as a good sorbent captured it from the circulating waters. The tungsten is genetically associated with the Krušné hory metallogenic province containing Sn–W deposits. Tungsten may have been transported in the form of $WO_3 \cdot n\, H_2O$ compounds, which are weathering products of tungsten minerals. The different behaviour of W and Sn in supergene conditions is corroborated by the difference in the occurrence of the two elements in the sediments of the Cheb Basin: with regard to its small migration ability Sn is not accumulated appreciably in coal.

Adsorption and desorption of W on peat, telinite, humic acids, and cation-exchange resins have been studied by Eskenazi (1977). The adsorption of W on peat decreases with increasing pH from 2 to 6 (maximum at pH 2 after 24 hours). It has been suggested that the binding of W on coals is due to the presence of hydroxycationic forms of W^{6+} in the solutions capable of forming covalent organometallic complexes. This idea may contribute to the explanation of tungsten accumulation in the coals of some basins. The adsorption of W on coal sorbents is neither physical adsorption nor a simple cation exchange. Adsorption of W by humic acids from solutions reaches up to 10 %.

Y — yttrium and rare earth elements

Mukherjee (1950) identified REE in ash of some Indian coals. Goldschmidt and Peters (1933) recorded contents of Y, Ce, Nd, Sm, Gd, Dy, Er, and Yb in ash of the German and British coals — on the average 100 ppm Y and maximum 800 ppm Y. According to Ershov (1961) Y and Yb in the Kiselovsk Basin are bound prevalently to organic matter. Eskenazi (1965) determined in the ash of gagate from Plevno (Bulgaria) the following amounts (in ppm): La — 103, Ce — 206, Pr — 32, Nd — 196, Sm — 53, Eu — 14, Gd — 81, Tb — 18, Ho — 31, Er — 28, Yb — 38, Lu — 6 and Y — 312. Eskenazi's assumption of a relationship between REE and phosphates has been borne out by the decrease of REE, Sc, Ca, P and Sr from the internal part of the gagate layer outwards. Schonfield and Haskin (1964) analysed lanthanides from two coal samples from West Pennsylvania and South Illinois (U.S.A.) and obtained the following average contents (in ash): 500 ppm $\sum Y$, La–Lu (Y 4.7 ppm, La 4.1 ppm, Ce 11.5 ppm, Pr 2.2 ppm, Nd 4.7 ppm, Sm 1.6 ppm, Eu 0.7 ppm, Gd 1.6 ppm, Tb 0.3 ppm, Ho 0.3 ppm, Er 0.6 ppm, Tm 0.1 ppm, Yb 0.5 ppm, Lu 0.07 ppm); $\sum Y$, La–Lu in coal $\cong$ 34 ppm. Coal is relatively enriched in light lanthanides.

Zn — zinc

Mott and Wheeler (1927) found 0.5 % Zn in clarain and durain ashes and Otte (1953) maximum 0.8 % Zn in ash of the organic coal matter. According to Leutwein and Rösler (1956) there is 800 ppm Zn in the Brandov anthracite. The average content of zinc in ash from the main seam of the Kladno coal district is 0.02 % and 0.014 % from the basal seam (Schejbal et al., 1972). In the ash of coal from the Ostrava-Karviná coalfield there is 48—1,108 ppm Zn on the average (Kessler et al., 1965). Čadková (1971a) gave 720 ppm Zn as an average value for the coal of the Radvanice seams.

Fortescue (1954) mentions minor zinc contents in the samples of American coals and Swanson et al. (1966) stated 100 ppm as an average Zn content in coal. Zinc may be derived from plant bodies or adsorbed in a later stage; part of it may be associated with pyrite or bound directly to sphalerite. Additional quantitative data are listed in Table 38.

Table 38
Zinc contents in peat and coal

Number of samples	Locality	Content (in ppm)	Average content (in ppm)	Authors
peat (76)	G.D.R.	10—1 000*	163*	1
brown coal (33)	G.D.R.	< 10—500*	≦27*	1
brown coal (6)	Victoria, Australia	70—900	300	2
brown coal (43)	South Australia	< 80—200	100	2
coal (68)	New South Wales, Australia	< 25—300	50	2
coal (53)	Queensland, Australia	< 30—1 000	200	2
coal (475)	E part of Int. coal prov., U.S.A.	<100—600 7 800*	44 590*	3
coal (376)	Appalachian Basin, U.S.A.	<100—230 5 000*	7.6 160*	4
coal (228)	SW part of Int. coal province, U.S.A.	<100—2 100 13 000*	108	5
coal (221)	N part of Great Plains, U.S.A.	<200—1 000 7 000*	59 560*	6
coal	W. Virginia, U.S.A.	up to 1 900	430*	7

* Contents in ash.
Authors: 1. Leutwein, 1956, 2. Swaine, 1971, 3. Zubovic et al., 1964, 4. Zubovic et al., 1966, 5. Zubovic et al., 1967, 6. Zubovic et al., 1961, 7. Headlee and Hunter, 1955.

Zr — zirconium

The occurrence of zirconium is referred in some papers. Goldschmidt gave 0.5 % Zr as the upper limit in coal ashes. According to Otte (1953), the vitrain component from the Katharina seam (Ruhr Basin) contained 0.7 % Zr (in ash).

There are no detailed data in the literature on other elements present in coal. They were either determined only qualitatively or occur in traces, such as Cd, In and others.

GEOCHEMICAL CONCENTRATION AND DISTRIBUTION OF TRACE ELEMENTS IN COALS

The concentration of some rare trace elements in coal is of geochemical importance. Two questions arise: In which stage of coal formation does the concentration of elements take place? Of what forms are the bonds between elements and organic coal components?

Coalification proceeds from complex organic molecules to simpler ones and up to coal under favourable conditions. Whilst the essential elements of coal are C, H, O, N, S and/or P, after the volatiles have escaped on its burning, the following elements accumulate in ash:

(1) elements that derived from original plants (primary ash matter);

(2) elements forming the mineral admixture of coal (secondary ash matter, elements alien to organic matter), which were brought into the developing organic sediment by water, wind and other external agents;

(3) elements that were physico-chemically captured from circulating water and bound to organic coal matter during coalification (mainly from the overlying beds, along tectonic fractures and margins of the basins);

(4) elements attached to coal at the contact with igneous intrusions, as a result of exhalations or hydrothermal activity.

The amounts of trace elements in coal ash, in the Earth's crust and in pelitic sediments are listed in Table 39, modified from Krejci-Graf (1972).

According to Goldschmidt (1950) and Headlee and Hunter (1951), who analysed 596 coal samples for 38 elements, the contents of Na, Rb, Ca, Mg, Si, Cr, and Mn are lower in coal than in the Earth's crust; in contrast Li, Sr, Ag, As, Bi, B, Ga, Ge, La, Hg, Pb, Sb, Sn, Zn and Zr are enriched in coal ash 10 to 185 times. Deul and Annell (1956) found unusually high concentrations of Sn, Cu, Zr, B, Ba and Sr in brown coal from the Milam County in Texas.

The results of quantitative spectral analyses of microelements in some coal seams, which are recorded in Tables 40 and 41, are taken from Leutwein and Rösler (1956) and from the summarizing Table of Kessler et al. (1965).

The trace elements in coal and coal ash were not identifiable by classic chemical analyses. Their presence was revealed when spectrographic analysis had come into

Table 39
Enrichment of elements in coal ash (ppm) after Krejci-Graf (1972)

Element	Earth's crust (average)	Pelitic sediments	Max. contents in coal ash
Li	65	46	960
Be	6	<4	2 800
B	10	310	8 600
Sc	5	6.5	400
Ti	4 400	4 300	20 000
V	150	120	11 000*
Cr	200	550	1 200
Mn	1 000	620	22 000*
Co	40	8	2 000
Ni	100	24	16 000
Cu	70	192	4 000*
Zn	40	200–1 000	10 000*
Ga	19	50	6 000
Ge	7	7	90 000
As	5	~5	8 000
Rb	280	300	33
Y	28	28	800
Zr	220	120	5 000
Nb	20	–	2
Mo	2.3	–	2 000*
Ag	0.02	0.05	5–10
Cd	0.18	0.3	80
In	0.1	0.5	2
Sn	40	40	6 000
Sb	1	3	3 000
I	3	0.3	950
Cs	3.2	12	4
La	18	18	31
Ta	2.1	–	0.1
Pt	0.005	–	0.7
Au	0.001	–	0.2–0.5
Hg	0.5	0.3	50
Tl	1.3	2	25
Pb	16	20	1 000*
Bi	0.2	1	200
U	4	1.2	600*

* Higher values are probably due to secondary enrichment.

Table 40

Contents of some elements (ppm) in ashes of coals in the G.D.R. (Leutwein and Rösler, 1956)

Geological age after Horst (1952)	Occurrence		Number of samples	Ash in %	Ge	Cu	Pb	Zn	Ag	As	Sn	Ga	Be	Co	Ni	Mo	V
Mesozoic coal from various occurrences			12	8.5	650	1 000	2 000	1 000	1.2	400	10	300	(70)	100	180	12	150
Lower Permian	Manebach stage	Manebach	8	22.3	4	60	2 600	400	1.3	380	10	160	14	100	110	58	190
		Freithal	53	17.3	87	36	750	1 300	1.2	850	21	110	85	35	72	91	170
	Gehren stage	Lauchagrund	5	14.1	13	640	1 300	400	3.0	1 700	4	tr.	60	120	42	12	38
		Crock	8	32.4	6	50	4 800	5 100	7.7	640	6	51	—	10	30	21	59
		Ilfeld	7	32.8	tr.	tr.	470	190	1.7	400	14	36	—	80	81	18	60
		Meisdorf	3	20.0	23	800	3 400	370	2.2	100	7	23	33	330	92	14	240
		Stockheim-Neuhaus	4	7.3	8	25	130	240	0.8	280	—	tr.	78	—	20	28	10
Upper Carboniferous	Stephanian	Ohrenkammer	4	10.0	10	3 200	80	1 400	12.0	3 000	5	50	—	260	170	110	240
		Dölau	6	22.4	3	170	1 400	260	1.8	70	42	40	12	5	42	17	100
		Wettin	8	12.1	24	140	2 800	1 500	7.8	770	13	88	140	25	53	44	150
		Löbejün	5	13.5	14	1 400	1 800	460	5.2	740	4	36	—	18	68	27	44
		Plötz	27	16.2	4	40	820	250	0.6	800	8	230	—	17	110	39	220
	Westphalian-B-D	Zwickau	134	8.6	340	190	1 300	1 600	3.7	880	32	250	310	190	370	69	120
		Mülsengrund	24	8.3	73	100	370	690	1.5	2 000	14	210	15	92	190	48	78
		Oelsnitz	108	7.7	420	325	1 200	2 100	2.3	930	25	470	100	290	550	54	170
		Flöha	8	21.1	14	25	160	65	1.5	25	11	370	220	20	71	23	150
		Schönfeld	7	14.6	90	160	240	660	1.6	160	10	200	260	96	140	24	190
		Zaunhaus	5	23.6	6	40	60	160	1.3	2 400	16	32	130	30	20	6	200
Lower Carboniferous	Viséan	Ebersdorf	6	16.6	68	100	100	250	1.0	2 400	10	240	(50)	130	140	70	250
		Berthelsdorf	4	18.2	71	150	500	(50)	1.3	1 700	10	50	(25)	33	45	28	160
		Doberlug	32	16.1	9	130	290	770	3.1	90	15	530	tr.	93	120	29	160

Table 41
Contents of elements in ashes of coal in the Ostrava-Karviná coalfield
(Kessler et al., 1965)

Elements	Contents in ppm in coal ash				
	Ostrava region				Karviná region
	Ostrava Formation				
	Petřkovice Member	Hrušov Member		Jaklovec Member	Jaklovec Member
		lower	upper		
Li	200	160	90	90	165
Na	15 660	13 840	13 690	17 000	16 800
K	16 350	16 390	24 110	12 900	12 370
Rb	99	127	214	86	63
Cs	30	36	57	32	15
Cu	282	280	190	239	310
Ag	6	4	3	3	3
Be	38	42	57	55	56
Ba	2 800	2 285	2 090	2 750	980
Zn	162	201	55	48	92
B	267	341	240	307	608
Ti	7 130	9 610	10 050	11 470	8 920
Ge	13	10	8	17	33
Sn	5	5	4	5	12
Pb	155	191	132	110	95
V	235	257	268	315	342
P	6 420	5 090	2 140	6 780	2 500
As	137	244	132	100	83
Cr	247	263	233	211	213
Mo	14	12	15	13	11
Mn	1 470	1 085	534	1 670	2 508
Co	66	53	63	98	101
Ni	166	128	116	195	147

use. Goldschmidt (1930, 1937, 1944) who was the first to apply spectrographic analysis to the study of the trace elements in coal ash, discovered jointly with Peters (1933) a high concentration of germanium and other elements in ashes of the European coals. Subsequent studies have shown that the concentration of trace elements is quite similar in coals of the same age throughout the world; the variations, rather quantitative than qualitative, are not great and depend on local conditions. Bibliography concerning this question was published in papers of Otte (1953), Stadnichenko et al. (1953), Leutwein and Rösler (1956), Bouška and Havlena (1959a), Bouška (1960, 1977) and Yudovich (1978).

Table 41 (continued)

	Contents in ppm in coal ash						
	Karviná region					Average	
Ostrava Formation	Karviná Formation						
Poruba Member	Anticlinal Seams Member	Suchá Member		Doubrava Member	Ostrava Formation	Karviná Formation
		lower	upper				
109	167	121	145	141	135	145
15 360	18 450	20 700	21 170	22 200	15 390	20 630
13 390	12 570	16 130	19 920	16 590	15 920	16 300
63	71	110	142	108	108	107
20	18	30	29	30	32	27
260	210	230	170	150	260	190
3	4	4	5	6	4	5
57	19	31	32	29	50	28
860	1 500	1 820	2 540	2 620	1 960	2 120
139	1 108	226	268	258	160	465
1 113	1 683	871	1 316	1 017	480	1 220
9 470	10 920	10 740	8 200	4 580	9 440	8 610
27	23	5	5	4	18	9
3	4	5	5	4	6	5
166	202	168	145	134	140	160
265	192	349	362	263	280	290
2 040	830	1 840	5 280	7 360	4 160	3 830
300	100	91	130	118	165	110
217	258	405	449	290	230	350
28	18	29	33	29	15	27
2 439	2 255	754	483	105	1 620	900
104	102	110	134	108	80	115
136	243	197	250	133	150	205

Goldschmidt assumed that the trace elements had accumulated in three stages:

(1) Concentration of trace elements during the life of plants. Such biological concentration is known in many recent plants.

(2) Concentration of elements during disintegration and decay of organic substances. This stage seems to be of much importance. The process occurs at the Earth's surface and the decay of plant matter produces humus and peat, under the extraction and removal of most water soluble compounds, chiefly alkali salts, Mg and Ca salts, and of anions of strong acids. On the other hand, humus and peat

entrap many trace elements. Soil colloids are known to sorb cations but with different capacity. In this stage various organo-metallic compounds are formed and arsenic probably accumulates.

(3) Subsurface concentration of elements in decomposed plant matter buried by sediments, when circulating water solutions function as an enriching factor. The reaction of thermal springs may be another contributive factor. A high boron content in coal ash from New Zealand (48,000 ppm) has been explained as due to the activity of thermal springs by Rankama and Sahama (1950).

Breger (1958) summarized more recent data on the geochemical effects of organisms on the accumulation of elements and derived the following distribution:

(1) accumulation of elements as a result of life activity of organisms: C, N, P, S, Fe, Si, Ca, Ba, Mn, I (Cu, V), Zn;
(2) accumulation after the death of organisms:
   (A) in a chemical way by:
      (a) incorporation into organic molecule: V, Ga, Ge,
      (b) precipitation as sulphides: Fe, Cu, Pb, Zn, etc.,
      (c) reduction: Ag;
   (B) by physical adsorption: V, Ag, Th, U and others.

It may be summed up that very complex physico-chemical processes are involved in the accumulation of trace elements in coal. Since the anions are more important for the life of plants (they are necessary for a good function of cells) they are taken preferentially. Some elements are known to be of direct importance for plants, but the role of some others remains unknown. There are also elements the concentration of which the plant cannot prevent. Adsorption and reduction play an important role in accumulation of some elements by plants. Silberminz and Rusanov (1936) explained the accumulation of germanium in coal by sorption from the circulating waters; Rankama and Sahama (1950) also advocated the theory of element accumulation by reduction and sorption. They believed that Ge occurs in coal in a bivalent form, probably as germanium monosulphide (but this opinion has not been confirmed).

The influence of geological and petrological factors on the content of trace elements in coal and the dependence of their accumulation on the geochemical environments is discussed comprehensively by Leutwein and Rösler (1956). Of Czech authors, Kudělásek (1959) deals with some of these questions in great detail, and Švasta et al. (1955) referred on adsorption and reduction with respect to element cumulation.

The problem can be resolved into several basic points:

(1) Concentration of trace elements in coal in relation to the geological age of coal seams. This aspect was best examined for germanium. Stadnichenko et al. (1953) reported brown coals of Cretaceous and Pleistocene age (Maryland, Columbia,

U.S.A.), which contain up to 7.5% Ge in ash, and Cretaceous coal from Montana with 4.7% Ge. In Czechoslovakia increased Ge has also been determined in the Cretaceous coals. Leutwein and Rösler (1956) published a profile through coal seams of different ages, which shows that the Upper Carboniferous and Mesozoic coals are richest in Ge (up to 1,500 ppm in ash of Jurassic coal).

(2) Dependence of trace elements in coal on facies and their pH and Eh. The formation of organic sediments and their specific facies is caused mainly by the changes of oxidation-reduction potential of environment. The changes of Eh affect not only the origin of typical petrographical coal facies but also the accumulation or dispersion of trace elements. The change in the oxidation or reduction degree results either in the increase of solubility and thus in dispersion of elements, or in the decrease in solubility and thus their concentration. The oxidation-reducing processes are controlled by the value of pH and the concentration of oxidized or reduced ions. The redox potential is limited by the fact that chemical reactions of supergenic processes occur prevalently in water solution and are defined theoretically by the reactions:

$$2 H_2O = O_2 + 4 H^+ + 4 e; \quad E_0 = 1.23 \text{ V},$$
$$2 H^+ + 2 e = H_2; \quad E_0 = 0.00 \text{ V}.$$

In extreme cases the water in volcanic areas may have pH = 1, and in areas containing sodium carbonate pH = 10 and even more. The value of pH of these waters usually ranges from 4 to 9, and is most frequently near 7. For pH = 7 the potential of the first reaction will be $+0.82$ V and for the second -0.41 V. These two numbers define the range 1.23 V. The values of oxidation potentials are somewhat different when pH value is lower or higher. Organic matter is the commonest and also most widespread reducing agent of sedimentary rocks. According to Krauskopf (1967), the reducing capacity of organic compounds is due to photosynthesis, when in reaction

$$6 CO_2 + 6 H_2O \rightarrow C_6H_{12}O_6 + 6 O_2$$

carbon is evidently reduced. The strongly reducing agent (saccharide) and strongly oxidizing agent (O_2) formed from weakly reducing (H_2O) and weakly oxidizing (CO_2) agents. Such reaction cannot be spontaneous. The necessary energy is received from an external source – solar radiation. The carbon compounds obtained in photosynthesis undergo further alterations, both in plant bodies and animal bodies and also during later decomposition processes. Many of these alterations produce again reduction and oxidation. The reducing environment of organic substances accumulation has Eh = -0.1 to -0.5 V. The pH values of solutions that come into contact with decaying plant matter are usually lower than 7; their minimum value is 4 when decomposition occurs under aerobic conditions. In this environment the reducing force of decomposing and decomposed organic matter affects the circulating water.

Besides Eh and pH factors, sorption plays a substantial role in the zone of supergenic processes. In streams, lakes and seas the elements are transferred in true solutions, as hydrosols and solid substances. The specific properties of hydrosols are due to their large surface relative to their volume. Two types of sorption are distinguished according to the nature of forces that attach the particles to the surface of colloids; when these forces are identical with van der Waals cohesive forces, the adsorption bond is relatively loose and only physical adsorption is involved; besides these unspecific attractive forces the valences may also be responsible for sorption, and in this case it is a chemical absorption or chemisorption with a much firmer bond. These two types of sorption usually co-act and many transitions exist between them. The cause of chemical absorption is the unsaturation of electrostatic charges of the ions. Sorption occurs between ions of opposite charge, which implies that negative colloids absorb cations from the solutions and positive colloids absorb anions. In water solutions there are important colloids of clay minerals, e.g. of kaolinite. These colloids originate from the hydrated silicic acid, which combines with the hydrate of aluminium oxide to form the nucleus of colloidal kaolinite under concomitant wrapping of bases. The nucleus is charged negatively and thus attracts positively charged ions. The negative charge of the nucleus is due to boundary anions, predominantly of the silicic acids and hydroxyls, which are concentrated densely on the nucleus.

In the organic component humic substances, bitumen and substances resistant to biochemical decomposition are strong sorbents. Plant waxes, resins and oil substances may be cited as examples. Humic substances are capable of binding 6—7 times more bases than the clay colloids.

In a reducing environment such elements accumulate most frequently which hydrolyse better at a lower valency (are less mobile) and are sorbed on highly dispersive substances of the opposite charge. Among these elements there are copper, molybdenum, uranium, vanadium and cobalt; in a reducing environment they can be reduced to metals and concentrate as such. Reduced forms of S, Se and Te concentrate in the compounds with cations and form numerous sulphides, less often selenides and tellurides. Sulphides originate mainly from elements having a close affinity to sulphur and those which under natural conditions migrate in water solution in a bivalent form. This group includes iron, lead, zinc, copper, nickel and others.

(3) Of interest is the relationship between the trace elements and the petrography of coal. According to Svoboda and Beneš (1955), sapropelites, i.e. cannels and bogheads, form in subaqueous conditions of a reducing anaerobic environment, in the bottom part of a peat bog. True peat is deposited in the upper parts in anaerobic and mixed environment. In the areas with tree growths the aerobic environment predominates over anaerobic conditions, and woody peat develops. The Tertiary forms of it are hemixylites, xylites and metaxylites. The marginal sediments of recent peat bogs represent terrestrial growths, the organic residue of which is

humus. These decomposition processes are purely aerobic and give rise to liptobioliths. Durains may develop as sediments of transitional aqueous and subaqueous environments (in purely oxidizing medium) or as sapropelitic sediments (of typical reduction character). An extremely oxidizing environment is postulated for the formation of fusains. The origin of banded coal is a result of the alternation of facies. Otte (1953) divided the coal into a light and a heavy fraction on the basis of density. According to this authoress the light fraction ($\varrho = 1.14 \text{ g} \cdot \text{cm}^{-3}$) contains mainly organic matter and the heavy fraction ($\varrho = 1.8 \text{ g} \cdot \text{cm}^{-3}$) contains chiefly alien substances. She differentiated then the elements typical of the light fraction and characteristic of the coal of a higher specific gravity. The relationship between the germanium content and the specific gravity of coal was studied by Zahradník et al. (1958). The separation of the petrographic components in a mixture of carbon tetrachloride and xylene has shown that germanium accumulates preferentially in vitrain. Clarains have lower contents of germanium and durains and fusains contain the least amounts.

On the basis of the study of trace elements in Upper Cretaceous coals of Transbaikalia Admakin (1974) has found that the concentrations of Be, Co, Cr, Cu, Ga, Ge, Mo, Ni, Pb, V, W and Zn in colinite exceeded 3—10 times their amounts in telinite, in which the Ge content, for example, only approached the clarke value. This implies that the elements tend to concentrate rather in coal of colinitic types, where the decomposition of plant material is well advanced.

(4) Kudělásek (1959) studied the contents of nickel and cobalt in relation to the stratigraphic position of the coal seams in the Lower Silesian Basin. He assessed that their contents increased with age (the Žacléř seams displayed the highest concentrations). In most cases nickel greatly prevailed over cobalt. After Goldschmidt (1937) the Co : Ni ratio in coal is 1 : 2 to 1 : 3. According to Schejbal et al. (1972) this ratio in the main Kladno seam (Zápotocký mine) is 1 : 10 and 1 : 19 in the basal seam. Otte (1953) and Leutwein and Rösler (1956) have proved that nickel predominates over cobalt in humitic coal. Typical sapropelites, in contrast, show the Co : Ni ratio = 2 : 1. The characteristic Co : Ni ratio in boghead or boghead aleuropelite is 1 : 1. In some cannels and bogheads (dy, gyttja deposits) the Co : Ni ratio may be 1 to 1 : 10. In plants it generally equals 1 : 5 to 1 : 10. Leutwein and Rösler (1956) also referred to the relationship between Ag and Pb and between Ga and V. Bouška and Havlena (1959a) found that in the coal of the Žacléř seams the greater part of copper is concentrated in fusain. Certain relations have been recognized between Ba and Sr and V and Ga in coal ash there. The Sr content increased with higher Ba. Gallium was identified only in samples containing increased vanadium. Čadková (1971a) determined major anomalies in Pb, Zn and Cu in the coal from the Radvanice seams, which may be accounted for by the presence of sulphides (galena, sphalerite, chalcosite, bornite). The Cu, Pb and Zn metals exhibit the highest average contents in coal and the lowest in siltstones. Vanadium, nickel and cobalt occur in largest amounts in claystones. Čadková

(1971b) established the same distribution in the Permian sediments of the Intra-Sudetic Basin.

Mathe (1961) stated relations between the V, As, Pb, Mo and U contents in the bituminous coal from the Permian Döhlen Basin in the G.D.R. A clear-cut relationship exists between vanadium and uranium: vanadium increases with the rise of uranium. The contents of arsenic, lead and particularly molybdenum vary. Davies and Bloxam (1974) observed the relationship between Ti and V in both light (i.e. organic) and heavy (inorganic) coal fractions.

(5) The study of germanium and its contents in coal has shown that it is predominantly restricted to the limnic basins; paralic basins contain very little of this element. This finding proved true also for Czechoslovak basins. The bituminous coals of central Bohemia are high in germanium whereas the paralic sediments of the Ostrava Formation are poor in it. Considering that some trace elements are associated essentially with only some coal constituents, it is obvious that the changes in petrographical composition of coal produce changes in the content of trace elements. The contents of trace elements in coal change both laterally and vertically (Bouška and Havlena, 1959).

Leutwein and Rösler (1956) stated that iodine is higher in marine than in continental sediments. Wilke and Römersperger (1930) studied the iodine contents from brown coal up to bituminous coal (0.05–11.17 ppm) and came to the conclusion that the iodine content rises with the geological age.

(6) The concentration of trace elements depending on their concentration in plants is feasible but probably of no great importance. It seems more likely that the trace elements present in plant bodies were liberated during the decomposition processes and enriched the waters of the coal-forming basin to be re-sorbed in the early stages of organic matter coalification. This indirect process is more reasonable than a direct relationship between the contents of trace elements in plants and in coal, which has not been borne out.

(7) The concentration of trace elements in coal may be influenced as by contact metamorphism so by regional metamorphism. Contact metamorphism is usually produced by intrusions of younger volcanic rocks and is often accompanied by mineralization. In the Ostrava coalfield many examples have been observed. Regional metamorphism pronouncedly affects the contents of many trace elements (B, Ge, etc.), which decrease with the rise of metamorphism. Medvedev (1971) maintains that during coalification mineral components (carbonates, oxides and other salts) are liberated from humates and other organic complex compounds and the soluble compounds like GeO_2 or Na_2GeO_3 are washed away from the seam by circulating water. Free organic radicals deficient in metals combine into macromolecular organic coal substances.

(8) The effects of mineral springs and genesis of ore veins on the contents of trace elements in adjacent coal seams have been assessed by Vnukov et al. (1971).

(9) Our knowledge of the influence of palaeogeography on the content of trace

elements in coal is so far insufficient. In this context, the distribution of some trace elements within or along the margins of coal seams would demand more attention. The chemistry of developing coal may be changed or at least influenced by the intensity of denudation in the surrounding area, i.e. by the material supplied to the coal-forming basin.

Yudovich (1972) pointed out the function of coal lenses and coalified logs as concentrators of trace elements. The highest Ge concentrations (6—7% in ash), for example, are invariably found in coalified tree stems, which have a larger sorption surface, are more accessible to circulating solutions and may attach more elements than a coal seam relative to their volume.

Brown coals and gagate inclusions are the richest. Germanium, molybdenum, zirconium, arsenic and antimony increase in amount in the gelification series from lignite to bright gagate. The elements Mo, Zr, As and Sb are correlative with Ge. In the Upper Permian sediments of the Pechora valley Zr was of higher concentration in coal lenses enclosed in sandstones than in claystones. Vanadium reaches the highest concentration in thin and small coal seams. Vanadium is correlative with Ge, Ti, Cr, Ni and Co. Copper, zinc, tin and lead are rarely enriched in thin and small coal seams and coalified woods, occasionally they are even lower than in coal seams of great thickness.

The elements Ba, Be, REE and Sc are not very characteristic of small coal seams, but they may be locally strongly enriched and in this case they usually occur together.

Coal lenses and coalified woods are generally deficient in B, Ba, Mg and Sr relative to coal seams. Boron and magnesium may be removed from a coal environment jointly with alkalies.

In the ash of the thin coal seams of southern Yakutsk Pavlov (1967) determined the following contents of rare elements: 20 ppm Nb and Ag, up to 0.005% W, about 0.001% Cd and Bi, up to 0.01% Hf, up to 0.003% Li and up to 0.01% Rb.

Uranium is usually evenly distributed in the coal lenses and does not form any minerals.

The coal lenses or small and thin seams and coalified tree trunks and branches have been recognized as extraordinary concentrators of a large group of elements in all coal basins of the world. The Ge, U and Mo contents can exceed their clarke amounts in sediments 100—1,000 times and the amounts of V, Cr, Ti, Ni and Co 10—100 times.

This circumstance is easily understood if we realize that a coal seam contacts the neighbouring rock with two surfaces, along which the enrichment is the highest. In a coal seam of great thickness the enriched contact zones are 10—15 cm thick. In a thin coal seam or lens and in coalified wood the two contact zones coalesce and, additionally, the coal body contacts the rock also on the sides. This explains such a high enrichment and suggests a relationship between the thickness and contact surface of the coal seam.

The enrichment of small and thin coal seams or isolated coalified woods in trace elements has been summarized by Yudovich (1972) as follows:

(1) enrichment in most cases: Ge, Mo, U, As, Sb, Zr, V, Cr, Ti, Ni, Co;
(2) frequent enrichment: REE, Sc, Ga, Sn, Cu, Pb, Zn (Nb, In, Cd, Hg, Ag);
(3) rare enrichment, occasionally deficit: B, Ba, Sr, Be, Mn.

In my opinion, the conclusion on Be is not quite substantiated by literary data.

Since germanium and uranium are of economic importance, much attention has been centred to their study during the investigations of coal seams. The data obtained made it possible to recognize the principle governing their distribution and concentration in a coal seam and the type of their bonding in coal. The problem of their presence in coal and of their complex bonding to the organic coal compounds has been satisfactorily solved.

Germanium. In the past different explanations have been proposed for the possibility and causes of the concentration of Ge (and other trace elements), but two theories predominate in the present literature: one says that germanium is derived from coal-forming plants which absorbed it from the soil. The alternative hypothesis is that germanium was concentrated in coal by sorption via the humic substances. I believe that sorption plays the principal role in the concentration of elements in the coal.

Goldschmidt was the first who expressed the idea that Ge accumulated in coal comes from the plant bodies. He supported his opinion by the find of at least traces of germanium in the ash of many recent plants, and by the close affinity between GeO_2 and SiO_2, which the plants are able to solve and take in. (Germanium was thought to be present in plants as GeO_2 or GeS_2, but recent studies have shown that the prevalent part of Ge is bound to organic compounds.)*
In some plants silica is an important structural element as, for example, in horsetails, which had so much contributed to the formation of coal in Carboniferous basins. Germanium, although in minute or trace amounts, is also found in conifers and other plants, which allows to explain its concentration also in younger coals. The concentration of germanium in low-ash coals is additional evidence that it is not associated predominantly with the inorganic coal component. The relationship between the Ge content in ash and the ash content of Czech Cenomanian coals is shown in Fig. 64 (Bouška and Honěk, 1962).

Stadnichenko et al. (1953) have studied the germanium content in sulphides and ashes of several coal seams. Their results have proved that it is not bound to sulphides. In most cases it was lacking in sulphides although it was present in coal

* Ge^{4+} has the ionic radius 0.053 nm, which is very near to Si^{4+} ion with a radius 0.042 nm (Ahrens, 1952). Therefore Ge often occurs in the silicates. Upon their weathering it is released into solutions and migrates in ground water, from which it is captured by organic substances forming at the decay of organic material.

ash. In several sulphides maximum 0.01 % Ge was determined but the ash from the coal in close vicinity gave 0.8 % Ge.

Germanium and other trace elements are concentrated particularly in vitrain. If we disregard the petrographical composition of coal and examine average samples, the result is necessarily affected by the ratio of the coal components.

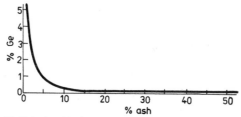

64. Relationship between the ash and germanium contents in the Cenomanian coal of Czechoslovakia (Bouška and Honěk, 1962).

Headlee and Hunter (1951) and Headlee (1953) pointed out that germanium in coal is usually concentrated mainly along the roof and floor of the seam. Šantrůček (1961) and Čadek et al. (1961) made the same observation in the Czechoslovak mines. Hak and Babčan (1967) reported an increase in Ge and Be near the footwall of coal seams and along the faults in the Sokolov Basin. The diagram in Fig. 65 shows the enrichment of germanium in the Sokolov Basin (Bouška and Honěk, 1962). The highest Ge concentrations have been established at the footwall in two boring profiles. The accumulation might be interpreted as due to a later migration or the influence of the adjacent rocks, but in several cases the rocks were found to be deficient in germanium.

The question how the plants could concentrate relatively large amounts of a finely dispersed element thus remains open. Rotter (1952) put forward a theory

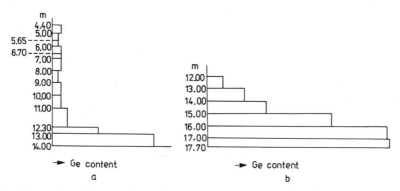

65. Enrichment of coal in germanium; data from two boring profiles (a, b) in the Sokolov Basin (according to Bouška and Honěk, 1962).

201

that the increased germanium contents in plants were raised additionally by forest fires and the growth of new plant generations in soils enriched in this element. This is a plausible interpretation but it can by no means be generalized, the less so if the proportion of coal-forming plants and the volume of coal is taken into consideration. Otte (1953) argued that the Ge content in ash if recalculated to the coal volume with due regard to the compression of the material, corresponds to its content in the Earth's crust. It is possible that after the death and decay of plants and trees the soil in swampy basins was enriched in germanium and other elements from the plant bodies, and the new generation of vegetation took them from the enriched humus and also from the primary substratum.

Rotter (1952) ascribed the accumulation of germanium in coal to the biochemical processes in *Equisetales*. He took the amount $10^{-4}\%$ Ge which had accumulated in the Carboniferous equiseta as a basic value. After forest fires their ash is assumed to contain 10^{-3} Ge and this value increased to 10^{-2} in plants of the new generation which grew on the ash; upon coalification the Ge concentration amounted to 10^{-1}. Rotter rejected the adsorption theory on the ground of the fact that fusain contains less Ge than an average coal sample or none at all, although it has the maximum sorption capacity of all coal components. He thought fusain to be an incompletely incinerated relic of the first-generation equiseta, from which Ge was abstracted by the plants of second generation. In his opinion the brown coal contains less germanium than bituminous coal because the plants from which it is derived were not able to accumulate this element sufficiently. Rotter's theoretical explanation has been substantiated by Švasta et al. (1955).

Although fusain has a low sorption capacity for germanium, it is a good sorbent of other substances but it does not concentrate the trace elements. A possible explanation is that the cell walls of the wood had been so damaged by fires that they did not undergo a normal series of alterations during the coalification process. No humic substances with a high sorption capacity were formed. The fusain cells were often filled with inorganic admixture, which penetrated easily into intercellular spaces. The inorganic admixture of fusains is well discernible in the microscope. According to the above theory, the cellular structure of the arborescent plants was so disturbed by forest fires that it lost its sorption capacity altogether. The cell walls and the non-activated environment of fusains let through all elements capable of migration; they only served as a passage way for solutions. Their cells were filled only with fine inorganic material to which additionally supplied elements may have been bound. On the contrary, the woods undergoing a normal coalification process filtered the circulating solutions and caught anions and cations on the cell walls. The high sorption activity of the developing humic substances contributed to the enrichment of the developing coal matter by trace elements.

It is not surprising that in an adequate environment, such as was frequent in Cretaceous times (coalification of isolated woods in separate basins with a rich material supply; solutions enriched in trace elements: preceding intensive weathering), the coals and coalified woods of Cretaceous age are highest in germanium. The contents and variety of the remaining trace elements are also relatively greater than in other coal types.

One relevant circumstance is to be mentioned in this context. The claystones and clays close to the coal seams or coalified logs high in germanium contain but trace amounts of this element. This differentiation cannot be explained even if we take into consideration that the ligneous matter of coal has an about 200 times higher sorption capacity than the clay minerals. It appears that the clay minerals sorb cations to a far greater extent than anions; it is probably only montmorillonite that can sorb anions such as OH and F up to an equivalent level. In contrast, the plants sorb anions very intensely. This may suggest that germanium was then present in the solutions as anions which could not be (or only to a limited extent) bound to clay material; this would also account for the differences in the Ge content in the individual petrographical coal components. This conception, however, is at variance with the preferential sorbing of cations by humic acids, owing to their acidic character. The anionic components can probably be bound under concurrent formation of complex compounds. During weathering processes, when, for example, germanium is most probably released from the silicate structures and passes into water solutions, it migrates as a complex anion or as an oxide dissolved in molecular form. Untrue solutions with germanium oxide dissolved in colloid form may also be formed, but this process has not yet been fully explained.

The accumulation of germanium in coal by absorption from the circulating waters was dealt with by Silberminz and Rusanov (1936). Egorov and Kalinin (1940) on studying the Ge contents in coals of different ages and ranks from the Kazakhstan coal basins have found that the largest amounts of Ge are in the youngest coal, mainly in thin coal seams and lenses enclosed in sandy and clayey rocks. They interpreted this enrichment in terms of adsorption from the circulating waters through pervious (particularly sandy) rocks. Aubrey (1952), Headlee (1953) and Stadnichenko et al. (1953) also presume the sorption of germanium from ground waters with regard to its highest concentrations in the hanging and foot walls.

The sorption theory is favoured by Zahradník et al. (1959a, b). They studied in detail the conditions in the basins of central Bohemia, particularly the relationship between the Ge contents in coal and the permeability of the adjacent rocks. They have established that the Ge contents are higher at the contact with sandstones than along the boundaries with claystones.

In the Plzeň Basin (Bohemia) the primary source of germanium is thought to be the granite pluton or its sphalerite-bearing vein suite. The most plausible explanation seems to be that the Ge concentration in solutions was higher by several orders of magnitude at the time of the most intensive mineralization (formation of ore veins), which there coincided with the stage of most intensive coalification of the Nýřany and Kounov seams. This presumption would also explain the relatively lower Ge contents in the Radnice coal seams. In the Plzeň Basin large amounts of germanium are in tectonically disrupted coal seams, which allow for easy percolation of water.

Germanium contents are dependent additionally on the distance from the coal/ /rock contact. They increase from the centre of a seam towards its margins, and thus the seams of great thickness are relatively lower in Ge. This general rule was expressed by Headlee and Hunter (1951) and Stadnichenko et al. (1953). Ammosov (1953) established the existence of a regular distribution of petrographical components in a coal seam: vitrain predominates at the margins and the heavier components such as durain and clarain occur in the centre.

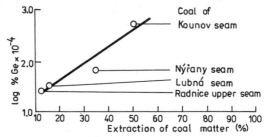

66. Interrelations between the germanium content, percentage of extracted coal matter (mainly humic substances) and the stratigraphical position of the coal seams in the Plzeň district (according to Zahradník et al., 1959).

The study of Zahradník et al. (1959a, b, 1960a, b) concerning the relationship between the Ge content in coal and the petrography of over- and underlying rocks has shown that Ge is higher at the contact with sandstones than at the contact with claystones. The authors have also stated that the Ge content decreases with the stratigraphic age of the coal seam, which is due to the increase of coal rank. This relationship is seen in coal of seams differing in age but belonging to one basin (Fig. 66). The association of Ge in coal with the free portion of the humic constituent is shown in Fig. 67. Since this portion decreases with the increase in

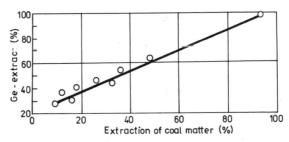

67. Relationship between the extracted germanium amount and the percentage of extracted coal matter from the Plzeň coal (Zahradník et al., 1959).

coalification, the identifiable amounts of Ge are the highest in the youngest coals. The petrography of the over- and underlying beds is also influenced by the coalification process. The supply of matter and elements in solutions is controlled by the filtration effect of the neighbouring rocks depending on two factors:

(a) internal — porosity of rocks, elementary concentration in the solution, size of the particles;
(b) external — temperature and pressure.

In case the coalification degree is not the same at all places of the seam, germanium is most frequently concentrated in the youngest, i.e. uppermost layers, where the lowest temperature and pressure are presumed.

In their paper of 1959 Zahradník et al. interpreted increased germanium contents in places of wedging-out of coal seams in terms of a lower pressure. Coalification occurred at a lower rate, and as a result, the existence of the active gels of humic acids persisted longer.

Many investigations have shown that Ge decreases in coal with the increase in ash content, and vice versa. Germanium thus differs from gallium, which is highest in coal richest in inorganic components. Gallium obviously accompanies aluminium present mainly in clay minerals; it is usually concentrated in claystone partings whilst germanium is contained directly in coal. In sandstones and arkoses gallium is lacking; in coal it is associated with secondary ash substances. The Ga concentration decreases markedly with the increase in silicates and rises with the increase in alumosilicates.

Germanium in coal is associated with organic compounds. The lightest gel of humic acids had already accumulated in the first humification phase along the margins of the basin and part of germanium was sorbed in it. The remaining part of germanium is probably bound by physical forces. Shirokov et al. (1971) maintain that the bulk of germanium had concentrated in coal during the early diagenesis and was subsequently transferred by circulating solutions.

The accumulation of germanium in coal by chemical sorption is most plausible; only a very small part of it was concentrated by physical adsorption. The substantial amount of germanium was cumulated by the gel of humic acids during the most intensive humification process. According to Ratynskiĭ (1946) those coals and coal components are highest in germanium which had passed through the longest and most complete humification stages. Therefore, Ge is most abundant in vitrain, the most completely altered coal constituent. Ratynskiĭ has also proved that coals developed without the presence of water and at the access of air have very little germanium. Figure 68 shows the sorption of Ge by humic acids at the destruction of lignin structures according to the concept of Manskaya and Kodina (1975).

Gordon et al. (1958) presume that Ge is present in coal:
(1) most probably as germanium humates,

(2) as silicates, germanates or silicogermanates, in association with the mineral portion of coal.

In dull coal, germanium is more associated with inorganic matter. According to these authors, fusain was practically deficient in germanium, which is in agreement with Rotter's observation. On the other hand, vitrain was high in Ge, but its mineral portion (secondary ash matter) contained a negligible amount of this element. Durain does not show any great difference in the Ge content between the organic and inorganic portions (see Table 42).

As follows from Table 42, germanium is predominantly associated with the organic matter and its content depends of the petrographical composition of coal. As concerns the relationship between the Ge content and the light coal fraction, opinions differ whether it is sorbed (Headlee, 1953) or present in germano-organic compounds. Travin (1957) examined this question with the use of electrodialysis. He plotted the relation between the Ge content and vitrinite portion (coal from the Kiselovsk basin, U.S.S.R.) on a graph, from which a direct proportionality of these two parameters is clear. During the electrodialysis of coal from the Kiselovsk Basin the following elements were trapped on the anode or cathode: Be,

68. Sorption of germanium by humic acids during the decomposition of lignin structures (Manskaya and Kodina, 1975).

Table 42
Germanium distribution in coal (Gordon et al., 1958)

	In organic matter (%)	In inorganic matter (%)
Brown coal	83.3	16.7
Vitrain	79.0	21.0
Durain	55.2	44.8
Concentrate from the above brown coal	97.3	2.7
Bituminous gas coal	87.2	12.8

Table 43
Sorption capacity of coal components (Kudělásek, 1960)

Locality	Coal type	Volatile matter	Humic acids	mg sorb. GeO$_2$
Donets Basin	anthr. coal	14.2	0	0
OKR*, Gottwald mine	vitrain	26.9	0.2	0
Handlová	metaxylite	40.2	8.0	4.8
locality unknown	xylite	46.9	58.3	37.6
J. Žižka mine	xylite	49.8	60.0	37.8
Skado mine (G.D.R.)	fossil wood	64.4	38.1	32.0
OKR*, Gottwald mine	durain	27.5	0.3	0
locality unknown	brown coal	41.8	10.2	2.0
Nová Ves	brown coal	50.2	16.3	6.0
locality unknown	brown coal	51.2	30.1	22.1
Kyjov	lignite	58.6	75.4	38.6
Hungary	pyropissite	71.7	65.0	20.0
Regis Geiseltal (G.D.R.)	pyropissite	69.1	72.0	27.2
Sokolov, Anežka seam	cannel	51.4	75.0	16.3
Slatina	hemixylite			95
Hnojné	lignite			89
Hnojné	lignite			87
Sokolov, Jiří seam	brown coal			60
Most	metaxylite			17
Obit (borehole)	metaxylite			15
OKR*, Čs. Pionýr mine	durain			0
OKR*, Jeremenko mine	vitrain			0
OKR*, Šverma mine	fusain			0
Žacléř	fusain			0
Most	fusain			0
Sokolov	fusain			0

* OKR — Ostrava-Karviná coalfield.

Na, Mn, Al, Si, P, Ca, Ti, V, Cr, Mg, Fe, Co, Ni, Cu, K, Sr, Y, Sn, Ba, Yb, Pb and Bi. Germanium has not been identified either in a control sample of brown coal (100 ppm Ge). As it was not determined in the complex of elements sorbed in coal, the author inferred that it occurs in coal in the form of a germano-organic compound.

Kurendová (1967) when studying the sorption properties of coal has recognized that Be is sorbed more intensely than Ge. Beryllium was sorbed up to 100 % by peat and brown coal orthotype. The maximum sorption of Ge is at pH 2 and pH 7.5—11, and minimum at pH about 5.5. The maximum sorption of Ge is in the range pH 4—12 and minimum at pH 2.5.

Kudělásek (1960) recorded a relationship between the sorption capacity and the amount of acidic components. Sorption declines with the increasing rank of coal. Vitrains and durains from various localities did not show sorption capacity in any case, whereas coals of lower rank displayed a rising sorption capacity with the increase in acidic components (Table 43).

Zhou (1974) mentions two likely types of Ge concentration in coal. He thinks that either the fundamental Ge pattern in peat beds has been preserved in coal beds, or that Ge was redistributed when the peat beds had been buried under different geochemical conditions.

Germanium was concentrated in coal most probably by chemical sorption. The concentration by organic way cannot be entirely rejected but it certainly plays only a minor role.

Uranium. The brown coal of Liassic age near the granitic complexes in Hungaria (in the proximity of the Meczek and Velence Mts.) proved to be high in uranium. Szalay (1954), who reported on this finding, observed that radioactive and non-radioactive layers of coal and claystones are arranged in a regular succession. There was up to 0.01 % U in coal ash, maximum in thin coal beds near the granite massif. The author presumed that uranium had been washed from the weathered granite into the basin where it accumulated concomitantly with the deposition of organic matter. Szalay (1974) explained the enrichment in U by the cation-exchange capacity of the insoluble humic acids. He proved experimentally that accumulation of uranium and other cations from very small concentrations in waters into the humic acids of peat exceeded a factor of 10,000. Subsequent experiments have demonstrated that some microelements such as V and Mo, which occur in anion migrating forms, are first reduced to cations by the humic acids and then fixed in the cation form. Humic acid particles transported by the rivers probably deplete water of its heavy cation content and humic acids might thus be highly responsible for the deficiency of the cation balance in the oceans. In 1954 Davidson and Ponsford published a report on the uranium occurrence in the European and American coals. They also explained the presence of U by the ground water supply from the disturbed granite massifs or from the overlying uranium-bearing beds. Uranium in the coal of the Warwickshire coalfield near Coventry (England) derived from the overlying Triassic sandstone. No uranium minerals have been identified in the polished sections of the coal itself, which contains 0.08 % U in ash. The Lower Permian coal from Freithal near Dresden (G.D.R.) is high in uranium (maximum 0.18 – 1.00 % U in ash), which is thought to be supplied from the granite of the Krušné hory and Lužické hory mountain ranges.

The uranium-bearing coals and xylites in the western part of the United States are associated with coeval or younger acid lavas and tuffs. The authors have found out that in South Dakota only the uppermost beds of brown coal lying directly under the volcanic rocks have an appreciable U content. In Idaho the coal beds at the top of the seam are higher in U than beds about 30 cm lower. This suggests again that percolating water is an important transporting agent for uranium. Pipiringos (1955) studied the Early and Middle Eocene uranium-bearing coal in the central part of the Great Divide Basin (Sweetwater County, Wyoming); he believed that uranium had been washed from the uranium-bearing tuffaceous sandstone of the Browns Park Formation containing the uranium mineral schroeckingerite. From the fact that coal associated with aleuropelites was lower in uranium than that deposited in sandstones or conglomerates Masursky and Pipiringos (1957) inferred that uranium might be brought from the overlying beds and that it is capable of lateral movement. Masursky (1955) considered the hydrothermal solutions along the fractures to be one of the possible uranium sources. Dybek (1962) described the leaching of elements from tuffs and their transport by ground water flow into a coal seam. Christoph (1963) presumed that the source of uranium and other elements in the Döhlen Basin (G.D.R.) was the Teplice quartz porphyry and granite porphyry, which reach from the eastern Krušné hory Mts. to a distance of 10 – 20 km from the basin. The increased U and the highest radioactivity were determined in the collinitic coal type.

Breger and Schopf (1955) correlated the uranium contents with the ash contents in the Devonian coal from Chattanooga:

Ash (%)	U in ash (%)	Ge in ash (%)
4.23	0.42	0.6
3.80	0.55	2.0
2.45	0.57	1.5
1.69	1.32	3.0

From the results showing that the lower the ash content the higher is the uranium content, the authors concluded that uranium had been adsorbed from water (during sedimentation and coalification) on organic compounds of the humic series, giving rise to urano-organic compounds.

Breger, Deul and Rubinstein (1955) wrote a detailed study on the uranium-bearing xylite from the Mendenhall strip mine (Harding County, North Dakota). The mineral-free part of xylite was separated using heavy liquids. Xylite contained

13.8 % ash and 0.31 % U in ash. The separated minerals (69 % gypsum, 10 % jarosite, 2 % quartz, 19 % kaolinite and other clay minerals, traces of calcite, no uranium mineral) yielded only 7 % from the total uranium content in coal, which indicates that uranium is associated with the organic coal components. Using 1 N HCl as much as 88.5 % U was extracted from the brown coal, and up to 98.6 % U was dissolved by continuing the extraction with hot 6 N HCl. This shows that the bulk of uranium is contained in coal in an organo-uranium complex, which dissolves at pH < 2.18. Only 1.2 % U from the total uranium content is probably present in the ionic state. The authors admit, although with reservation, that an ionic organo-uranium compound may be present. The elements Mn, Sr, Fe, Ni, Co, Be, Mo, Ti, Li, Pb, V, Zr, Cr, Sn and Mg were not present as ions. Jorscik et al. (1963) presumed that uranium is bound to humic acids in a ratio of one U atom to two monomers of the humic acid. Breger and Chandler (1960) studied the Triassic and Jurassic coal beds in Colorado, which are derived from the roots, stems and branches of *Araucarixylon* and are high in uranium (7.28 % in ash, average of 8 samples). In one sample the U content in ash exceeded 30 % and 7 % in coal. The coals studied had higher H and organic sulphur, less N and were impoverished in volatiles in relation to the coals of the same rank from Kentucky.

Breger, Deul and Meyrowitz (1955), who also favoured the theory of uranium transport by water, presumed that one of the components of ground waters is the alkaline or alkaline-earth metal uranyl-carbonate complex. This complex is unstable and in an acidic environment it gives uranyl ion UO_2^{2+}, which reacts with the organic coal components, particularly humic acids, to form uranyl-organic compounds insoluble at pH > 2.2.

Lepka (1967) reported that in the Kladno-Rakovník Coal Basin the migration of uranium is hindered by the reducing properties of the rocks, which cause the entrapping of uranium from the circulating ground water and its transition from the liquid to solid phase. In the permeable rocks of the upper and lower red beds, which have a higher oxidation potential, migration of uranium in ground water is considerable. In the Kladno-Rakovník Basin the U anomalies conspicuously follow the contact between the red and grey beds. The rocks having reducing properties, i.e. the coal seams of grey bed complexes, show increased uranium contents of thousandths to hundredths of one per cent.

It appears that uranium cumulates in coal predominantly by sorption, which may result in organometallic complexes or a physical bond (Davidson and Ponsford, 1954). Uranium minerals form when the sorption capacity of a system is lower than the U concentration in solution. The U minerals identified in coal are pitchblende and coffinite, and in the oxidation zone of coal carnotite, autunite, metaautunite, metatyuyamunite, metauranocircite, torbernite, metatorbernite, zeunerite and metazeunerite. Uranium in organometallic complexes or in ion organic compounds is soluble at pH 3—6 (Breger et al., 1955). The uranium concentrations are higher in coal seams of a smaller thickness. Nekrasov (1957)

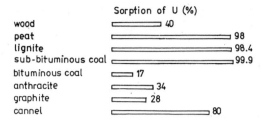

69. Sorption of uranium from solution by various natural materials (Moore, 1954b).

maintained that U mineralization is not associated with peatification processes but is a result of U assimilation from the ground waters in the existing coal seams.

Humic acids form uranyl humates within pH 4–5; fulvic acids at pH about 7 form uranyl fulvates.

The experiments of Moore (1954) have shown that peat and brown coals are most effective in trapping U from the solution (98–99.9%). The increase in coalification causes the reduction of the adsorption of metals on coal (in anthracite only 34%). Uranium may be present in organic matter as U^{6+} to U^{4+}. Depending on the degree of polymerization and on pH of solutions the humic acids may transfer uranium or bind it in uranium-organic complexes.

In this context the paper of Cameron and Leclair (1975) should be mentioned. The authors refer that coke, lignite and subbituminous coals are most effective in extracting uranium from solutions and that fusinite fragments proved more effective than the fragments rich in vitrinite. The washing process was strongly influenced by the acidity of the uranium solution in which the samples were immersed:

at pH < 5 → less uranium was washed out,

at pH > 5 → more uranium was washed out.

The sorption of uranium from solutions by various natural materials was studied by many authors. Moore (1954b) quoted some instances (see Fig. 69). Figure 70 shows the relationship between uranyl absorption by humic and fulvic acids and the pH values.

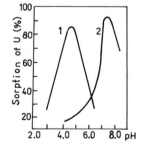

70. Absorption of uranyl by humic acids (1) and fulvic acids (2) in relation to pH of solutions (Manskaya and Kodina, 1975).

The role of humus in the geochemical uranium enrichment in coal and other bioliths was assessed by Szalay (1957). He believes that uranium is replaced in coal by H^+ at pH 1—3 and by some cations of high valency as, for example, Th^{4+} and La^{3+}. He got a series:

$$Ag^{1+} < Ni^{2+} < Cu^{2+} < Ba^{2+} < Pb^{2+} < UO_2^{2+} < Fe^{3+} < Th^{4+}.$$

Otte (1953) made an attempt to distinguish between the elements of the organic coal matter and those of external ash matter. She presumed that the homogeneous organic substances are lighter than components containing a varying portion of inorganic substances. On the basis of specific gravity she separated the coal components and determined their element contents. According to Otte several coal components can thus be differentiated: vitrain ($\varrho = 1.3$), clarain (1.3), durain (1.35), fusain (1.5), coaly claystone (1.7 and more). This division is clearly not quite reliable. Otte presented the following scheme:

vitrain, clarain $\}$ Ge, Ga, Be (V, Ni, Zr, Cr)

durain

fusain $\}$ Ti, Cu, Mo

coaly claystone, coal with mineral admixture

$\}$ Co, Pb, Zn, Sn

$\}$ Mn

In order to assess which elements belong to the organic coal matter, Manskaya and Kodina (1975) studied their assigment to coal fractions according to density:

$$< 1.40\,g \cdot cm^{-3} \quad \text{and} \quad > 1.60\,g \cdot cm^{-3}.$$

They determined that 92.7 to 99.1 %, on the average 95.9 % of organic substances are included in fraction below $1.40\,g \cdot cm^{-3}$. The fraction above $1.60\,g \cdot cm^{-3}$ comprises mainly mineral components of coal. The fraction 1.40 to 1.60 is transitional. The assignment of an element to the organic coal substances is expressed by coefficient

$$F = \frac{M_1}{M_2},$$

where F — coefficient of enrichment in the element, M_1 — content of the element in the fraction $< 1.40\,g \cdot cm^{-3}$ of the sample examined, M_2 — content of the element in coal fraction $> 1.60\,g \cdot cm^{-3}$ of the same sample.

If $F > 1$, the element is associated chiefly with organic substances, if $F < 1$, the element is related mainly to the mineral coal components.

Manskaya and Kodina (1975) gave the following F values:

Ge	2.6	Be	1.1	Sc	0.4	Zn	0.03
W	2.2	Nb	0.7	Y	0.2	Pb	0.02
Ga	1.7	Mo	0.6	La	0.2		

Deul (1956) believed that except for U, La, Ce and such, no element concentrated in the organic coal matter is dominantly lithophile; V, Fe, Co, Ni, Cu, Zn, Mo, Ag, Pb and Sn constitute characteristic complex substances, which may react with organic material to form organic complex compounds. In contrast, the mineral fraction shows an exceptional concentration of Si, Al, Ca, Mg, Na, K, Ba, Sr, Ti and Zr. The elements Be, Sc, Cr, Ga, Mn, Y and Yb are not markedly concentrated in either fraction. Deul has not classified Bi and Ge, because he had not found them in the samples examined. His scheme differs from Otte's classification in several points (e.g. the assignment of Be and Zr).

In the geochemical study concerning the trace elements in the Ostrava-Karviná coalfield Macháček (1957) placed Be among the typical elements of the organic coal matter.

This type of element classification is more or less a statistical problem and conclusions can be drawn only from a large number of analyses. Therefore, the division proposed by Leutwein and Rösler (1956) and refined by Bouška and Havlena (1959a) is more adequate. Issuing from the idea that the purely organic coal matter will contain little ash and that coal with a larger amount of inorganic matter will be higher in ash, they divided the elements into four groups (see in the summarizing remarks).

Their scheme may be refined according to the results of further investigations. For example, several studies (e.g. Deul, 1956) have shown that U, La and Ce should be grouped with the elements contained predominantly in the organic coal matter. Elements as Sc, Y, Yb, Ga and V have a varying position which has been only broadly defined. Other elements determined in coal are present in very small to trace amounts, which do not allow to decide whether they are associated with the organic or inorganic matter.

According to Zubovic et al. (1961a, b), the elements Ge, V, Be, Ti, Ga, B, Ni, Cr, Co, Y, Mo, Cu, Sn, La and Zn form with organic substances metallo-organic complexes of the chelate type. The metal is most probably attached to nitrogen. The bond of these complexes is firmer than that of complexes in which metal is bound to oxygen. The pH value of environment plays an important role in the formation of metallo-organic complexes.

In his paper of 1966a Zubovic discussed the relationship between chelate stability and organic affinity of several minor metallic elements in coal. In the decomposition of plant material, chlorophyll and amino acids having nitrogen as the donor element would be good chelating agents for trivalent vanadium (tetravalent vanadium forms chelates rather with oxygen, i.e. also with the decomposition products of lignin), nickel, copper and possibly iron. Lignin, in which oxygen is the donor element would complex with Be, Ge, Ga, Ti, Co, Al and Si. One of the possible important chelating agents derived from lignin is 4-hydroxy-3-methoxyphenyl group. The chelate stability of bivalent metals is $Be > Cu > Ni > Co > Zn > Fe$; except for copper it is the same as that of the organic affinity of these metals in coal. Gallium, yttrium and lanthanum have the same order of organic affinity and chelate stability. Copper may have been in a reduced state already in the coal-forming swamps or may have been precipitated as a sulphide. The relationship between organic affinity of the metallic elements and their ionic potential is plotted in Fig. 71.

Manskaya and Kodina (1975) also conclude that the formation of chelates is controlled by the ionic potential of the metal. Metals having a high ionic potential have a higher capacity to binding with the organic matter of coal, and metals with a low ionic potential show a lower capacity of producing chelates with the organic coal substances. Germanium and vanadium attain maximum concentrations in coalified wood. Vanadium forms vanadium porphyrins in the form of vanadyl VO^{2+} bond with four pyrrole nuclei, in which nitrogen is the donor element, and vanadium phenols with oxygen as the donor element. Porphyrin pigments are more abundant in oil and oil shales than in coal. Magnesium in them was replaced by vanadium, nickel or other elements. The predominant part of the substances of this type are organo-metallic complexes with mixed ligands containing oxygen and sulphur as additional donor atoms besides nitrogen.

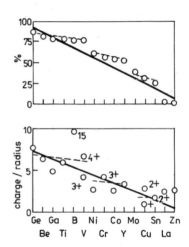

71. Decreasing affinity of elements for organic coal matter (Zubovic, 1966a).

From the study of the trace element distribution in some coals of the U.S.A. Zubovic (1966a) recognized that of 13 elements it was Be, B, Ge and Ga that had concentrated near the source areas of a coal basin. These elements form small, highly charged ions having a high ionic potential. The other elements have shown variable distribution patterns which may be due to the factors of geological transport. The author observed anomalous increases in trace elements in the samples of weathered coal.

The investigation results can be summarized as follows:

(1) The accumulation and concentration of elements in coal is the result of different processes, which may be divided into two groups:

(A) accumulation of elements as a result of living activity of plants and partly of animals: C, N, P, S, Fe, V, I, Ca, Mg, Al, Sn, Zn, Pb and others;
(B) accumulation of elements after the death of organisms:
    (a) mechanical — minerals and rock fragments supplied during the coal formation: Si, Al, Ca, Mg, Fe, Mn, Na, K and others;
    (b) chemical — by incorporation into organic compounds — Ge, U, V, etc.;
        — by precipitation as sulphides — Fe, Cu, Zn, Pb, etc.;
        — by reduction — Ag and others;
    (c) by physical adsorption — U, Th, Ge, V and others.

(2) The available data allow for the following classification of elements on the basis of their assignment to the organic coal matter or to the inorganic coal components:
    (a) elements associated almost completely with the organic coal matter: Be, Ge, U, Zr (part may be derived from zircon), (Cu), (Cr);
    (b) elements associated predominantly with the organic coal matter: Cu, Cr, Pb, Ag, Au, Sb, Mo, (Ga), (V), (Ni), (Sn), (As), (Ti), (B);
    (c) elements bound predominantly to the external ash substances: Ga, V, Ni, Sn, Co, As, Ti, B, (Li);
    (d) elements bound almost completely to external ash substances: Mn, Ba, Sr.

In this classification the hydrothermal mineralization processes which could have enriched even low-ash coals in elements typical of inorganic coal components were not taken into consideration.

The typical organogenic elements C, O, H, N, S, P and elements forming the essential part of external ash substances, such as Si, Al, Fe, Ca, Mg, Na and K have not been included either. Their contents are given nearly to a full extent by the mineral admixture in coal. The assignment of these elements is quite clear. The scheme is based on the generalization of several instances and may be locally influenced by an increased mineral admixture. The assignment of the elements in parentheses is not unequivocal. Many other elements could not be classified for a lack of reliable data.

(3) The concentration processes of trace elements should be considered from several points of view. The enrichment by living processes in plants is negligible. The elements liberated during the decay of plant bodies are rather dispersed in the coal-forming environment than concentrated, even if their influence on the total enrichment of the solutions circulating in the coal basins cannot be denied. It can also be assumed that the trace elements had been leached from the allochthonous plant material earlier than the material capable of coalification was transported into the coal-forming environment.

The fundamental process of trace element concentration in coal is sorption, both chemical and physical. Humic acids and humic substances with a high sorption capacity play an important role, particularly in the early stage of coal formation. The majority of trace elements present in coal is bound to them.

Another decisive factor is time, during which the solutions transporting the trace elements circulate in the developing or existing coal seam. The accumulation of trace elements in a coal seam is controlled by the following factors: (1) duration of supply, (2) possibility of solution circulation, (3) pH and Eh of environment, (4) concentration of the supplied constituents, (5) degree of coalification, (6) type of prominent sorbents [depending on (5)] — humic substances, ligneous matter, etc., (7) ash content, (8) size and quality of inner surface, (9) thickness, position and surface extent of the coal seam, (10) porosity of the neighbouring rock, and (11) overall geological structure, recent and original depths of the coal seam, alteration grade, jointing and other parameters.

(4) The principal source of enrichment of the coal seams in chemical elements were predominantly the rocks of granitic character, strongly kaolinized or lateritized, during which processes a number of elements were released from the minerals to be transported by water solutions. This circumstance provides evidence of a connection between the formation of coal and kaolinization processes, which complement each other. Additional sources of elements may be nearby ore deposits, a wider source area having an anomalous concentration of elements, or an overlying complex of volcanic and volcanodetrital rocks. Of importance is often a superjacent layer of permeable rocks (e.g. sandstones), which makes possible an easy water circulation and supply of metals into the coal seam.

The elements liberated by intensive weathering were transported into the basin during its development and in the early coalification stages they were brought into coal by circulating water. The transport was mediated by surface water, ground water, hydrothermal water, water of post-volcanic and perhaps also of metamorphic origin.

During the proceeding coalification the trace elements become to be distributed more continuously in the individual coal components, in the organic and clayey constituents, and their interrelations are gradually defined more sharply. The elements forming soluble compounds are washed out from the coal seams by percolating water.

(5) Of the petrographical coal components, xylite and vitrain and their transitional members are most enriched in trace elements, because they had the highest sorption capacity during the coal formation. Fusain is low in trace elements and its sorption capacity is extremely low. It has cells usually filled with clay material.

MECHANISM AND TIME SCALE OF THE ENRICHMENT OF COAL IN TRACE ELEMENTS

The trace elements such as Ge, U and V are present in coal mostly in organometallic complexes of chelate type (Zubovic, 1966b), whereas others may be bound in a functional group or sorbed on the organic coal matter.

There is no doubt that the coal seams were enriched in trace elements for the most part in post-sedimentary time. The reduction of the plant material volume in the course of coalification and selective leaching of ash substances were only of local significance, the supply of elements from the neighbouring rock complexes being an essential process. The elements were sorbed from the solutions. The role of the oxidation-reduction potential of the environment must not be omitted. The question remains in which development stage the elements were entrapped in a coal seam.

Sorption begins roughly in the stage of early diagenesis, as soon as the plant material has been covered with sediments, and ends broadly at the transition from the brown coal to the bituminous coal stage. It thus starts with the initiating gelification of the plant tissues and ends in the bituminous coal stage, when a decrease of metal elements is observed. For example, germanium is more regularly distributed, both laterally and vertically, in the higher coal ranks but it is less abundant than in the lower coalified types. The coal is enriched mainly by physicochemical sorption (Bouška et al., 1963), the chemical part of this process predominating in the initial stages and the physical sorption in the later stages, prevalently in the coalified plant material. Breger and Chandler (1960) quote the uranium contents from the Triassic and Jurassic coal seams in Colorado, into which U was brought by percolating solutions as late as in the Cretaceous, when the plant material had already been fully coalified.

Our knowledge of the sorption process itself on the organic material is still meagre. It is unquestionable that the elements were supplied into the developing coal seam by solution percolating from the neighbouring rocks. The sorption of elements in a sedimentary rock occurs by filtration or molecular diffusion. The mixed filtration-diffusive mechanism functions in convective diffusion.

Yudovich (1972) and Ryazanov and Yudovich (1974) applied these processes to characterize the mechanism of element transport in the enrichment of coal seams and to determine the time required for the enrichment of a coal seam having a given volume. Yudovich based his calculations on the following parameters: permeability of the neighbouring rocks 10^{-2} to $10^2\ d$, porosity of the neighbouring rocks 5 to 30%, temperature of the environment 0 to 70°C, coefficient of diffusion 10^{-7} to $10^{-9}\ cm^2 \cdot s^{-1}$. The results achieved are reliable to a certain extent and applicable to further considerations. Yudovich (1972) considers the enrichment processes intentionally as stationary.* This simplifaction was necessary because of the lack of knowledge of many input data and of the sorption process itself. No consideration is taken of the changes of physical properties in time and of such factors as are sorption of the transported component during migration, osmotic transfer of the solvent and ion exchange; thermal factors, and the mode and rate of absorption are neglected.

1. Progressive filtration

The laminar flow of ground water is filtered in passing through a coal seam of a certain shape and arrangement. The mathematical model of this filtration may be expressed by the equation of continuity

$$-\mathrm{div}\,(c\mathbf{u}) - v = \frac{\delta c}{\delta t},$$

where c — concentration of the dissolved element, v — velocity of sorption, $\mathbf{u}$ — vector of linear velocity of filtration, t — time.

Boundary conditions must be determined for the solution of this equation. Let us assume that the concentration of metal in solution is C_0, and porosity of the coal seam n. Under the condition that at the surface $x = 0$ and $C = nC_0$, the change of metal concentration in the solution at a distance x from the surface of the coal seam will be

$$nC_0 - c(x) = nC_0[1 - \exp(-Kx/u)],$$

where K — constant of sorption velocity and x/u — time of migration of the dissolved metal in coal seam until its complete sorption.

The metal quantity (q) sorbed over an area S in time t will be

$$\frac{\delta q}{\delta t} = nC_0 u \varrho_0 S,$$

where ϱ_0 — specific gravity of the solution. The time T required for the sorption

---

* A non-stationary model has been presented by Ryazanov and Yudovich in their later paper (1974).

of a certain amount of metal in coal seam with volume V and constant cross-section S will then be

$$T = \frac{Q}{nC_0} \cdot \frac{\varrho}{\varrho_0} \cdot \frac{V}{nS},$$

where ϱ — specific gravity of a coal seam having porosity n, and Q — weight concentration of the sorbed metal in the coal seam.

In compiling the two last equations, the author issued from the assumption that the constant of sorption K, which may be evaluated from Herzfeld's equation as $K \approx 2.5 \times 10^8$ per year at temperature $t \approx 25\,°C$, is appreciably higher than the filtration velocity and may be regarded as practically infinite.

2. Diffusion

Diffusion in a three-dimensional space is expressed by the second Fick's law $D\,\Delta c = \delta c/\delta t$, where D is diffusion coefficient and $\Delta c = \delta^2 c/\delta x^2$. For simplification we can visualize the coal seam as a spherical body with radius r and the surrounding environment as homogeneous and unlimited. At the beginning the metal concentration on the sphere surface is the same as in the environment and equals nC_0. At the moment when the sorption of metal ions on the seam surface begins, a concentration gradient will appear and diffusion from the surrounding medium into the seam will start. At any point of the environment and at any moment, with the exception of the initial, concentration c of the dissolved metal will be the function of time and distance x of this point from the seam surface; in accord with the second Fick's law it can be expressed using the Gauss integral as

$$c = nC_0 \left[1 - \frac{r}{x} + \frac{r}{x} \cdot \mathrm{erf}\,\frac{x-r}{2\sqrt{Dt}} \right],$$

where D — diffusion coefficient (in $\mathrm{cm}^2 \cdot \mathrm{s}^{-1}$).

Using the first Fick's law for the determination of transfer rate of the metal to the spherical seam surface with area S we get

$$\frac{\delta c}{\delta t} = -DS \cdot \frac{\delta c}{\delta x}.$$

The quantity of metal q sorbed by the seam surface can be calculated as the function of time t from the equation of concentration and Fick equation:

$$\frac{dq}{dt} = 4\pi DrnC_0\varrho_0 \left(1 + \frac{r}{\sqrt{\pi Dt}} \right).$$

Since Yudovich (1972) studied only a stationary instance, the second member in parentheses which depends on time t may be neglected and after integration and simple modification the time T required for the sorption of a given metal amount in a coal seam with volume V will be

$$T = \frac{Q}{nC_0} \cdot \frac{\varrho}{\varrho_0} \cdot \frac{V}{4\pi Dr}.$$

3. Convective diffusion

The metal brought by filtration flow into the area of a coal seam moves by diffusion to the reactive surface of the seam. The equation for the metal transfer will be

$$D \Delta c = \frac{\delta c}{\delta t} + \text{div}(c\mathbf{u}).$$

This equation has not a simple solution. First, a stationary process should be presumed (i.e. $\delta c/\delta t = 0$) and, secondly, an approximate expression should be used for the rate of filtration flow near the seam surface. Assuming a spherical shape for the seam, the metal amount q sorbed on its surface can be found as a function of time t (according to Golub'ev and Garibyants, 1968):

$$\frac{dq}{dt} = 7.9 n C_0 \varrho_0 D^{2/3} u^{1/3} r_0^{4/3}.$$

Time T necessary for the accumulation of metal amount Q in volume V will be

$$T = \frac{Q}{nC_0} \cdot \frac{\varrho}{\varrho_0} \cdot \frac{V}{7.9 D^{2/3} u^{1/3} r_0^{4/3}}$$

From this relation it follows that the smaller the coal seam, the less time is necessary for its saturation.

Geologically, the two instances are not equivalent. For example, a gelified coal matter makes the filtration process virtually impossible; it may undergo mineralization, at the most. Mineralized wood, however, in contrast to coalified wood does not contain major concentrations of trace elements.

The diffusion mechanism seems more plausible. The selection between pure diffusion and convective diffusion models depends on the ratio of filtration rate of the ground water to the diffusion rate. In case the filtration velocities are high, diffusion is suppressed and acts only close to the seam surface, and at a small filtration rate diffusion is the only process of substance transfer. For the assessment of the transfer mechanism of metals the dimensionless quantity (hydrodynamic criterion of Péclet) may be applied; it expresses the relation of filtration rate to the diffusion coefficient and depends on the coefficient of permeability (K_{pr}, cm^2):

$$Pe = \frac{u \cdot \sqrt{K_{pr}}}{D},$$

where u is the rate of filtration. The Pe value above 1 indicates the predominance of filtration over diffusion. The Pe value of 0.01 to 1.0 is characteristic of the convective diffusion and the values below 0.01 represent pure diffusion. The relations of u (filtration rate) and D (diffusion coefficient) in the rocks with different coefficients of permeability (K_{pr}) are shown in Fig. 72.

It is apparent that pure filtration is not characteristic of the enrichment of coal seams: it is virtually impracticable in clays and in sandstones it occurs only at a filtration rate of tens of centimetres per year. The same holds for convective diffusion, which may function perhaps in sands. There are so far very few data relating to the real filtration rates and to the coefficient of ion diffusion in the solutions filling the pores in rock. According to Gurevich (1969), the motion velocity of mineralized ground waters of the order of $1-10$ cm/year is unrealistic, since it implies that the amount of removed salts would exceed the weight of the rocks involved. He demonstrates that the filtration rate is much lower than that calculated on the basis of Darcy's law; in his opinion, it ranges from 10^{-4} to 10^{-2} cm per year.

Yudovich (1972) presented concrete calculations for enrichment of coal in germanium by diffusion, which may be applied also for other elements (Table 44).

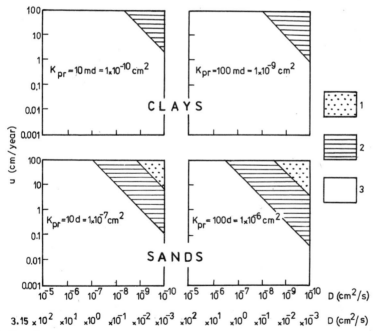

72. Transfer of substances as related to the coefficient of permeability (K_{pr}), coefficient of diffusion (D) and the rate of filtration (u) (after Yudovich, 1972). 1 — field of filtration ($Pe > 1$), 2 — field of convective diffusion ($0.01 < Pe < 1$), 3 — field of diffusion ($Pe < 0.01$).

Table 44

The period of coal enrichment in germanium by diffusion under various conditions (in m.y.); after Yudovich (1972)

Initial conditions: $V = 5$ cm^3, $r = 1.06$ cm

Ge concentration in water	6×10^{-8} g/l				6×10^{-6} g/l			
Ge concentration in coal seams (in %)	0.005	0.010	0.03	0.3	0.005	0.010	0.03	0.3
D (cm^2/s)								
				Room temperature				
				Porosity of environment 30%				
10^{-7}	0.155	0.310	0.925	9.25	0.002	0.003	0.009	0.093
10^{-8}	1.55	3.10	9.25	(92.5)	0.016	0.031	0.093	0.925
10^{-9}	15.5	31.0	(92.5)	(925)	0.155	0.310	0.925	9.25
				Porosity of environment 20%				
10^{-7}	0.230	0.460	1.39	13.9	0.002	0.005	0.014	0.139
10^{-8}	2.30	4.60	13.9	(139)	0.023	0.046	0.139	1.39
10^{-9}	23.0	46.0	(139.0)	(1 390)	0.23	0.460	1.39	13.9
				Porosity of environment 5%				
10^{-7}	0.925	1.85	5.55	(55.5)	0.009	0.09	0.055	0.555
10^{-8}	9.25	18.5	(55.5)	(555)	0.93	0.185	0.555	5.55
10^{-9}	(92.5)	(185)	(555)	(5 500)	0.925	1.85	5.55	(55.5)
				Temperature 50–70°C				
				Porosity of environment 20%				
10^{-7}	0.077	0.155	0.460	4.60	0.001	0.002	0.005	0.046
10^{-8}	0.77	1.55	4.60	(46.0)	0.008	0.016	0.046	0.46
10^{-9}	7.7	15.5	(46.0)	(460)	0.077	0.155	0.460	4.6
				Porosity of environment 5%				
10^{-7}	0.31	0.615	1.85	18.5	0.003	0.006	0.019	0.185
10^{-8}	3.10	6.15	18.5	(185)	0.031	0.062	0.185	1.85
10^{-9}	31.0	(61.5)	(185)	(1 850)	0.310	0.615	1.85	18.5

Values in parentheses are improbable.

He thinks that filtration rates of at least 10^{-2} cm in a year are possible. Diffusion is accelerated by higher temperatures; an increase of 50°C doubles the intensity of diffusion. The values in parentheses seem to be unreal. The duration of enrichment of coal seams in trace elements may be estimated reasonably at 1 million or maximum a few tens of millions years from the beginning of coalification. Table 44 should be understood as an example of calculations. The germanium concentrations in coal seams with $3-7\%$ in ash would be difficult to explain, although they are found in the nature. We have to presume the participation of additional factors, which are either still unknown (e.g. local enrichment of coal seam) or surmised, e.g. increased concentration of elements in ground water in the initial stage of coal formation.

Yudovich (1972) summarized the results of his study as follows:

(a) The calculations of the enrichment of coal seams in trace elements have shown that the most plausible procedure is diffusion or convective diffusion. Diffusion probably predominates in nature. Convective diffusion requires higher filtration rates, which may occur in the early coalification stages, perhaps until the brown coal stage. Enrichment by progressive filtration seems to be out of question.

(b) The concentrations of trace elements that exceed their clarke contents in the neighbouring rocks 10–100 times may be explained satisfactorilly by diffusion if the ground waters have clarke contents of elements and the physical parameters of rocks are most convenient (permeability from 10^{-2} to 10^2 d, porosity $5-30\%$, temperature $0-70$°C, coefficient of diffusion $10^{-7}-10^{-9}$ cm$^2 \cdot$ s^{-1}). Under such conditions the enrichment in trace elements up to the observed concentrations may be completed within several thousands up to ten million years. If the ground waters themselves are enriched in element ions, the enrichment of coal occurs at a higher rate.

The transfer of metals by diffusion occurs in two stages (Ryazanov and Yudovich, 1974). The first, which is of a higher rate, takes place in the early stage of diagenesis, when the seam is in a relatively close contact with its environs. The second is slower and the mechanism involved is characterized by very low coefficients of diffusion, of the order 10^{-12} to 10^{-14} cm$^2 \cdot$ s^{-1}, which is million times lower than in the first case. Diffusion in a colloidal system is distinguished by analogous parameters.

The formation of the contact zone of a coal seam may be far more complicated. For the time being we have considered only an irreversible sorption process, i.e. the supply of metal from outside. In the early coalification stage the sorption of trace elements by peats and all organic substances attains a maximum in strictly defined intervals of pH and possibly also of Eh, which are interdependent in natural conditions. In the environments of higher acidity or basicity, however, the sorption process may be replaced by desorption, i.e. the elements may be washed out from coal. Germanium, for example, shows optimum conditions for being sorbed by coal matter at pH $5.5-6.5$ and gallium at pH $3-6$. If these conditions change, the enrichment of coal may change into impoverishment in trace elements as a result of desorption. It should be remarked that some humates of metals are unstable in alkaline or neutral conditions, in which case the trace elements are actually removed from the coal seam. Beryllium and gallium, for example, are readily hydrolysed and both of them require alkaline or strongly acidic environment. Therefore they are occasionally less concentrated in the contact zone of a coal seam, in the sense of Ryazanov and Yudovich (1974).

Ryazanov and Yudovich (1974) expressed the time necessary for the enrichment of coal in an element by the equation:

$$\tau = \frac{(h_1 + 0.3h_2 - 0.5l)^2 \, x_{max} \varrho}{D_0 n c_0 k} \quad \text{(years)}.$$

In taking the maximal Ge content in coal seam as $x_{max} = 120$ ppm, i.e. 1.2×10^{-4} g/g, the thickness of contact zone as 10 cm ($h = h_1 + h_2$; h_2 evaluated as one third), and the parameters $l \cong 0$ (thickness of the thin contact layer), density $\varrho = 1.4$, $D_0 = 10^{-6}$ cm$^2 \cdot$s^{-1} (coefficient of diffusion), $n = 0.3$ (porosity of the medium 30%), $c_0 = 6 \times 10^{-11}$ g $\cdot$ cm^{-3} (volume concentration of metals in solution), $k = $ coefficient of converting the number of seconds to years (strictly 3.1536×10^7), by substituting these values into the above equation they obtained the time required for enrichment as

$$\tau = \frac{(0.3 + 0.3 \times 9.7)^2 \times 1.2 \times 10^{-4} \times 1.4}{1 \times 10^{-6} \times 0.3 \times 6 \cdot 10^{-11} \times 3.1536 \times 10^7} = 3.05 \text{ million years}.$$

Generally speaking, the formation of a contact zone covers several millions to a few tens million years. The coefficient of diffusion in the coal seam is much lower, varying between 10^{-12} and 10^{-14} cm$^2 \cdot$s^{-1}. These values are characteristic of diffusion in colloidal systems. The process is to be visualized as the absorption of metal by a thin layer at the contact of the seam with the neighbouring rocks, followed by slow diffusion in the seam itself.

In Czechoslovakia, Hokr (1975a) estimated from the geological-petrographical and geochemical data the accumulation of sulphur in the coal of the North Bohemian Brown-coal Basin at a period of the order of eight million years from the end of basinal sedimentation.

APPLICATION OF THE GEOCHEMICAL DISTRIBUTION OF ELEMENTS IN COAL SEAMS TO THE IDENTIFICATION, CORRELATION AND STRATIGRAPHY PURPOSES

The possibility of identifying the coal seams with the use of analysis of coal ash is mentioned in every article on the correlation of coal-bearing series and identification of their seams (e.g. Zhemchuzhnikov, 1948; Beneš, 1954), but concrete examples are very rarely quoted. These articles likewise lack any information on the advantages and drawbacks of these identification methods.

The geochemical papers devoted to the study of elements in coal ashes and their distribution in a seam treat the identification problem but indirectly. For example, Otte (1953) drew attention to the differences in the chemical element association within one seam, or Leutwein and Rösler (1956) pointed out the regularity of the lateral distribution of elements in one seam or of their vertical distribution within a seam series.

Except for the cited paper of Leutwein and Rösler (1956), most of the geochemical studies do not give a statistical evaluation of analyses of a major number of samples from one seam, which is prerequisite for identification purposes. Since a coal seam cannot be characterized adequately from a few samples, the geochemical works can be used only as a basis for considerations on the vertical and lateral distributions of elements. The conclusions drawn from these considerations show that the practical application of the analysis of coal ashes is strongly limited by the geological factors and the deficiency of statistical data. Theoretically, however, it provides many possibilities for the identification of coal seams. The results depend above all on the number of analyses and their correct evaluation. One total sample or a sample of one fraction may be characteristic at most of the close vicinity of the sampling site. A coal seam was forming under conditions that made possible the accumulation of plant matter during a long period and over a large area, but it can hardly be presumed that the set of these conditions would have remained unchanged. In other words, a coal seam as a whole is a product of particular genetic conditions but the share and effect of each of these conditions were not the same in all parts of the coal-forming basin.

The most important parameters are the rate of sinking of the partial area, the type of substratum, the composition of the floral association, and the geographical position of the individual areas. These were not separated by a sharp boundary but passed one into another. Accordingly, the chemistry of the coal-forming environment changes gradually in the horizontal direction. Both the development and chemistry of the environment also changed with time, i.e. in the vertical direction.

In a coal seam the gradual vertical and lateral changes of genetic conditions are shown today by the changes in the lithology of its sectors. These changes make it possible to recognize, to a certain extent, the genetic conditions of the individual parts of the coal-forming basin.

On the other hand, the chemical composition of a coal-forming environment (particularly the content of trace elements) may have been somewhat effaced by the supply of foreign elements into the organic matter during diagenesis or by the removal of some elements by percolating water. The supply and removal may be shown both quantitatively and qualitatively. As a result, the present-day coal seams do not reflect the pattern of element contents as it was in the primary environment. It has been preserved in the individual seam segments only in broad outlines but the detailed distribution of elements is generally distorted by their migration during diagenesis. The chemical composition of a seam segment cannot therefore be regarded as analogous to the chemistry of the corresponding original area of coal formation.

This circumstance, however, is not an unsurmountable obstacle to the identification or correlation works.

Theoretical possibilities of correlation and identification of coal seams have been discussed in detail by Bouška and Havlena (1959b). They argued that correlation and identification of seams is only apparently a problem that can be solved by correlation of seams in the vertical direction alone. For the correlation of seams in the vertical direction on the basis of one characteristic feature it is necessary to know its behaviour in the horizontal direction. In practice two inseparable points of view have to be considered:

(A) lateral variability (quantitative and qualitative) of the element content in a coal seam;

(B) the content of elements (quantitative and qualitative) in the seams in vertical direction.

In case (A) several possibilities of the distribution of elements may occur; they are enumerated below, arranged according to the applied method of spectral analysis:

1. Semiquantitative or quantitative analyses:
    (a) The quantitative representation of an element is strikingly uniform within the whole seam.
    (b) An element is more abundant in one seam segment (e.g. at the margin of the basin) than in another (e.g. in the interior of the basin) as, for example, boron in the Kladno-Rakovník Basin (Somasekar, 1969). These two possibilities may be also valid for a group of elements.

(c) The concentration of one element or of a group of elements in the seam by far prevails over the amount of the remaining ones.

(d) There is a relationship or ratio between the contents of two or more elements. These two parameters either remain constant in all segments of the seam or change regularly (decrease or increase) in a particular direction.

(e) The contents of all elements are distributed randomly in the seam in lateral direction. This theoretical possibility in itself is not useful for correlation or identification; it can be applied only when other seams are distinguishable by a regularity in the content of one element or a group of elements, but this case is very improbable.

2. Qualitative analyses:

(a) An element or a group of elements is contained in all samples of one coal seam. Identification or correlation is possible only if a characteristic trace element or a group of such elements is present that is absent from other seams. This eventuality is possible when the seam developed under very stable conditions. No positive result can be inferred from the presence of common elements.

(b) An element or a group of elements is present only in some samples, i.e. is characteristic of particular seam segments such as segments richer in partings or with cannel lenses. In this case their presence cannot be used for correlation or identification. Only if the distribution of the individual segments shows some regularity (e.g. areas of quiescent development or places where streams entered the basin), the results could be applied to the comparison of principal chemical features in the basinal parts of different genesis.

In case (B) concerning the correlation or identification in vertical direction the following possibilities come into consideration: Similarly as ad (A), whatever possibility must first be valid over the whole area of a seam or of a group of seams, or at least within an exactly defined part of the seam, e.g. in a particular facies.

1. Semiquantitative or quantitative analyses:

(a) One from the group of seams is characterized by a high concentration of one trace element, whilst the other trace elements recede to the background or are absent altogether. In the remaining seams the contingent trace element occurs in the amount not exceeding the average content of other elements or is lacking.

Several instances of unusually high concentrations of one element in coal have been quoted in the literature. For example, some trace elements occur in strikingly high amounts in ashes from the seams of the Ruhr Basin. Otte (1953) recorded 0.2% Co (in the Dickebank seam), 0.1% Ga (in Katharina seam), and up to 1.6% Ni in the vitrain of the latter seam. In the vitrain ash from Northumberland 7.0–24.3% TiO_2 has been determined by Jones and Miller (1939). The uranium content of 0.051% was found in the ash from a Miocene seam at Creston Ridge (Masursky, 1956). Increased concentrations of Zr, Mn, Cr, V and other elements have been reported in a number of papers, but they have never been used for identification purposes and their lateral distribution within the seam has not been examined.

Worth of notice is the observation of Katchenkov (1952) that the amount of strontium in the sterile Permian complexes of the Urals varies conspicuously, consistently with the boundaries between the stratigraphical units. The author thinks it feasible to correlate the stratigraphical units in deep borings on this basis.

(b) The circumstances given ad (a) can be assessed also for a group of elements, especially a group of physically and chemically allied elements.

(c) The concentrations of two chemically and physically allied elements occur in one seam at a ratio which is widely different from that ascertained in other seams. This can hold for more than two elements. All the various possibilities cannot be discussed here; they have to be considered individually and, in essentials, they represent only more or less complicated alternatives. Katchenkov (1952) correlated in the Permian sediments of the Urals two sedimentary complexes on the basis of a prominent concentrations of Ca : Sr and Fe : Mn ratios. Leutwein and Rösler (1956) observed in the coal seams several pairs of elements (e.g. Ag : Pb, Co : Ni, Ga : V, Fe : Mn, Ca : Sr, Ba : Sr) and recorded regular quantitative relations between them.

2. Qualitative analyses:

(a) The seam is distinguished by the presence of one element or a group of elements (regardless of quantity) which is absent from other seams.

No instance of such correlation or identification has been mentioned in the literature. On the contrary, it has been proved that there are invariably more seams in a group of seams that contain a given element. Leutwein and Rösler (1956) compared twelve successive seams (Zwickau, Oelsnitz), but they did not reveal the presence of an element in only one seam.

(b) The seam is characterized by a variety of elements, the other seams being, in contrast, qualitatively poor. This case is not known from practice. It is commonly said in the summarizing papers on the chemical elements in coal that in the Palaeozoic coals there occurs repeatedly an association of about 40 elements which are invariably present in the samples, at least in traces.

Of the theoretical possibilities given above Bouška and Havlena (1959b) regard B (1a) as acceptable for the identification or correlation of coal seams, and to a less extent the alternatives (1c) and (2a). The other variants are either purely theoretical or do not furnish unequivocal results.

The decision whether to use the granular or componental fraction for identification purposes depends on the macro- and micropetrographical composition of the coal seam. In the seams showing a quiet development no differences would appear between the essential elements of secondary ash substances in the individual seams, but changes may occur in the presence and content of elements associated with the organic coal matter. In this case the componental fraction and sampling at places ca. 100—200 m apart may be used for investigation. The preliminary results of such research in the seams of the Ostrava-Karviná coalfield seem promising (Bouška and Krejčí, 1959, 1965).

Where the development of the basin was disturbed, for example, by synsedimentary movements, inflows of streams, volcanic activity in close vicinity, the elements bound to the organic matter of coal would provide very ununiform results. Since in this case the coal is usually contaminated by secondary ash substances, the granular fraction is more suitable for the identification of seams, as the elements of both the organic substances and secondary ash substances are available. It would be most desirable to use a ground granular fraction and the corresponding componental, e.g. fusain, fraction and analyse them separately.

The comparable results may be obtained only under the same working conditions, i.e. identical grain-size distribution of the fraction, the same temperature of burning (a temperature not exceeding 500 °C is generally thought to be most adequate, as practically all elements except for the highly volatile, remain preserved in the sample) and identical experimental conditions. The ash content is decidedly an important basis for the correlation of results. Huszka and Láda (1956) managed to identify a coal seam to a distance of 100–150 m according to the silicate analyses of the ash of the vitrain component.

With regard to the lateral and vertical changes in the genetic conditions and chemistry of the individual segments, the identification demands the study of a sufficient number of samples so as to record the variability of lithology and chemistry both in space (lateral) and time (vertical). This would imply taking samples in a sufficiently dense grid and from the whole seam profile at every sampling site and examining every coal band. Since this procedure would be very laborious and since the aim of identification is, as a matter of fact, to place a seam in a succession of seams, the vertical variability in the genetic conditions of a coal seam and thus in the chemistry of coal is usually "concentrated" into a channel sample. By grinding and quartering a channel sample we obtain a "concentrate" of the vertical variability of the coal seam at the sampling site.

In doing so, we interrupt the continuity of the development of the primary areal segment but the simplification does not distort the essential relations. For identification it is important to define one characteristic feature of the seam and not a series of features of the coal bands. If, however, the changes in genetic conditions are expressed by changes in the macropetrographical composition of the seam, the channel samples are taken from the segments differing in macropetrography, as for example, from a sapropelite bed or a bed rich in dull coal. Channel samples proved suitable also for other identification methods, particularly for palynological analysis.

Several examples of a complete or at least partial solution of the correlation and stratigraphical problems are given below.

Yudovich (1972) recorded an example from the Sangara region of the Lena Basin in the U.S.S.R., where the Upper Cretaceous Chechum series contains thin coal seams and coalified logs in sand. These deposits were difficult to distinguish from the sandy sediments of the Sangara series of Early Cretaceous age, which also contains coal interseams. The study of heavy minerals (appearance of epidote in the upper parts of the profile) did not fit with the boundary defined by spore and pollen analyses. The problem was solved unequivocally using geochemical methods.

The characteristic elements of coal in the Upper Cretaceous Chechum series are Cr, Ga and Mo. Increased Ga is characteristic only of these coals and has never been found in Lower Cretaceous coal. (Chromium, up to 0.5% in amount, and molybdenum were found in the underlying Jurassic coal beds.)

The Lower Cretaceous coals of the Batylykh series contain V, Ni, Co, As and Sr as specific elements. The medium concentrations of ubiquitous Pb, Ti, Nb and REE are characteristic of the Lower Cretaceous coal seams of the Eksenyakh series.

A similar case is known from southern Bohemia (Bouška, 1961). Germanium shows there a remarkable distribution in caustobioliths since the Permian to the Recent, which might be applied to discriminating between the Cretaceous and

Tertiary basinal sediments. Germanium has not been identified in any sample of anthracite (Permian-Carboniferous) from Lhotice and Úsilné, from the Miocene coal seams near Mydlovary (see also Vlášek, 1956 and Bouška, 1966) and a small seam from the Veselí nad Lužnicí, or in samples of peat and woods extracted from the peat bog Borkovická blata near Veselí nad Lužnicí (lowmoor bog, transitional and highmoor bogs Jitra and Kozohrudky), lowmoor bogs Ruda near Horusice, Vimperka near Třeboň and lowmoor bog at Schwarzenberský pond. It was, however, regularly present in the coal seams and coalified logs in the Upper Cretaceous sediments, even if in very small to trace amounts, e.g. in coaly clays. Germanium is a characteristic element of coal matter in the freshwater Cenomanian sediments in Bohemia and Moravia (Bouška et al., 1963).

Other examples can be found in the detailed study of Kessler et al. (1965) on the importance of trace and minor elements for the correlation and identification of seams in the Ostrava-Karviná coalfield. The paper expanding the former knowledge (Králík et al., 1960; Kudělásek, 1960) provided information on the Ostrava and Karviná Formations. Some of the seams or groups of seams strikingly differ in the content of elements from the nearby seams. The enriched seams are more frequent in the Ostrava Formation, which is associated with its paralic character and with the presence of marine bands. The alkaline metals, particularly the Na : K ratio have proved most suitable for correlation and identification purposes. Phosphorus, boron and beryllium are second in importance. The Li content in coal ash from the seams of the Upper Hrušov Member (90 ppm on average) greatly differs from its average content 200 ppm in the Lower Hrušov and Petřkovice Members. The Na : K ratio changes with the development of megacycles; it is highest at the base of every cycle, decreasing upwards (to < 1) with the increase in K. The contents of boron are markedly higher in ash of coal from the Karviná Formation than from the Ostrava Formation.

Králík et al. (1960) determined increased germanium in some coal seams of the Ostrava-Karviná coalfield. Macháček (1957) recorded greater facies changes in the elements of the iron and other metal groups than are their vertical variations in the seams of the Lower Hrušov Member.

Many papers refer on the element pairs which occur in the coal seams in constant ratios, such as Ba and Sr, V and Ga, less frequently Ni and Co, Pb and Ag. The first element invariably predominates; if it is present in a small amount, the paired element occurs only in trace amount or is absent altogether. These relations have been observed by Leutwein and Rösler (1956) and already in 1937 Goldschmidt determined the Co : Ni ratio in coal at 1 : 2 to 1 : 3. Otte (1953) reported that Ni prevails over Co in humitic coal; in typical sapropelitic coals the Co : Ni ratio equals 2 : 1, in boghead or boghead claystone even 1 : 1. The V : Ga ratio has been studied in the anthracite ash from Lhotice and Úsilné (Bouška, 1966) and the V : Ga and Ba : Sr ratios in the ash of coals from the Žacléř group of seams (Bouška and Havlena, 1959a).

Hokr (1975a) defined the chronostratigraphical boundary in the North Bohemian Brown-coal Basin from the changes in the content and composition of ash, particularly of potassium, SiO_2, Al_2O_3 and in the Si:Al ratio. He placed the stratigraphical boundary between the so-called upper and middle bench, where a sudden change of the association of clay minerals is observable. All samples from the seam and from the underlying clays, regardless of the ash content, have only kaolinite; SiO_2 varies about 50—55 % and Al_2O_3 about 35—40 % at the $SiO_2:Al_2O_3$ ratio equal to ca. 1.5. The K_2O contents are more varying but in the stratigraphical succession they are relatively constant for every drill hole. The upper coal bench and the Lom seam possess the same association of clay minerals: kaolinite, illite and a minor amount of montmorillonite. The SiO_2 contents rise almost to 60 % and those of Al_2O_3 drop to 30 % so that their ratio is about 2. The TiO_2 shows a general decreasing tendency with many deviations down to or near the base of the upper coal bench. The K_2O content is everywhere higher above than below the boundary.

For identification purposes also other than purely geochemical methods may be used. A good indentification tool is, for example, the character and amount of ash matter and its distribution. Of importance are particularly ash substances of syngenetic origin, as they might reflect the conditions of the coal formation. Mineralization of the whole or part of the seam can be often assessed macroscopically right in the mine or, if the ash substances are finely dispersed, microscopically in the laboratory. An example of a mineralized seam serving as an index horizon was recorded by Beneš (1954) from the Ostrava-Karviná coalfield. According to a bed of highly pyritized dull banded coal the seam Milan II in the A. Západocký mine has been correlated with the upper coal layer of seam 39 in the Žofie mine. Malán et al. (1966) used pyrite and inertinite and their ratio for correlation in the same coalfield. Ferrari (1948) correlated a seam in the Ruhr Basin with the aid of a vitrain band containing a large amount of finely dispersed quartz. Hoehne (1954) recorded several similar examples from the Ruhr Basin, where a silicified bed or the character of the interlaid or overlying clay was usually an index horizon. Ratajczak (1974) studied samples from 14 drill holes in the Ljubljana Coal Basin. The results of mineralogical, geochemical and petrographical examinations were usable for the interpretation of lateral and vertical changes in the sediment type.

The quoted examples indicate that geochemistry can be justifiably regarded as a promising means for the solution of correlation, identification and stratigraphic problems, but it has its limitations as have other methods.

The geochemical distribution of elements in lateral or vertical direction within a coal seam should thus not be overrated and it must be kept in mind that the geochemical correlation methods are above all statistical methods.

REFERENCES

ABELSON, P. H. (1957): Some aspects of paleobiochemistry. Ann. N. Y. Acad. Sci., 69, 276–285, New York.
ADAMS, A. S. and WEAVER, Ch. E. (1958): Thorium-to-uranium ratios as indicators of sedimentary processes: example of concept of geochemical facies. Bull. Amer. Assoc. Petrol., Geol., 42, 387–430, Tulsa.
ADMAKIN, L. A. (1974): Metallonosnosť geneticheskikh tipov ugleĭ nekotorykh mestorozhdeniĭ Zabaĭkaľya. Doklady AN SSSR, 217, 919–922, Moskva.
AHRENS, L. H. (1952): The use of the ionization potentials, I. Ionic radii of the elements. Geochim. Cosmochim. Acta 2, 155–169, Oxford.
A ĽBOV, M. N. and KOSTAREV, I. I. (1968): Copper in the peat bog formations of the Middle Ural. Sov. Geol., 11 (2), 132–139.
ALEKSEEV, F. A. and LEBEDEV, V. S. (1977): Izotopnyĭ sostav ugleroda uglya, CH_4 i CO_2 upoľ nikh mestorozhdeniĭ yugo-zapadnoĭ chasti Donetskogo basseĭna. Geokhimiya, 2, 306–310, Moskva.
ALLAN, J., BJORØY, M. and DOUGLAS, A. G. (1975): Variation in the content and distribution of high molecular weight hydrocarbons in a series of coal macerals of different ranks. In: Advances in organic geochemistry 1975, ed. R. Campos and J. Goni, pp. 633–654, Enadimsa.
ALLAN, J. and DOUGLAS, A. G. (1977): Variations in the content and distribution of n-alkanes in a series of carboniferous vitrinites and sporinites of bituminous rank. Geochim. Cosmochim. Acta, 41, 1223–1230, Oxford.
AMMOSOV, I. I. (1953): Khimiya i genezis tverdogo topliva. Sb. 1, 1–252. Izd. Akad. Nauk SSSR, Moskva.
— (1961): Stades du métamorphisme des roches sédimentaires et rapports paragénétiques des combustibles minéraux. Soviet Geol., 4, p. 117.
— (1967): Petrologiya ugleĭ i paragenez goryuchikh iskopaemykh. Nauka, Moskva.
AMMOSOV, I. I., GORCHKOV, V. I., GRECHICHNIKOV, N. P. and KAIMYKOV, G. S. (1975): Vitrinite: an indicator of paleotemperatures and thermal genesis of petroleum formation zones. 8ᵉ Congrès du Carbonifère, Moskva (in print).
ANDERSON, P. C., GARDNER, P. M., WHITEHEAD, E. V., ANDERS, D. E. and ROBINSON, W. E. (1969): The isolation of steranes from Green River oil shale. Geochim. Cosmochim. Acta, 33, 1304–1307, Oxford.
ARAMU, F. and URAS, I. (1957): Alcuni risultati su uno analisi di radioattività naturale dei carboni del Sulcis. Rend. Soc. mineral. ital., 13, 114–119, Pavia.
AUBREY, K. V. (1952): Germanium in British coals. Fuel, 31, 429–437, London.
BABIČKA, J. (1937): Analysis of the ashes of Lycopodium clavatum and L. annotinum, and Bibliography of Aluminium in Plants. Věstník Král. čes. spol. nauk, tř. II, 3, 18 p., Praha.

BACHMAN, G. O. and READ C. B. (1952): Trace elements reconnaissance investigations in New Mexico and adjoining States in 1951. U. S. Geol. Survey TEM-443-A, 1 – 22, Oak Ridge, Tenn., USA.

BACHMAN, G. O., VINE, J. D., READ, C. B. and MOORE, G. W. (1957): Uranium-bearing coal and carbonaceous shale in the La Ventana Mesa area, Sandoval County N. Mex. U. S. Geol. Survey Bull., 1055 – Washington.

BAJOR, M. (1960): Amines, amino acids and fats as facies indicators in Lower Rhenish brown coals and their analytical determination. Braunkohle, 12, 472 – 478, Halle.

BALME, B. E. (1956): Inorganic sulphur in some Australian coals. J. Inst. Fuel, 29, 21 – 25, London.

BANERJEE, N. N., RAO, H. S. and LAHIRI, A. (1974): Germanium in Indian coals. Indian Journ. Techn., 12, 353 – 358.

BARGHOORN, E. S. and SCHOPF, J. W. (1966): Microorganisms three billion years old from pre-Cambrian of South Africa. Science, 152, 758 – 763, Washington.

BAUMEISTER, W. (1952): Mineralstoffe und Pflanzenwachstum. 176 p., Jena.

BEĽKEVICH, P. I., VERKHOLETOVA, G. P., KAGANOVICH, F. L. and TORGOV, I. V. (1963): β-Sitosterol from peat wax. Izv. Akad. Nauk SSSR, Otd. Khim. Nauk 1963, 112 – 115. Chem. Abstr., 58, 10011 (1963), Moskva.

BENEŠ, K. (1954):Význam uhelně petrografických metod pro otázky identifikace slojí. Přír. sborník Ostrav. kraje, 15, 471 – 490, Ostrava.

BENEŠ, K. and DOPITA, M. (1954): Makropetrografická klasifikace černého uhlí a grafická dokumentace uhelných slojí. Uhlí, 4, 144 – 148, Praha.

BENEŠ, K., FOJTÍK, Z. and KRÁLÍK, J. (1964): Předběžná zpráva o zjištění zvýšených obsahů bóru v handlovském uhlí. Sb. věd. prací VŠB, Ostrava, 10, 201 – 204, Ostrava.

BERGIUS, F. (1913): Die Anwendung hoher Drücke bei chemischen Vorgängen des Entstehungsprozesses der Steinkohle, 58 p., Halle.

BERGMANN, W. (1963): Geochemistry of lipids. In Organic Geochemistry, I. A. Breger (ed.), 503 – 542. Oxford: Pergamon Press.

BERKNER, L. V. and MARSHALL, L. C. (1965): The history of the growth of oxygen in the earth's atmosphere. Progress in Radio Science, III. The Ionosphere. Elsevier, New York.

BERNARD, J. H. and PADĚRA, K. (1954): Bravoit aus dem Kladno-Rakonitzer Steinkohlenbecken. Zeit. f. angew. Geologie, B. 3, 2, 155 – 169, Berlin.

BERNER, R. A. (1964): Iron sulfides formed from aqueous solution at low temperatures and atmospheric pressure. J. Geol., 72, 293 – 306, Chicago.

BERONI, E. P. and MCKEOWN, F. A. (1952): Reconnaissance for uraniferous rocks in northwestern Colorado, southwestern Wyoming and northwestern Utah. U. S. Geol. Survey TEI - 308 - A, 1 – 42. Oak Ridge, Tenn., USA.

BERTETTI, J. (1955): Sulla presenza del gallio e del germanio in alcuni carboni fossili Italiani. Atti Accad. Ligure Sci. e Lettere, 11, p. 53, Genova.

BEUS, A. A. (1966): Geochemistry of beryllium. Freeman & Co., 401 p., London.

BLAYDEN, H. E., GIBSON, J. and RILEY, H. L. (1944): An X-ray study of the structure of coals, cokes, and chars. Proc. Conf. Ultra-fine Structure of Coals and Cokes, Brit. coal utilization. Res. Assoc. 176 – 231.

BLOM, L., EDELHAUSEN, L. and VAN KREVELEN, D. W. (1957): Chemical structure and properties of coal. XVIII. Oxygen groups in coal and related products. Fuel, 36, 135 – 153, London.

BLOSS, F. D. and STEINER, R. L. (1960): Biogeochemical prospecting for manganese in north-east Tennessee. Bull. Geol. Soc. Am., 71, 1053 – 1065, New York.

BLOXAM, T. W. (1974): Geochemical parameters for distinguishing palaeoenvironments in some Carboniferous shales from the South Wales coalfield. In The Upper Palaeozoic and post-Palaeozoic rocks of Wales, ed. by R. T. Owen. Cardiff (Univ. Wales Press), 263 – 284.

BOCTOR, N. Z., KULLERUD, G. and SWEANEY, J. Z. (1976): Sulfide minerals in Seelyville Coal III, Chinook Mine, Indiana. Min. Deposita, 11, 249 – 266.

BOGDANOVA, M. V. (1969): Khimicheskie osobennosti i genezis burykh ugleï Kazakhstana. Nedra, 1—166. Moskva.
BOGVAD, R. and NIELSEN, A. H. (1945): Vanadinindholdet i en Raekke danske Bejergarter. Medd. Dansk Geol. Foren., 10, p. 532, København.
BOHOR, B. F. and GLUSKOTER, H. J. (1973): Boron in illite as an indicator of paleosalinity of Illinois coals. Journ. Sedim. Petr., 43, 945—956.
BOROVIK-ROMANOVA, M. F. (1939): Spectroscopic determination of barium in the ash of plants. Tr. Biogeokhim. Lab. Akad. Nauk SSSR, 5, p. 175, Moskva.
BOROVSKIÏ, S. A. and RATYNSKIÏ, V. M. (1944): Olovo v uglyakh Kuznetskogo basseïna. Doklady AN SSSR, 45, 120—121, Moskva.
BOUŠKA, V. (1956): Železité konkrece z třetihorních sedimentů třeboňské pánve. Čas. Nár. muzea (NM) odd. přírod., č. 1, 329—340, Praha.
— (1959): Mineralogický a geochemický výzkum některých rašelinišť třeboňské pánve. Reseach report. Přírod. fak. UK v Praze, 1—29., Praha.
— (1960): Geochemie pevných kaustobiolitů uhelné řady. Diploma work, pt. 1. Katedra mineralogie, geochemie a krystalografie UK v Praze, 1—221, Praha.
— (1961): Geochemie pevných kaustobiolitů uhelné řady některých oblastí ČSSR,. Ph. D. thesis, pt. 2. Katedra mineralogie, geochemie a krystalografie UK v Praze, 1—207, Praha.
— (1962a): Zpráva o geochemickém výzkumu stopových prvků v pevných kaustobiolitech uhelné řady z pánevních sedimentů v jižních Čechách. Zprávy o geologických výzkumech v r. 1961. Geofond, Praha, 231—233.
— (1962b): Goethit ze slatiniště Ruda u Horusic. Acta Univ. Carolinae, Geologica, Nos. 1, 2, 39—53, Praha.
— (1966): Stopové prvky v antracitu ze Lhotic u Českých Budějovic. Sborník Jihočeského muz. v Č. Budějovicích. Přír. vědy, 6, 1—8, České Budějovice.
— (1972): Geochemické srovnání mořských a sladkovodních horizontů OKR. Report for Geol. průzkum n. p., Ostrava-Hrabová. ÚGV UK v Praze, 1—32. Unpublished, Praha.
BOUŠKA, V. and HAVLENA, V. (1959a): Sloje dolu Jan Šverma (Lampertice u Žacléře) a geochemický výzkum jejich stopových prvků. Rozpravy ČSAV, 69, 3, 1—64. ČSAV, Praha.
— — (1959b): O možnosti použití spektrální analýzy popelů jako metody k identifikaci uhelných slojí. Čas. pro mineral. geol., 4, 2, 189—194, Praha.
BOUŠKA, V., HAVLENA, V. and ŠULCEK, Z. (1963): Geochemie a petrografie cenomanského uhlí z Čech a Moravy. Rozpravy ČSAV, řada mat. přír. věd, 73, 8, 1—79, Praha.
BOUŠKA, V., HAVLENA, V. and TACL, A. (1964): Geochemický a petrografický výzkum svrchních slojí doubravských vrstev OKR. Čas. pro mineral. geol., 9, 1, 57—67, Praha.
BOUŠKA, V. and HONĚK, J. (1962): O koncentracii elementov primeseï v tverdykh kaustobiolitakh ugolnogo ryada nekotorykh oblasteï Chekhoslovakii. Min. Sb. Ľvovskogo geol. Obschch., 16, 334—342, Lvov.
BOUŠKA, V. KLIKA, Z. and PEŠEK, J. (1975): The geochemical role of boron in the sediments of the Carboniferous basins of Central Bohemia. Folia musei rerum naturalium Bohemiae occidentalis, Geologica, 5, 1—25. Západočeské muzeum, Plzeň.
BOUŠKA, V. and KREJČÍ, B. (1958): Zpráva o geochemickém sledování stopových prvků v uhelných slojích spodního sušského pásma za účelem využití výsledku k identifikačním účelům. Geofond, Unpublished., Praha.
— — (1959): Předběžná zpráva o geochemickém výzkumu stopových prvků v uhelných slojích spodního sušského pásma karvinského revíru. Zprávy o geol. výzkumech v r. 1959, 77—78. ÚÚG, Praha, 1960.
— — (1965): Geochemický výzkum stopových prvků v uhelných slojích sušského pásma v OKR. Acta Mus. Silesiae, A, 14, 57—64, Opava.

BOUŠKA, V. and NOVÁK, J. (1963): Cristobalit z limnokvarcitu od Sejkova. Acta Univ. Carol., Geologica, 3, 149—157, Praha.
BOUŠKA, V. and PEŠEK, J. (1976): The geochemical role of boron in the Carboniferous sediments of Czechoslovakia. Seventh Conference of Clay Mineralogy and Petrology, Karlovy Vary, 203—209, Přírodovědecká fakulta UK v Praze.
BOUŠKA, V., POVONDRA, P. and STREIBL, M. (1979): Distribuce boru a aminokyselin v mořských a sladkovodních horizontech Ostravsko-karvinského revíru. Čas. pro mineral. geol., 24, 1, 51—64, Praha.
BOWEN, H. J. M. (1966): Trace elements in biochemistry. Academic Press, New York.
BRADFORD, H. R. (1957): Fluorine in western coals. Mining Eng., 9, 78—79, New York.
BRANDENSTEIN, M., JANDA, I. and SCHROLL, E. (1960): Seltene Elemente in österreichischen Kohlen und Bitumengesteinen. Tcherm. Mineral. Petr. Mitt., 7, 260—285, Wien.
BREGER, I. A. (1951): Chemical and structural relationship of lignin to humic substances. Fuel, 30, 204— —208, London.
— (1958): Geochemistry of coal. Econ. Geol., 53, 823—841, Lancaster.
BREGER, I. A. and DEUL, M. (1956): The organic geochemistry of uranium. U. S. Geol. Survey. Prof. Paper, 300, 505—510, Washington.
BREGER, I. A., DEUL, M. and MEYROWITZ, R. (1955): Geochemistry and mineralogy of a uraniferous subbituminous coal. Econ. geol., 50, 610—624, Lancaster.
BREGER, I. A., DEUL, M. and RUBINSTEIN, S. (1955): Geochemistry and mineralogy of a uraniferous lignite. Econ. geol., 50, 206—226, Lancaster.
BREGER, I. A. and CHANDLER, I. C. (1960): Extractability of humic acid from coalified logs as a guide to temperatures in Colorado Plateau sediments. Econ. geol., 55, 1039—1047, Lancaster.
BREGER, I. A. and SCHOPF, J. M. (1955): Germanium and uranium in coalified wood from upper Devonian black shale. Geochim. Cosmochim. Acta, 7, 287—293, Oxford.
— — (1956): Geochemical studies of coalified wood from the Colorado Plateau. Abstract. Geol. Soc. Am. Bull., 67, pt. 2, p. 1675, New York.
BROOKS, C. E. P. (1950): Climate through the ages. 356 p, London.
BROOKS, J. D., GOULD, K. and SMITH, J. W. (1969): Isoprenoid hydrocarbons in coal and petroleum. Nature, 222, 257—259, London.
BROOKS, J. D. and SMITH, J. W. (1967): The diagenesis of plant lipids during the formation of coal, petroleum and natural gas — I. Changes in the n-paraffin hydrocarbons. Geochim. Cosmochim. Acta, 31, 2389—2397, Oxford.
BROOKS, R. R. and KAPLAN, I. R. (1972): Biogeochemistry. In R. W. Fairbridge (ed.): The Encyclopedia of Geochemistry and Environmental Sciences. Van Nostrand Reinhold Company, New York, p. 74—82.
BROWN, J. K. (1959): Infrared spectra of solvent extracts of coals. Fuel, 38, p. 55, London.
BROWN, J. K. and HIRSCH, P. B. (1955): Recent infra-red and X-ray studies of coal. Nature, 175, 229— —233, London.
BROWN, H. R. and SWAINE, D. J. (1964): Inorganic constituents of Australian coal. I. Nature and mode of occurrence. J. Inst. Fuel, 37, 422—440, London.
BRUS, Z. (1965): Výskyt auripigmentu v uhlí severočeské hnědouhelné pánve. Čas. Nár. muz., odd. přírodov., 134, 192—194, Praha.
BURKART, E. (1953): Moravské nerosty a jejich literatura. NČSAV, Praha.
BUTLER, J. R. (1953): Geochemical affinities of some coals from Svalbard. Kong. Ind., Handverk, Skipsfartdept. Norsk Polarinst., Skr. 96, 1—26, Oslo.
CAMERON, A. R. and LECLAIR, G. (1975): Extraction of uranium from aqueous solutions by coals of different rank and petrographic composition. Paper Geol. Surv., Canada, 74—35, 11 p.
CAMPBELL, D. D., JORY, L. T. and SAUNDERS, C. R. (1977): Geology of the Hat Creek coal deposits. Can. Mining Metall. Bull., 70, 99—108.

CANNON, H. L., SHACKLETTE, H. T. and BASTRON, H. (1968): Metal absorption by equisetum (Horsetail). U. S. Geol. Surv. Bull., 1278-A, Washington.

CARTZ, L., DIAMOND, R. and HIRSCH, P. B. (1956): New X-ray data on coals. Nature, 177, p. 500, London.

CASAGRANDE, D. J. and ERCHULL, L. D. (1976): Metals in Okefenokee peat-forming environments: relation to constituents found in coal. Geochim. Cosmochim. Acta, 40, 387–393, Oxford.

− − (1977): Metals in plants and waters in the Okefenokee swamp and their relationship to constituents found in coal. Geochim. Cosmochim. Acta, 41, 1391–1394., Oxford.

CASAGRANDE, D. J., SIEFERT, K., BERSCHINSKI, C. and SUTTON, N. (1977): Sulfur in peat-forming systems of the Okefenokee Swamp and Florida Everglades: origins of sulfur in coal. Geochim. Cosmochim. Acta, 41, 161–167, Oxford.

CHAKRABARTTY, S. K. and BERKOWITZ, N. (1976): Non-aromatic skeletal structures in coal. Fuel, 55, p. 362., London.

CHENERY, E. M. (1948): Aluminium in the plant world. Pt. 1, General survey in dicotyledons. Royal Botanical Gardens (Kew, England), Kew Bull., 2, 173–183.

CHINOVNIKOV, A. S., POGREBINSKAYA, M. L., FILIPPOVA, A. A. and KLEPCHINA, Z. B. (1964): The content of arsenic in solid fuels. Koks i khim., 4, p. 17, Moskva.

CHRISTOPH, H. J. (1963): Untersuchungen an den Kohlen des Döhlener Beckens mit besonderer Berücksichtigung der radioaktive Substanzen enthaltenden Kohlen. Dissertation, Mineral. Inst. d. Bergakademie, Freiberg.

CLARK, M. C. and SWAINE, D. J. (1962): Trace elements in coal. I. New South Wales coals. C.S.I.R.O. Div. Coal Res. Tech. Commun. 45, London.

COBB, J. C. and KULP, J. L. (1957): Age of the Swedish kolm. Abstract. Program 1957 Annual Meeting, p. 41., The Geol. Soc. Am., Atlantic City, N. Jersey, USA.

CODY, R. D. (1970): Anomalous boron content of two continental shales in eastern Colorado. Journ. Sedim. Petrology, 40, 750–754.

COLOMBO, U., GAZZARRINI, F., GONFIANTINI, R., KNEUPER, G. TEICHMÜLLER, M. and TEICHMÜLLER, R. (1970): Carbon isotope study on methane from German coal deposits. In Advances Org. Geoch. Proceedings of the Third Intern. Cong. Ed. G. D. Hobson and G. C. Speers. Pergamon Press. (In Earth Science. Ed. D. E. Ingerson, vol. 32.) 1–26, London.

CONNAN, J. (1974): Diagenèse naturelle et diagenèse artificielle de la matière organique à éléments végétaux prédominants. In Advances in Organic Geochemistry, ed. B. Tissot and F. Bienner, 73–96. Technip. Paris, 1973.

COOPER, M. (1955): Bibliography and index of literature on uranium and thorium and radioactive occurrences in the United States. Bull. Geol. Soc. Amer., 66, 257–326, New York.

COOPER, B. S. and MURCHISON, D. G. (1969): Organic geochemistry of coal. In Organic Geochemistry, editors G. Eglinton and M. T. J. Murphy, 699–726., Springer Verlag.

CORRENS, C. W. (1956): The geochemistry of the halogenes. In: Physics and Chemistry of the Earth, vol. 1, p. 181, Pergamon Press, London/New York.

CRAIG, H. (1953): The geochemistry of the stable carbon isotopes. Geochim. Cosmoch. Acta, 3, 53–92, Oxford.

− (1957a): The natural distribution of radiocarbon and the exchange time of CO_2 − atmosphere and sea. Tellus 9, 1–17, Stockholm.

− (1957b): Isotopic standards for carbon and oxygen and correction factors for mass-spectrometric analysis of carbon dioxide. Geochim. Cosmoch. Acta, 12, 133–149, Oxford.

CROSSLEY, H. F. (1946): Inorganic constituents of coal. Inst. of Fuel Wartime Bull., 57, London.

ČADEK, J., MACHÁČEK, V. and VÁCL, J. (1961): Zjištění obsahů germania a dalších stopových prvků ve spodní hnědouhelné sloji hrádecké části Žitavské pánve. Sborník ÚÚG, 28, odd. geol., 581–600, Praha.

ČADKOVÁ, Z. (1971a): Distribuce kovů v sedimentech radvanického souslojí. Časopis pro min. a geol., 16, 147–157, Praha.

- (1971b): Distribuce kovových prvků v permských sedimentech vnitrosudetské pánve. Sb. geol. věd, ř. LG, 14, 91–126, Praha.
ČECH, F. and PETRÍK, F. (1972): Klasifikačné zatriedenie a popis minerálnej prímesi v slojoch handlovsko-nováckej oblasti. Mineralia Slovaca, 4, 257–265, Spišská Nová Ves.
DALTON, I. M. and PRINGLE, W. J. S. (1962): The gallium content of some Midland coals. Fuel, 41, 41–48, London.
DANYUSHEVSKAYA, A. A. J. (1959): The chromatography of hydrocarbon found in the bitumen of the Sangarsk coal. Tr. Nauchn. Issled. Inst. Geol. Arktiki. Min. Geol. i Okhrany Nedr. SSSR 98, 120– –129; Chem. Abstr., 55, 9836g. Moskva.
DANNENBERG, A. (1915): Geologie der Steinkohlenlager. Berlin.
DAVIDSON, C. F. and PONSFORD, D. R. A. (1954): On the occurrence of uranium in coals. Mining Mag. (London), 91, 265–273.
DAVIES, M. M. and BLOXAM, T. W. (1974): The geochemistry of some South Wales coals. In The Upper Palaeozoic and post-Palaeozoic rocks of Wales, ed. by T. R. Owen. Cardiff (Univ. Wales Press), 225–261.
DEGENS, E. T. (1965): Geochemistry of sediments. A brief survey, Prentice-Hall, Inc., Englewood Cliffs. New Jersey.
DEGENS, E. T. and BAJOR, M. (1960): Die Verteilung von Aminosäuren... Glückauf, 96, 1525–1534, Essen.
– – (1962): Die Verteilung der Aminosäuren in limnischen und marinen Schiefertonen des Ruhrkarbons. Fortschr. Geol. Rheinld. u. Westf., 3, 429–440, Krefeld.
DEGENS, E. T. and MOPPER, K. (1975): Early diagenesis of organic matter in marine soils. Soil Sci., 119, 65–72.
DEKATÈS, Y. G. (1967): Tungsten occurrences in India and their genesis. Econ. Geol., 62, 556–561, Lancaster.
DENSON, N. M., BACHMAN, G. O. and ZELLER, H. D. (1967): Uranium-bearing lignite and its relation to the White River and Arikaree formations in northwestern South Dakota and adjacent States. U. S. Geol. Survey Bull., 1055-B, New York.
DENSON, N. M. and GILL, J. R. (1955): Uranium-bearing lignite and its relation to volcanic tuffs in eastern Montana and the Dakotas. Contr. to the Internat. Conf. on peaceful uses of atomic energy, Geneva, Switzerland, August 1955. U. S. Geol. Survey Paper, United Nations No. P/57, Washington.
DEUL, M. (1956): Concentration of minor elements in carbonaceous fractions mechanically segregated from some sedimentary rocks. Abstract. Program 1956 Annual Meetings, p. 41. The Geol. Soc. of Am., etc. Minneapolis, Minnesota, USA.
DEUL, M. and ANNELL, C. S. (1956): The occurrence of minor elements in ash of low-rank coal from Texas, Colorado, North Dakota and South Dakota. U.S. Geol. Survey Bull. 1036-H, 155–172, Washington.
DEVIS, D., DZHOVANELLI, DZH. and RIS, T. (1966): Biokhimiya rastenii. „Mir", Moskva.
DEWIS, F. J., LEVINSON, A. A. and BAYLISS, P. (1972): Hydrogeochemistry of the surface waters of the Mackenzie River drainage basin, Canada. IV. Boron-salinity clay mineralogy relationship in modern deltas. Geochim. Cosmochim. Acta, 36, 1359–1375, Oxford.
DIACONESCU, E. and DOMA, M. (1960): Relation between the geological age of coal and exchange capacity of ions derived from coal. Bull. Inst. politeh. Iasi, 6, 139–156, Iasi.
DILCHER, D. L. (1967): Chlorophyll in der Braunkohle des Geiseltales. Natur. Mus., 97, 124–130, Frankfurt a. M.
DILCHER, D. L., PAVLICK, R. J. and MITCHELL, J. (1970): Chlorophyll derivatives in Middle Eocene sediments. Science, 168, 1447–1449, Washington.
DOBROLYUBSKII, O. K. (1962): Ob uchastii soedinenii skandiya v okislitelno-vostanovitelnykh protsessakh proiskhodyashchikh v rasteniyakh. Dokl. AN SSSR, 144, 1174–1177, Moskva.

DOHNAL, Z., KUNST, M., MEJSTŘÍK, V., RANČINA, Š. and VYDRA, V. (1965): Československá rašeliniště a slatiniště. Academia, Praha.
DOPITA, M. (1955): Nález sideritizovaných kmenů ve slojích petřvaldské pánve (OKR). Přírodov. sb. Ostr. kraje, 16, 313–317, Opava.
DOPITA, M. and KRÁLÍK, J. (1977): Uhelné tonsteiny ostravsko-karvínského revíru. OKD Ostrava, 1–213.
DOUGLAS, A. G. and EGLINTON, G. (1966): in „Comparative Phytochemistry" ed. T. Swain, Academic Press, London, 57–77.
DRECHSLER, M. and STIEHL, G. (1977): Stickstoffisotopenvariationen in organischen Sedimenten. I. Untersuchungen an humosen Kohlen. Chem. Erde, 36, 126–138.
DUCK, N. W. and HIMUS, G. W. (1951): Arsenic in coal and its mode of occurrence. Fuel, 30, 267–271, London.
DUDYKINA, A. F. and SEMENOV, E. I. (1957): Lovozerskiĭ i Khibinskiĭ massivy – redkometaľnaya biogeokhimicheskaya provintsiya. Trudy Inst. Mineralog. Geokhim. i Kristallokhim. Redkikh Elementov, AN SSSR, 1, 35–37, Moskva.
DVORNIKOV, A. G. (1967): Some features of the mercury content of coal from the eastern Donbas (Rostov Region). Dokl. Akad. Nauk SSSR, 172, 199–202.
– (1971): Increased mercury concentrations in certain coal fractions. Coke Chem. USSR, 9, 4–5.
DYBEK, J. (1962): Zur Geochemie und Lagerstättenkunde des Urans. Clausthaler Heft zur Lagerstättenkunde und Geochemie, 1, Berlin.
EAGAR, R. M. C. and SPEARS, D. A. (1966): Boron content in relation to organic carbon and to paleosalinity in certain British upper carboniferous sediments. Nature, 209, 5019, 177–181. London.
ECHLIN, P. (1970): Primitive photosynthetic organisms. Advances in organic geochemistry. Proc. Third Intern. Congr., Ed. G. D. Hobson a G. C. Speers. Pergamon Press. (In: Earth Sciences. Ed. D. I. Ingerson, vol. 32.) 523–537, London/New York.
EGLINTON, G. (1968): Advances in Organic Geochemistry. Pergamon Press, London.
EGUTI (1955): Zapasy germania v burykh uglyakh v raïone Tohoku (Japan). Tohoku kenhkju, 5, 4, 26–27. Ref. zhurnal 1957, 11, 125, Moskva.
EKLUND, J. (1946): Urantillgänger och energiförsörjning. Kosmos 24, p. 74, Stockholm.
ENDERS, C. (1943): Wie entsteht der Humus in der Natur. Chemie, 56, 281–285, Berlin.
ERNST, W., KREJCI-GRAF, K. and WERNER, H. (1958): Parallelisierung von Leithorizonten im Ruhrkarbon mit Hilfe des Borgehaltes. Geochim. Cosmochim. Acta, 14, 211–222, Oxford.
ERNST, W. and WERNER, H. (1964): Anwendung der Bor-Methode in den geologischen Formationen zu ihrer besseren Unterteilung in wissenschaftlichem und praktischem Interesse sowie Untersuchungen über Bindung und Festlegung des Bors in natürlichen und künstlichen Sedimenten. Forsch.-Ber. Land Nordrhein-Westf., 1433, 1–27.
ESKENAZI, G. (1965): Redki zemi i skaniĭ v gagatnite vglishcha ot Plevensko. Godishn. Sof. univ., geol.--geogr. fak., t. 58, kn. I. Geologia, Sofia.
– (1974): Experimental study on the form of silver fixation in coal. Yearbook Univ. Sofia, Geol., 66, 279–284, (in Russian).
– (1977): On the binding form of tungsten in coal. Chem. Geol., 19, 153–159.
FEARON, W. R. (1933): A classification of the biological elements, with a note on the biochemistry of beryllium. Sci. Proc. Roy. Dublin Soc., 20, p. 531, Dublin.
– (1947): An introduction to biochemistry. New York.
FEIGEĽMAN, KH. E. (1949): Distribution of phosphorus in Fan'yabnobsk coal. Doklady Akad. Nauk USSR, 12, 30–33, Moskva.
FEIGEĽMAN, KH. E. and VOĬNALOVICH, M. V. (1955): Kharakter raspredeleniya i formy soedinenii fosfora v fan'yabnobskikh uglyakh. Tr. AN Tadzhik SSR, 41, 17–25. Dushanbe.
FERM, J. C. (1955): Radioactivity of coals and associated rocks in Beaver, Clearfield, and Jefferson Counties, Pa. U. S. Geol. Survey TEI - 468, 52 p., Washington.

FERRARI, B. (1948): Flözgleichstellung im Osten des Ruhrkohlenbezirkes auf Grund neuartiger Leitschichten. Glückauf, 81/84, 216—221, Essen.
FISCHER, F. and SCHRADER, H. (1921): The origin and chemical structure of coal. Brennstoff-Chemie, 2, 213—219, Essen.
FLAIG, W. (1966): Chemistry of humic substances in relation to coalification. In Coal Science, Am. Chem. Soc., Adv. Chem. Ser., 55, 58—68, Washington.
— (1968): Origin of nitrogen in coal. Chem. Geology, 3, p. 485.
FOLDVARI, A. (1952): The geochemistry of radioactive substances in Mecsek Mountains. Acta Geol. Acad. Sci Hungaricas, 1, 37—48, Budapest.
FORTESCUE, J. A. C. (1954): Germanium and other trace elements in some western Canadian coals. Am. Mineralogist, 39, 510—519, Washington.
FRANCIS, W. (1961): Coal, 2nd ed. E. Arnold Ltd., London, 567 s.
FRASER, G. K. (1955): In; „Chemistry of the Soil", Bear, F. E. et al., 149—176. Reinhold, New York.
FREDERICKSON, A. F. and REYNOLDS, R. C. (1960): Geochemical method for determining paleosalinity. Clays and clay minerals. Proc. Nat. Conf. Clays and Clay Minerals, 8 (1959), 201—213., Washington.
FRIEDEL, R. A. and QUEISER, J. A. (1959): Ultraviolet-visible spectrum and the aromaticity of coal. Fuel, 38, p. 369., London.
FRIEDMAN, S., KAUFMAN, M. L., STEINER, W. A. and WENDER, I. (1961): Determination of hydroxyl content of vitrains by formation of trimethylsilyl ethers. Fuel, 40, 33—46, London.
FUCHS, W. (1931): Die Chemie der Kohle, Berlin.
— (1935): Rare elements in German brown-coal ashes. Ind. and Eng. Chem., 27, 1099—1100, Easton.
— (1942): Thermodynamics and coal formation. Am. Inst. Min. Met. Eng. Trans., 149, 218—223, New York.
GEILMANN, W. and BRÜNGER, K. (1935): Über die Aufnahme von Germanium durch Pflanzen. Biochem. Z., 275, 387—395, Berlin.
GEISSLER, CH. and BELAU, L. (1971): Zum Verhalten der stabilen Kohlenstoffisotope bei der Inkohlung. Z. angew. Geol., 17, 1—2, 13—17, Berlin.
GHIGHI, E. and FABBRI, G. (1965): —. Atti Accad. Sci Ist. Bologna, 12, 1. Cited in „Organic Geochemistry" ed. G. Eglinton a M. T. J. Murphy. Springer-Verlag, Heidelberg, 1969.
GIBSON, F. H. and SELVIG, W. A. (1944): Rare and uncommon chemical elements in coal. U. S. Bureau of Mines, Tech. P. 669, Washington.
GILL, J. R. (1953): Parts of Colorado, Wyoming and Montana; in: Search for and geology of radioactive deposits. Semiannual progress report. U. S. Geol. Survey TEI - 330, p. 118., Washington.
— (1954a): Ekalaka lignite field, Carter Country, Montana; in Geologic investigations of radioactive deposits. Semiannual progress report. U. S. Geol. Survey TEI - 440, 109—112, Washington.
— (1954b): Mendenhall Area, Slim Buttes, Harding County, South Dakota, in: Geologic investigations of radioactive deposits. Semiannual progress report. U. S. Geol. Survey TEI - 440, 113—117, Washington.
— (1954c): Northwestern South Dakota, southwestern North Dakota, and eastern Montana; in: Geologic investigations of radioactive deposits. Semiannual progress report. U. S. Geol. Survey TEI - 490, 149—155, Washington.
— (1957): Reconnaissance for uranium-bearing lignite in the Ekalaka lignite field, Carter County, Montana. U. S. Geol. Survey Bull. 1055-F, Washington.
GILL, J. R. and ZELLER, H. D. (1957): Results of core drilling for uranium-bearing lignite, Mendenhall area, Harding County, South Dakota. U. S. Geol. Survey Bull. 1055-D, Washington.
GILLET, A. (1948): The molecule of coal. Bull. soc. chim. belges, 57, 298—306, Bruxelles.
— (1955): The evolution of humins from woody materials: cellulose to anthracite. Brennstoff-Chemie, 36, 103—120, Essen.
— (1956): From cellulose to anthracite. The development of humins in ligneous matter. Brennst. Chemie, 37, 395—402, Essen.

GILLILAND, M. R., HOWARD, A. J. and HAMMER, D. (1960): Polycyclic hydrocarbons in crude peat wax. Chem. Ind. (London), 1357–1358.

GIPSH, A. A., KAPATURIN, G. G. and YUDOWICH, YA. E. (1971): Nekotorye voprosy raspredeleniya i genezisa radioaktivnosti v uglyakh Pechorskogo basseĭna. Izv. vyssh. uchebn. zavedeniĭ. Geol. i razvedka, 6, 61–70, Moskva.

GIVEN, P. H. (1960): The distribution of hydrogen in coals and its relation to coal structure. Fuel, 39, p. 147, London.

– (1962): Chemicals from coal. New Scientist, 14, 355–357, London.

– (1964): The chemical study of coal macerals. In Advances in Geochemistry, ed. U. Colombo a G. D. Hobson. Macmillan Comp., New York, 39–48.

GIVEN, P. H. and PEOVA, M. E. (1960): Investigation of CO-group in solvent extracts of coals. J. Chem. Soc., 1960, 394–400, London.

GLUSKOTER, H. J. and HOPKINS, M. E. (1970): Distribution of sulphur in Illinois coals. Ill. State Geol. Surv. Guideb., 8, 89–95, Washington.

GLUSKOTER, H. J. and Lindahl, P. C. (1973): Cadmium: mode of occurrence in Illinois. Coal Sci., 181, 264–266.

GOLDSCHMIDT, V. M. (1930): Über das Vorkommen des Germaniums in Steinkohlen und Steinkohlen--Produkten. Nachr. Ges. Wiss. Göttingen, Math.-Phys. Kl., 3, 398–401.

– (1935): Rare elements in coal ashes. Ind. Eng. Chem., 27, 1100–1102, Easton.

– (1937): The principles of distribution of chemical elements in minerals and rocks. Journ. Chem. Soc., London, 655–673.

– (1944): The occurrence of rare elements in coal ashes. Coal Research. Scient. and Techn. Reports of BCURA, London.

– (1950): Occurrence of rare elements in coal ashes. Progress in coal science. Interscience, New York, 238–247.

– (1954): Geochemistry. Oxford, 730 s.

GOLDSCHMIDT, V. M. and PETERS, C. (1932a): Zur Geochemie des Germaniums. Nachr. Ges. Wiss. Göttingen, Math.-Phys. Kl., 141–166.

– – (1932b): Zur Geochemie des Berylliums. Nachr. Ges. Wiss. Göttingen, Math.-Phys. Kl., 360– –376.

– – (1932c): Zur Geochemie des Bors. Nachr. Ges. Wiss. Göttingen, Math.-Phys. Kl., I, 402–407, II. 528–545.

– – (1933): Über die Anreicherung seltener Elemente in Steinkohlen. Nachr. Ges. Wiss. Göttingen, Math.-Phys. Kl., 4, 371–386.

GOLDSCHMIDT, V. M. and STROCK, L. (1935): Zur Geochemie des Selens. II. Nachr. Ges. Wiss. Göttingen, Math.-Phys. Kl., IV. N. F. 1, 11, 123–142.

GOLOVKO, V. A. (1960): K raspredeleniyu malykh elementov v kamennougoľnykh otlozheniyakh centraľnykh oblasteĭ. Dokl. AN SSSR, 132, 4, 911–914, Moskva.

GOLUBEV, V. S. and GARIBYANTS, A. A. (1968): Geterogennye processy geokhimicheskoĭ migratsii. Gosgeoltekh, Moskva.

GORDON, M. (1952): Trace elements in peat. Torfnachrichten, 3, 12; Chem. Abstr., 47, 4533i (1953).

GORDON, S. A., VOLKOV, K. YU. and MENKOVSKII, M. A. (1958): O formakh soderzhaniya germaniya v ugle. Geokhimiya, AN SSSR, 4, 384–388, Moskva.

GORHAM, H. R. (1949): Nitrates and nitrogenous compounds. In: Industrial Minerals nad Rocks (ed. S. H. Dolbear a O. Bowles), p. 643, New York.

GOROKHOVA, V. N. and POKROVSKAYA, V. L. (1962): Genetic types of rhenium deposits. Reñii. AN SSSR, Inst. Met., TR. 2-go Vses. Soveshch. Moskva.

GOTT, G. B., WYANT, D. G. and BERONE, E. P. (1952): Uranium in black shales, lignites and limestones in the United States. Selected papers on uranium deposits in the United States. U. S. Geological Survey, Circ. 220, Washington, 31–35.

GÖTTLICH, K. (ed.) (1975): Moor- und Torfkunde. E. Schweizerbart'sche Verlagsbuchhandlung, Stuttgart, 1–269.
GOUCH, L. J. and MILLS, J. S. (1972): Composition of succinite (Baltic amber). Nature, 239, 527–528., London.
GROPP, W. and BODE, H. (1932): Über die Metamorphose der Kohlen und das Problem der künstlichen Inkohlung. Braunkohle, 1, 6–17, Halle.
GUDE, A. J. and MCKEOWN, F. A. (1953): Results of exploration at the Old Leyden coal mine, Jefferson County, Colorado. U. S. Geol. Survey TEM-292, 14p., Washington.
GUILD, F. N. (1922): The occurrence of terpin hydrate in nature. J. Am. Chem. Soc., 44, p. 216, Washington.
GULYAEVA, L. A. and ITKINA, E. S. (1962): Galogenidy V. Ni i Cu v ugle. Geokhimiya, 4, 345–355, Moskva.
– – (1974): Mikroelementy ugleĭ, goryuchikh slantsev i ikh bituminoznykh komponentov. „Nauka". Moskva.
GUREVICH, A. E. (1969): Processy migratsii podzemnykh vod, nefteĭ i gazov. Leningrad, 109 p.
GÜNTHER, L. (1966): Obnovené procesy prouhelnění v některých evropských kamenouhelných pánvích. Uhlí, 8, 225–227, Praha.
HADAČ, E. (1953): Československé peloidy. St. zdrav. naklad., 246 p., Praha.
HADZI, D. and NOVAK, A. (1954): The heat of wetting of sulfurous coals of Istria and of some other Yugoslav coals. Věstnik Slovin. Kem. Drustva, 1, 175–184, Beograd.
HAIL. W. J. and GILL. J. R. (1953): Results of reconnaissance for uraniferous coal, lignite, and carbonaceous shale in western Montana. U. S. Geol. Surv. Circ., 251, p. 9, Washington.
HAK, J. and BABČAN, J. (1967): A contribution to the geochemistry of germanium and beryllium in coals of the Sokolov Basin. Geochemie v Československu. VŠB, Ostrava, 163–166.
HALLAM, A. and PAYNE, K. W. (1958): Germanium enrichment in lignites from the Lower Lias of Dorset. Nature 181, No. 4614, p. 1008., London.
HAVLENA, V. (1959): Typy usazenin z řady uhlí – jílovitá hornina. Návrh makro a mikropetrografické klasifikace. Acta Univ. Carol. – Geologica, 1–2, 111–123, Praha.
– (1963): Geologie uhelných ložisek 1. NČSAV, Praha, 342 p.
– (1964): Geologie uhelných ložisek 2. Academia, Praha, 437 p.
– (1965): Geologie uhelných ložisek 3. Academia, Praha, 382 p.
– (1974): Principy ložiskové geologie kaustobiolitů. III. postgraduální studium ložiskové geologie radioaktivních surovin. Přírodovědecká fakulta Univerzity Karlovy, 28–67, Praha.
HAWLEY, J. E. (1955): Germanium content of some Nova Scotia coals. Econ. Geol., 50, 517–532, Lancaster.
HEADLEE, A. J. W. (1953): Germanium and other elements in coal and the possibility of their recovery. Mining Engin., 6, 1011–1014, New York.
HEADLEE, A. J. W. and HUNTER, R. G. (1951): Germanium in coals of West Virginia. West Virginia Geol. Econ. Survey Rept. Inv., 8, 1–15, Morgantown.
– – (1953): Elements in coal ash and their industrial significance. Ind. Eng. Chem., 45, 548–551, Easton.
– – (1955): Characteristics of minable coals of West Virginia. P. 5. West Va. Geol. Econ. Surv. Bull., 13-A, p. 36, Morgantown.
HILT, C. (1873): Les rapports entre la composition des charbons et leurs propriétés industrielles. Ann. Ass. Ing. Liège, 387.
HIRSCH, P. B. (1954): X-ray scattering from coals. Proc. Roy. Soc. London, A 226, 143–169.
HLAVICA, B. (1924): Složení popelu z uhlí nýřanského. Hornický věstník (Paliva a topení), 6 (25), 137––139, Praha.
HOEFS, J. (1973): Stable isotope geochemistry. Springer-Verlag Berlin, 140 p.

HOEHNE, K. (1954): Zur Neubildung von Quarz in Kohlenflözen. N. Jb. Geol. Pal., Abh. 99, 209–222, Stuttgart.
HOFFMANN, J. (1943): Uran in Kohlen und Torf. Chem. d. Erde, 15, 277–282, Jena.
HOKR, Z. (1971a): Geneze síry v uhlí v severočeské hnědouhelné pánvi. Report on theme T-0-20-28/4/09. Archív ÚÚG, Praha.
— (1971b): Ke genezi síry v severočeské uhelné pánvi. Geol. průzk., 325–327, 351–352, Praha.
— (1975a): Změny složení popele v závislosti na vývoji sloje v severočeské hnědouhelné pánvi. Sb. II. uhel. geol. konference katedry ložiskové geologie přírod. fak. UK v Praze, 57–60, Praha.
— (1975b): Vznik a procesy hromadění síry v uhlí severočeské hnědouhelné pánve. Sb. geol. věd, LGM, 17, 95–123, Praha.
— (1977): Arsenic in the coal of the North Bohemian Brown Coal Basin. Věstník ÚÚG, 52, 267–273, Praha.
HOOD, A., GUTJAHR, C. C. M. and HEACOCK, R. L. (1975): Organic metamorphism and the generation of petroleum. Bull. Amer. Assoc. Petrol. Geol., 59, 986–996.
HRDÝ, V. (1961): Germanium v uhlí kounovské sloje. Geol. průzk., 3, 353–354, Praha.
HUBÁČEK, J. (1939): Jihočeský antracit a jeho popeloviny. Uhlí, 19, 7–9, 52–56, Praha.
— (1948): Tuhá paliva Československé republiky. Nákladem Matice hornicko-hutnické, 138 p., Praha.
— (1964): Pasportizace a klasifikace hnědých uhlí v ČSSR a jejich popelů. Práce ÚVP, 9, Praha.
HUBÁČEK, J., KESSLER, M. F., LUDMILA, J. and TEJNICKÝ, B. (1962): Chemie uhlí. SNTL, Praha a SVTL, Bratislava, 469 p.
HUBÁČEK, J. and LUSTIGOVÁ, H. (1962): Huminové kyseliny v československých hnědých uhlí a jejich určení. Práce Úst. výzk. paliv, 4, 357–380, Praha.
HUCK, G. and KARWEIL, J. (1955): Physikalisch-chemische Probleme der Inkohlung. Brennstoff-Chemie, 36, 1–11.
HURNÍK, S. (1972): Koeficient sednutí některých sedimentů v Severočeské hnědouhelné pánvi. Čas pro mineral. geol., 17, 365–372, Praha.
HUSZKA, L. and LÁDA, A. (1956): Köszentelepek azonosítása vitrithamu szilikátelemzése alapján. Magyar állami földtany inl. évk., 45, p. 95, Budapest.
HUTCHINSON, G. E. (1945): Aluminium in soils, plants and animals. Soil Sci., 60, 29–40, Baltimore.
IKAN, R. and MCLEAN, J. (1960): Triterpenoids from lignite. J. Chem. Soc., 1960, 893–894, London.
INAGAKI, M. (1957): J. Coal. Res. Inst. Tokyo, 8, 4, 97–108.
— (1958): Germanium in coal. J. Mining Inst. Kiushiu, 23, 555–563.
INAGAKI, M. et al. (1956): J. Coal. Res. Inst. Tokyo, 7, 193–200; 11, 1–8. Ref. zhurnal 1957, 11, p. 123, Moskva.
INAGAKI, M. and YAMAGUCHI, T. (1958): Gallium in Japanese coals. Tankem, 9, p. 161, Chem. Abstr., 53, 684 (1959), Easton.
ITKINA, E. S. (1955): Rasprostranenie ioda i broma v otlozheniyakh uglenosnogo gorizonta Saratovskogo Povolzhya. Doklady AN SSSR, 101, 521–523, Moskva.
IVANOV, G. A. (ed.) (1975): Metamorphizm uglei i epigenez vmeshchayushchikh porod. „Nedra", 1–256, Moskva.
JECZALIK, A. (1970): Geochemia uranu w uranonósnych węglach kamiennych w Polsce. Biul. Inst. geol., 224, 104–204, Warszawa.
JEDWAB, J. (1964): Coal as a source of beryllium. NASA Accession No. N 65-21075, Rept. No. EUR-2106f, p. 51.
JEFFERY, M., COMPSTON, W., GREENHALGH, D. and DELAETER, J. (1955): The carbon 13 abundance of limestones and coals. Geochim. Cosmochim. Acta, 7, 255–286, London.
JEGOROV, A. I. (1948): Rifeiskie uglistye tolshchi Kazakhstana. Bjulleten Mosk. isp. prirody, otd. geologii, T. 23 (6), 23–32, Moskva.
JEGOROV, A. I. and KALININ, S. K. (1940): Distribution of germanium in the coals of Kazakhstan. Doklady AN SSSR, 26, 925–926, Moskva.

— — (1941): Verteilung von Germanium in den Kohlen von Kazakhstan. Chem. Zentralbl., 1, p. 1405, Berlin, Leipzig.
JERSHOV, V. M. (1961): Rare earth elements in the coals of the Kazelovskii coal basin. Geochemistry (USSR), English Transl., 306—308, Moskva.
JONES, J. H. and MILLER, J. M. (1939): The occurrence of titanium and nickel in the ash from some special coals. Chem. Ind. 58, 237—245, London.
JONES, J. M., MURCHISON, D. G. and SALEH, S. A. (1973): Reflectivity and anisotropy of vitrinites in some coal seams from the Coal Measures of Northumberland. Proc. Yorks. Geol. Soc., 39, 515—526.
JORSCIK, I., UPOR, E., HOHMANN, E. and JUHASS, S. (1963): Fixation of radioactive elements on coal and the nature of the bond between uranium and humic acid. Acta chim. hung., 35 (2), 225—232, Budapest.
JURAIN, G. (1968): Les schistes uranifères d'âge westphalien du versant oriental des Vosges. Université de Nançy. Faculté des Sciences, Nançy.
KADAŇKOVÁ-VOTAVOVÁ, Z. (1960): Distribuce Ge v uhelných pánvích ČSSR. II. Germanium v pánvi rakovnické. III. Germanium v pánvi rosicko-oslavanské. Sborník VŠCHT v Praze, odd. fak. anorg. a org. technologie, 4, 1, 475—484, 485—491, Praha.
KAGANOVICH, F. L. and RAKUSKIĬ, V. E. (1958): Selective extraction of peat bitumens of low temperature. Vest. AN Belorus. SSR, Ser. Fiz. Techn. Nauk, 117—122. Chem. Abstr., 57, 3721c, Easton.
KARAVAEV, M. N., RUMYANTSEVA, Z. A. and POLYANSKAYA, V. S. (1965): Alcohol-benzene and alcohol extracts of fusinated brown coals. Izv. Akad. Nauk Tadž. SSR. Otd. Fiz-Techn. i Chim. Nauk, 1, 79—88. Chem. Abstr., 64, 10984e, Easton.
KARWEIL, J. (1973): The determination of paleotemperatures from the optical reflectance of coal particles in sediments. Colloque International: Pétrographie de la Matière Organique des Sédiments. Relations avec la Paléotemperature et le Potentiel Pétrolier. C. N. R. S., Septembre 1973, 195—203, Paris.
KASATOCHKIN, V. I. (1969): Problema molekuľarnogo stroeniya i strukturnaya khimiya prirodnykh ugleĭ. Khim. tverd. topl., 4, p. 33, Moskva.
KAŠPAR, J. (1939): Mineralogie kladenských uhelných slojí. Knihovna SGÚ Československé republiky, 20, Praha.
KAŠPAR, J., HUDEC, I., SCHILLER, P., COOK, G. B., KITZINGER, A. and WÖLFL, E. (1972): A contribution to the migration of gold in the biosphere of the humid mild zone. Chem. Geol., 10, 299—305, Amsterdam.
KATCHENKOV, S. M. (1952): Korrelyatsiya nizhnepermskikh otlozheniĭ po khimicheskim elementam, opredelyaemym metodom spektraľnogo analiza. Dokl. AN SSSR, 82, 961—964, Moskva.
KATCHENKOV, S. M. and FLEGONTOVA, E. I. (1955a): Malye elementy v devonskikh otlozheniyakh Volgo-Uraľskoĭ oblasti po dannym spektraľnogo analiza. R. Vses. nef. n.-i. geol.-razved. instituta, 83, 466—505, Leningrad.
— — (1955b): Raspredelenie elementov v osadochnykh porodakh mezozoya Severovostochnogo Kavkaza po dannym spektraľnogo analiza. Geol. Sb., 3, 90—98, Ľvov.
KEITH, M. L. and LEGENS, E. T. (1959): Geochemical indicators of marine and freshwater sediments. — In: P. H. ABELSON (editor): Researches in Geochemistry, 38—51, New York.
KENH, T. M. (1957): Selected annotated bibliography of the geology of uranium-bearing coal and carbonaceous shale in the United States. U. S. Atomic Energy Comm., U. S. Geol. Survey Bull. 1059-A, Washington.
KELLER, W. D. (1970): Environmental aspects of clay minerals. Journ. Sed. Petrology, 40, 788—854, Tulsa.
KERVELLA, F. (1958): Les gisements uranifères dans les formations sédimentaires en France et dans ľUnion Française. Rapport CEA, 911, Saclay.
KESSLER, M. F. (1974): Příspěvek k vyjádření stupně prouhelnění kaustobiolitů. Uhlí, 22, 153—156, Praha.

— (1972): Chemické složení macerálů ostravsko-karvínského revíru. Uhlí, 20, 507 – 511, Praha.
KESSLER, M. F. and DOČKALOVÁ, L. (1955): Stanovení sodíku a draslíku v uhelných popelech. Paliva, 35, 206 – 212, Praha.
— — (1960): Stanovení malých množství lithia v uhelných popelech. Geol. průzkum, 8, 244 – 246, Praha.
KESSLER, M. F., MALÁN, O. and VALEŠKA, F. (1965): Význam stopových a minoritních prvků pro korelaci a identifikaci slojí a komplexní využití uhlí. Rozpravy ČSAV, řada mat.-přír. věd, 75, 10, 1 – 123, Praha.
KESSLER, M. F., MALÁN, O. and VALEŠKA, F. (1967): Beziehungen der Alkalimetalle zur Stratigraphie und Flözidentifizierung der paralischen Kohlenbecken. Glückauf-Forschungshefte, 28, 149 – 154, Essen.
KESSLER, M. F. and VALEŠKA, F. (1957): Spektrografické stanovení křemíku, železa a hliníku v černém uhlí. Paliva, 37, 158 – 160, Praha.
KESSLER, J. V. and VEČEŘÍKOVÁ, V. (1954): Roentgenografický průzkum černých uhlí a studie jejich mikrostruktury. Paliva, 34, 97 – 105, Praha.
— — (1960): Roentgenometrická analysa grafitu, uhlíkatých materiálů a uhlí. Čes. čas. pro fysiku, sekce A, 10, 125 – 135, Praha.
KHALDNA, YU. L. (1958): O soderzhanii vanadiya v goryuchikh slantsakh Estonskoï SSR. Uch. Zap. Tartusk. Inst. 62, 219 – 225, Tallin.
KLETEČKA, A. (1933): Zajímavý vztah rašeliníku k obsahu Ca v půdě. Věda přírodní, 14, 199 – 205, Praha.
KNOCHE, H., ALBRECHT, P. and OURISSON, G. (1968): Organische Verbindungen in fossilen Pflanzen (Voltzia brongniarti, Coniferales). Angew. Chem., 80, 666 – 667, Weinheim.
KNOCHE, H. and OURISSON, G. (1967): Organische Verbindungen in fossilen Pflanzen (Schachtelhalm). Angew. Chem., 79, 1107 – 1108, Weinheim.
KOBLIC, J. (1950): Antracit ze Lhotic u Českých Budějovic. Věstník SGÚ ČSR, 25, 38 – 48, Praha.
KOCHETKOV, O. C. (1966): Koncentraciya malykh i redkikh metallov v biolitakh iz devonskikh otlozheniï Srednego, Severnogo Timana i poluostrova Kanin. Izv. AN SSSR, ser. geol., 3, 25 – 35, Moskva.
KOCHLOEFL, K., SCHNEIDER, P., ŘEŘICHA, R. and BAŽANT, V. (1963): Untersuchung über die Zusammensetzung der Braunkohlenteerfraktion vom SPD. Coll. Czech. Chem. Comm., 28, 3362 – 3381, Praha.
KOROLEV, D. F. (1957): Nekotorye osobennosti raspredeleniya molibdena v porodakh Bylymskogo ugoľnogo mestorozhdeniya (Severnyï Kavkaz). Geokhimiya, 5, 420 – 424, Moskva.
KOVÁCS, F., KRÁLÍK, J. and POLICKÝ, J. (1960): Příspěvek k výzkumu stopových prvků mostecké pánve v severočeském uhelném revíru. Sb. věd. prací VŠB, 6, 98 – 116, Ostrava.
KOVALEV, V. A. and GENERALOVA, V. A. (1969): Geochemical aspects of the movement of iron in recent peat bogs of Belorussia. Geokhimiya, 2, 210 – 220.
KOZENOV, A. B., ZINOV'EV, B. B. and KOVALEVA, S. A. (1965): Nekotorye osobennosti processa nakopleniya urana v torfyanikakh. Geokhimiya, 1., Moskva.
KRÁLÍK, J. (1959): Tonsteiny v ostravských vrstvách OKR. Sb. prací konf. o geologii OKR. VŠB v Ostravě, 103 – 107, Ostrava.
KRATOCHVÍL, J. (1939): Kladensko po stránce mineralogické a petrografické. Báňský svět, 11, 85 – 89, Praha.
— (1937, 1943, 1948, 1957, 1958): Topografická mineralogie Čech, I – IV a V. Dodatky. Archív pro přírod. výzkum Čech. Praha, I – II (992 p.), III – IV (993 – 2014 p.), V (646 p.).
KRAUSKOPF, K. B. (1956): Factors controlling the concentrations of thirteen rare metals in seawater. Geochim. Cosmochim. Acta, 9, 1 – 32 B, Oxford.
— (1967): Introduction to geochemistry. McGraw-Hill, New York, 721 p.
KREILEN, B. I. (1938): Osnovnye napravleniya khimii i sistematiki ugleï, 1 – 119, Charkov.
KREJCI-GRAF, K. (1972): Trace metals in sediments, oils, and allied substances. In The Encyclopedia

of Geochemistry and Environmental Sciences. Ed. R. W. Fairbridge. Van Nostrand Reinhold Com. New York, 1201 – 1209.

KREULEN, D. J. W. (1928): Über die durchšchnittliche quantitative Zusammensetzung von Ruhrkohlenaschen. Brennstoff-Chemie, 9, p. 399.
— (1935): Grundzüge der Chemie und Systematik der Kohlen. Centen's Uitgevers Maatschappij, Amsterdam, 179 p.
— (1938): Elements of coal chemistry. Nijh and van Ditmar, Rotterdam.
— (1952): Sulfur coal of Istria. Fuel, 31, 462 – 467, London.

KREULEN, D. J. W. and KREULEN-VAN SELINS, F. J. (1956): Humic acids and their role in the formation of coal. Brennstoff-Chemie, 37, 14 – 19.

KŘÍBEK, B. (1978) — personal communication.

KRÖGER, C. and BÜRGER, H. (1959): Die physikalisch-chemischen Eigenschaften der Steinkohlengefügebestandteile. X. Autoxydation und chemische Konstitution. Chem., 40, 76 – 85, Essen.

KRUŤA, T. (1949): O nerostech, nerostných surovinách a předních nalezištích ve Slezsku. Přír. sb. Ostrav. kraje, X/4, 281 – 319, Opava.
— (1951): O nerostech z ostravsko-karvinského revíru. Přír. sb. Ostrav. kraje, 12, 451 – 486, Opava.
— (1952): O nerostech z ostravsko-karvinského revíru. Přír. sb. Ostrav. kraje, 26, 402 – 408, Opava.

KUDĚLÁSEK, VL. (1959): Stopové prvky uhlí Dolnoslezské pánve, I, II. Sb. věd. prací VŠB v Ostravě, 5, 319 – 348; 457 – 481, Ostrava.
— (1960): Stopové prvky Dolnoslezské pánve, III. Sb. věd. prací VŠB v Ostravě, 6, 987 – 1013, Ostrava.

KUKHARENKO, T. A. (1959): Changes in structure and properties of humic acids during coalification. Genezis Tverd. Goryuch. Iskopaem. AN SSSR, Inst. Goryuch. Iskop., 319 – 337, Moskva.
— (1970): Khimicheskie aspekty preobrazovaniya organicheskogo veshchestva gumitov v processe litogeneza. Sb. uglenosn. formacii i ich genezis. Moskva, 103 – 108.

KUKHARENKO, T. A. and EKATERININA, L. N. (1960): The humic acid of weathered coals. Tr. Inst. Goryuch. Iskop. AN SSSR, 14, 58 – 72, Moskva.

KUHL, J. (1959): Chemiczno-mineralna budowa nieorganicznej substancji mineralnej w węglu brunatnym z Konina. I.: Badania popiołów. Kwartalnik geol., Warszawa, 3, 751 – 766.

KUNSTMANN, F. H. and BODENSTEIN, L. B. (1961): Arsenic content of South African coals. J. S. African Inst. Mining Met. 62, p. 234, Johannesburg.

KURENDOVÁ, J. (1967): Methods of investigation of the binding of some rare and trace elements in coal. Geochemie v Československu, 193 – 197, VŠB Ostrava.

KUTINA, J. (1956): O khimisme sfalerita v pustotakh sferosiderita iz Gnidousa bliz Kladna. Miner. sb. Ľvovskogo geol. obshch., 10, 165 – 170, Ľvov.

KUZNETSOVA, V. V. and SAUKOV, A. A. (1961): Possible forms of occurrence of molybdenum and rhenium in coals of Central Asia. Geokhimiya, 750 – 756, Moskva.

KWIATKOVSKI, A. (1965): Composition of resin components of peat bitumens. Valtion Tek. Tutkimuslaitos, Julkaisu, 93, p. 37; Chem. Abstr., 64, 13973e, Easton.

KYSHTYMOVA, L. T. (1976): Raspredelenie bariya i strontsiya v uglenosnoï tolshchi Sb. Materialy po geol. i polezn. iskopaemykh Sev.-Vost. Evrop. chasti SSSR, 8, Syktyvkar, 390 – 395.

LAZAROV, L. and ANGELOVA, G. (1976): Struktura kamennykh ugleĭ. Khim. tverd. topl., 3, p. 15, Moscow.

LEPKA, F. (1967): Distribuce uranu v sedimentech a podzemních vodách kladensko-rakovnické pánve. Geochemie v Československu, 199 – 208, VŠB Ostrava.

LESSING, R. (1922): The study of mineral matter in coal. Fuel, 1, 6 – 10, London.
— (1926): Coal ash and clean coal. Fuel, 5, p. 17, 65 and 117, London.

LEUTWEIN, F. (1956): Untersuchungen über das Vorkommen von Spurenmetallen in Torfen und Braunkohlen. Freiberger Forsch., C 30, 28 – 48, Freiberg.
— (1966): Geochemische Charakteristica mar:ner Einflüsse in Torfen und anderen quaternären Sedimenten. Geol. Rundschau, 55, 97 – 112, Stuttgart.

LEUTWEIN, F. and RÖSLER, H. J. (1956): Geochemische Untersuchungen an paläozoischen und mesozoischen Kohlen Mittel- und Ostdeutschlands. Freib. Forsch., C 19, 1 – 196, Freiberg.
LEYTHAEUSER, D. and WELTE, D. H. (1969): Relation between distribution of heavy n-paraffins and coalification in carboniferous coals from the Saar District, Germany. In: Advances in Organic Geochemistry 1968, eds. P. A. Schenck and I. Havenaar, 429 – 442. Pergamon Press, London.
LINSTOW, V. O. (1929): Bodenanzeigende Pflanzen. Abh. d. Preuss. Geol. Landesanstalt, N. F., 114, 2. ed., Berlin.
LLOYD, S. J. and CUNNINGHAM, J. (1933): The radium content of some Alabama coals. Am. Chem. Journ., 50, 47 – 51, Easton.
LOMASHOV, I. P. (1961): O nekotorykh zakonomernostyakh raspredeleniya germaniya v uglenosnykh otlozheniyakh. Dokl. AN SSSR, 137, 3, 692 – 694, Moskva.
LOPEZ DE AZCONA, J. M. and PIUG, A. C. (1948): Investigation of minor elements in the ash of Austrian coals. Inst. geol. min. España, 60, p. 393, Madrid.
LOTSPEICH, F. B. and MARKWARD, E. L. (1963): Minor elements in bedrock, soil, and vegetation at an outcrop of the Phosphoria Formation on Snowdrift Mountain, southeastern Idaho, U. S. Geol. Surv. Bull., 1181-F, Fl, Washington.
LOUNAMAA, J. (1956): Trace elements in plants growing wild on different rocks in Finland. Ann. Botanici Soc. Zool., Bot. Fennicae „Vanamo", 29, 4, 1 – 196, Helsinki.
LOVERING, T. S. (1959): Significance of accumulator plants in rock weathering. Bull. Geol. Soc. Amer., 70, 781 – 800, New York.
LUKASHEV, K. J., KOVALEV, V. A., ZHUKCHOVITSKAYA, L. A., KHOMICH, A. A. and GENERALOVA, V. A. (1971): Geokhimiya ozerno-bolotnogo litogeneza. Izd. „Nauka i tekhnika", Minsk 1971.
MACKOWSKY, M. T. (1945): Mikroskopische Untersuchungen über die anorganischen Bestandteile in der Kohle und ihre Bedeutung für Kohleaufbereitung und Kohleveredelung. Arch. Bergbau, Forschung, 4, 1, Essen.
– (1951): Inkohlung und Chemie der Kohle. Glückauf, 87, 551 – 557, Essen.
MÄGDEFRAU, K. (1953): Paleobiologie der Pflanzen. Jena, 438.
MACHÁČEK, V. (1957): Geochemický výzkum stopových prvků v ostravsko-karvínském revíru. Zprávy o geol. výzkumech v r. 1956, 107 – 108, Praha.
MACHÁČEK, V., ČADEK, J. and HÁJKOVÁ, J. (1955): Orientační geochemický výzkum stopových prvků a těžkých minerálů v karvinském souvrství OKR. Research report. Ústav pro průzkum uhelných ložisek. Geofond, Praha.
MACHÁČEK, V., ŠULCEK, Z. and VÁCL, J. (1962): Ke geochemii berylia v uhlí na Karlovarsku. Věstník ÚÚG, 37, 291 – 294, Praha.
– – – (1966): Geochemistry of beryllium in the Sokolov basin. Sbor. geol. věd, Techn., Geochem., 7, 33 – 66, Praha.
MALÁN, O. (1976): Geologický význam dispergovaných organických látek (MOD). Geol. průzkum, 18, 3, 70 – 73, Praha.
MALÁN, O., KESSLER, M. F. and VALEŠKA, F. (1966): Výskyt pyritu a inertinitu v OKR a možnosti jejich využití pro korelaci a identifikaci slojí. Uhlí, 8, 6 – 9, Praha.
MALECHA, A. (1959): Geologie okolí Měcholup u Žatce se zřetelem k ložiskům cenomanských jílovců. Ph. D. thesis ÚÚG, Praha, (unpublished).
MALKOVSKÝ, M. (1959): Jílové minerály ostravských vrstev. Acta Univ. Carol., Geologica, 1 – 2, 197 – 210, Praha.
MANSKAYA, S. M. and DROZDOVA, T. V. (1964): Geokhimiya organicheskogo veshchestva. Izd. Nauka, Moskva.
– – (1968): Geochemistry of Organic Substances, 1 – 345. Pergamon Press, London.
MANSKAYA, S. M., DROZDOVA, T. V., KRAVTSOVÁ, V. P. and TOBELKO, K. I. (1961): K biogeokhimii germaniya. Geokhimiya, 5, 433 – 439, Moskva.
MANSKAYA, S. M. and KODINA, L. A. (1975): Geokhimiya lignina. Izd. Nauka, Moskva, 229 p.

Mapel, V. J. and Hail, W. J. Jr. (1957): Tertiary geology of the Goose Creek District, Cassia County, Idaho; Box Elder County, Utah and Elko County, Nevada. U. S. Geol. Surv. Bull. 1055-H Washington.
Martin, A. and Garcia-Rossell, L. (1970): Uranio y renio en rocas sedimentarias. I. Lignitos de Arenas del Rey (Granada). Bol. geol. y minero, 81, 45 – 55, Madrid.
– – (1971): Uranio y renio en rocas sedimentarias. III. Lignitos de la depresión del Ebro. Bol. geol. y minero, 82, 62 – 69, Madrid.
Masursky, H. (1955): Trace elements in coal in the Red Desert, Wyoming; Contribution to the Internat. Conf. on peaceful uses of atomic energy, Geneva, Switzerland, August 1955. U. S. Geol. Survey Paper, United Nations No. P/56, Washington.
– (1956): Uranium-bearing coal in the Red Desert, Wyoming. Abstract. Program 1956 Annual Meeting, p. 73 – 74, Minneapolis, Minnesota. The Geol. Soc. of America, New York.
– (1962): Uranium-bearing coal in the eastern part of the Red Desert area, Wyoming. U. S. Geol. surv. Bull., 1099-B, Washington.
Masursky, H. and Pipiringos, G. N. (1957): Preliminary report on uranium bearing coal in the Red Desert, Great Divide Basin, Sweetwater County, Wyoming. U. S. Geol. Survey Bull. 1055-G, Washington.
Mathe, G. (1961): Geochemische und lagerstättengenetische Untersuchungen an erzführenden Kohlen des Döhlener Beckens. Diplomarbeit Mineral. Inst. d. Bergakademie, Freiberg.
Mazáček, J. (1960): Příspěvek k výskytu a získání vanadu z grafitových surovin. Rudy, 8, 415 – 417, Praha.
Mc Hugh, D. J., Saxby, J. D. and Tardif, J. W. (1976): Pyrolysis – hydrogenation – gas chromatography of carbonaceous material from Australian sediments. Part I.: Some Australian coals. Chem. Geol., 17, 243 – 259.
Mc Lean, J., Rettie, G. H. and Spring, F. S. (1958): Triterpenoids from peat. Chem. Ind., 1515-1516, London.
Medek, J. (1956): Kvantitativní spektrografické stanovení cesia v uhelných popelech. Paliva, 36, 370 – 374, Praha.
Medvedev, K. P. (1971): O roli makro- i mikroelementov v protsessach ugleobrazovaniya. Khimiya tverdogo topliva, 4, 3 – 11. Izd. Nauka, Moskva.
Mecháček, E. (1975): Die Geochemie des B, Ba und Sr in Kohlenflözen aus Tertiären Kohlenbecken der Westkarpaten. Geol. Zbornik, 26, 295 – 303, Bratislava.
Mecháček, E. and Petrík, F. (1967): Distribúcia stopových prvkov v uholných slojach handlovsko-nováckeho ložiska. Geol. průzkum, 9, 266 – 268, Praha.
Michalíček, M. (1967): Clay minerals exchangeable cations in Neogene sediments of the Danube Lowland and their significance as indicators of hydrochemical environment of deposition. – Geochemistry in Czechoslovakia. Transactions of the First Conference on Geochemistry. Ostrava. Sept. 20 – 24, 1965, 219 – 227, Ostrava.
Miller, H. P. (1949): The problems of coal geochemistry. Econ. Geol., 44, 649 – 662, Lancaster.
Minchev, D. and Eskenazi, G. (1966): Germanii i drugi elementi-primesi v vglishchata ot Vlchepolskoto nakhodishche – Iztochni Rodopi. God. Sof. Univ.. Geol.-Geogra. Fak.. 1964/5, 55, Sofija.
Minchev, D. and Eskenazi, G. (1969): Zonalno raspredelenie na germaniï v samostoyatelni vitrenovi fragmenti. Spis. Blg. geol. dr., 30, 2, Sofija.
Minguzzi, C. and Naldoni, K. M. (1950): Supposed traces of arsenic in wood; its determination in the wood of some trees. Atti. soc. toscana sci. nat. (Pisa), Mem. 57, ser. A, 38 – 48, Pisa.
Miropolskii, L. M. (1939): O rasprostranenii vanadiya v kaustobiolitakh Tatarii. Izv. AV SSSR, ser. geol., 3, 103 – 110, Moskva.
Moore, E. S. (1940): Coal. Wiley, 473 p., New York.
– (1954a): North Dakota. In: Geologic investigations of radioactive deposits, Semiannual progress report; U. S. Geol. Survey TEI-440, 102 – 109, Oak Ridge, Tenn., USA.

– (1954b): Extraction of uranium from aqueous solution by coal and some other materials. Econ. Geol., 49, 652–658, Lancaster.

MOORE, J. R. (1963): Bottom sediment studies, Buzzards Bay, Massachusetts. J. Sed. Petrology, 33, 511–558.

MORGAN, G. and DAVIES, G. R. (1937): Germanium and gallium in coal ash and flue dust. J. Soc. Chem. Industry, 56, 717–721, London.

MORISHIMA, H. and MATSUBAYASHI, H. (1978): ESR diagram: a method to distinguish vitrinite macerals. Geochim. Cosmochim. Acta, 42, 537–540.

MORITA, H. (1975): A phenolic palmitate in sphagnum peat. Geochim. Cosmochim. Acta, 39, 218–220, London.

MOROZOV, V. B. (1971): Raspredelenie i genezis sery v uglyakh Kizelovskogo basseïna. Sb. Geol. polezn. iskopaemye Urala, p. 82–84, Sverdlovsk.

MOSTECKÝ, J. and VČELÁK, VL. (1978): Některé nové možnosti zpracování a využití uhlí. Uhlí, 26, 4, 166–168. SNTL, Praha.

MOTT, R. A. and WHEELER, R. V. (1927): The inherent ash of coal. Fuel, 6, 416–420, London.

MRÁZEK, A. and VLÁŠEK, Z. (1958): Zjištění germania v sedimentech jihočeských pánví. Věstník ÚÚG, 33, 74–75, Praha.

MUELLER, G. (1972): Organic mineraloids. In: The Encyclopedia of Geochemistry and Environmental Sciences (Ed. R. W. Faibridge), Van Nostrand Reinhold Comp., New York, p. 823–829.

MUCHEMBLÉ, M. G. (1943): Sur la radioactivité élevée des roches marines du terrain houiller du Nord de la France. Compt. Rend. Acad. Sci. Paris, 216, 270–271.

MUKHERJEE, B. (1950): Detection of rare earths in the ashes of Indian coals. Fuel, 29, 264–266, London.

MUKHERJEE, B. and DUTTA, R. (1949): Germanium in Indian coals ash. Sci. a. Cult., 14, 538, Calcutta.

– – (1950): A note on the constituents of the ashes of Indian coals determined spectroscopically. Fuel, 29, 190–192, London.

MYERS, A. T. and HAMILTON. J. C. (1961): Rhenium in plant samples from the Colorado Plateau. U. S. Geol. Survey, Profess. Papers, 424-B, 286–288, Washington.

NEISH, A. C. (1964): In Biochemistry of Phenolic Compounds (ed. J. H. Harborne), p. 328. Academic Press.

NEKRASOV, Z. A. (1957): K voprosu o genezise uranovogo orudeneniya v uglyakh. Sb. Vopr. Geol. Urana. Atomizdat, Moskva, 37–54.

NĚMEC, A. (1947): Příspěvek k seznání biochemie tisu. Lesnická práce, 26, 1–11, Písek.

– (1948): Biochemie lesních dřevin. Publikace min. zemědělství RČS, 120, 1–310, Praha.

NĚMEC, A. and BABIČKA, J. (1934): Chlorosa rostlin způsobená kobaltem. Věstník Král. čes. spol. nauk, tř. II, 19, 1–29, Praha.

NĚMEC, A. and KÁS, V. (?): Studie o fysiologickém významu titanu pro organismus rostlinný. Reprint from Zemědělský archiv v Praze. "Politika", Praha.

NĚMEC, B., BABIČKA, J. and OBORSKÝ, A. (1936a): Výskyt zlata v přesličkách. Rozpravy Čes. akad., II. tř., 46, 1, 1–8, Praha.

– – – (1936b): Nové analysy popelu rostlin ze zlatonosných území. Rozpravy Čes. akad., 46, 13, 1 13, Praha.

– – – (1937): Gold im Körper einiger Säugetiere. Věstník Král. čes. spol. nauk, tř. mat.-fys., 1937, 18, 1–5, Praha.

NĚMEJC, F. (1959): Paleobotanika I. NČSAV, Praha, 402 p.

– (1963): Paleobotanika II. NČSAV, Praha, 529 p.

– (1969): Paleobotanika III. NČSAV, Praha, 478 p.

NOVÁČEK, R. (1931): Linnéit z dolu Prago u Kladna. Rozpravy Čes. akad., II. tř. 40, 37, 1–6, Praha.

– (1932): Granáty československých pegmatitů. Věstník Král. čes. spol. nauk, 1931, tř. II, 1–55, Praha.

OBORN, E. T. (1960): Iron content of selected water and land plants. U. S. Geol. Survey. Water-Supply Papers, 1459-G, Washington.

OELSCHLÄGER, W., WÖHLBIER, W. and MENKE, K. H. (1968): Über Fluor-Gehalte pflanzlicher, tierischer und anderer Stoffe aus Gebieten ohne und mit Fluor-Emission. III. Mitteilung, Landwirtsch. Forsch., 21, p. 82, Berlin.

O' GORMAN, J. V., SUHR, N. H. and WALKER, P. L. (1972): The determination of mercury in some American coals. Appl. Spectrosc., 26, 44—48.

O' GORMAN, J. V. and WALKER, P. L. JR. (1972): Mineral matter and trace elements in U. S. coals. Penn. State Univ. College Earth Mineral Sci Res. Develop. Rep. No. 61. Interim. Rep., 2, 183 p.

OKA, KANNO, AYUSAWA, NADA (1955): Rare elements in bituminous coal and lignite. 1. Germanium in lignites from North Miyagi and South Iwate prefectural areas. Bull. Res. Inst. Min. Dressing and Met., 11, 17—28, Sendai.

OKA. L., KANNO. T., HAGA, K. and SUZUUKI, S. (1956): Rare elements in lignite. Journ. Chem. Soc. Japan Pure Chem. Sec., 77, 7, 1026—1030. Ref. zhurnal, khem. 1958, p. 109, Moskva.

O' NEIL, R. L. and SUHR, N. H. (1960): Determination of trace elements in lignite ashes. Appl. Spectroscopy, 14, p. 45, Boston.

OTTE, M. U. (1953): Spurenelemente in einigen deutschen Steinkohlen. Chemie der Erde, 16, 239—294, Jena.

OTTENJANN, K., TEICHMÜLLER, M. and WOLF, M. (1974): Spektrale Fluoreszenz-Messungen an Sporiniten mit Auflichtanregung, eine mikroskopische Methode zur Bestimmung des Inkohlungsgrades gering inkohlter Kohlen. Fortschr. Geol. Rheinl. Westf., 24, 1—36.

OUCHI, K. and IMUTA, K. (1963): The analysis of benzene extracts of Yubari coal. II. Analysis by gas chromatography. Fuel, 42, 445—456, London.

PAPPAS, A. C., ALSTAD, J. and LUNDE, G. (1963): Determination of trace elements in opium by means of activation analysis. I. Methods of analysis and studies referring to the determination of the origin of opium. Radiochem. Acta, 1, 109—117, Frankfurt a. M.

PARK, R. and EPSTEIN, S. (1960): Carbon isotope fractionation during photosynthesis. Geochim. Cosmochim. Acta, 21, 110—126, Oxford.

PARSON, J. W. and TINSLEY, J. (1975): Nitrogenous substances. In J. E. GIESEKING (ed.): Soil components, 1. Organic Components. Springer Verlag, Berlin.

PATTEISKY, K. (1954): The thermal springs of the Ruhr area and their juvenile gases. Glückauf, 47, 1908— —1919, Essen.

PATTEISKY, K. and TEICHMÜLLER, M. (1960): Inkohlungs-Verlauf, Inkohlungs-Maßstäbe und Klassifikation der Kohlen auf Grund von Vitrit-Analysen. Brennstoff-Chemie, 41, 79—84, 97—104, 113— —137, Essen.

PATTERSON, E. D. (1954): Radioactivity of some coals and shales in southern Illinois. U. S. Geol. Survey TEI-466, Oak Ridge, Tenn., USA.

PAVLÍČEK, F. (1927): Chemie uhlí. Nakl. Prometheus, 178 p., Praha.

PAVLOV, A. V. (1967): Uglistye vklyucheniya yursko-melovykh uglenosnykh otlozheniĭ Aldano-Chuľmanskogo raĭona (Yuzhno-Yakutskiĭ kamennougoľnyĭ basseĭn). Litol. i polezn. iskop., 2, Moskva.

PAVLŮ, D. (1956): Obsah germania v severočeském hnědém uhlí. Diploma work. Katedra mineralogie VŠCHT, Praha.

PEDERSEN, K. R. and LAM, J. (1975): Organic compounds from the Rhaetic-Liassic coals of Scoresby Sund, east Greenland. Bull. Grønlands Geol. Unders., 117, 1—39.

PELÍŠEK, J. (1940): The occurrence of Ca in Moravian soils. Sb. Čs. akad. zeměd., 15, 61—64, Praha.

PETRÁNEK, J. and DOPITA, M. (1956): Prouhelnění slojí v ostravsko-karvínském revíru a jeho závislost na geologických činitelích. Sb. ÚÚG, 22, 593—634, Praha.

PIPIRINGOS, C. N. (1955): Uranium-bearing coals in the central part of the Great Divide Basin, Sweetwater County, Wyoming, Contribution to the Internat. Conf. on peaceful uses of atomic energy, Geneva, Switzerland. U. S. Geol. Survey Paper, United Nations No. p/287, Washington.

PIRSCHLE, K. (1939): Die Bedeutung der Spurenelemente für Ernährung, Wachstum und Stoffwechsel der Pflanzen. Ergebn. Biol., 17, 255—413, München.
PLUSKAL, O. (1972): Úvod do geologie uranových ložisek. Universita Karlova v Praze. St. pedag. nakl., 196 p., Praha.
POKORNÝ, J. (1954): Příspěvek ke geochemii vanadu. Ústav pro výzkum rud, res. ústav MHD, Výzkumné středisko Kutná Hora. Unpublished.
POLAŃSKI, A. (1961): Geochemistry of Isotopes. Wydaw. Geol. Warszava, 392 p. (English translation published in 1965 by the U. S. Dept. of Interior.)
POLICKÝ, J., TOMŠÍK, J. and KLIKA, Z. (1976): Bor jako indikátor paleosalinity sedimentačního prostředí. Sborník geol. průzkumu, Ostrava, 11, 35—73.
POTONIÉ, H. (1910): Die Entstehung der Steinkohle und der Kaustobiolithe überhaupt. 225 p., Berlin.
POTTER, P. E., SHIMP, N. F. and WITTERS, J. (1973): Trace elements in marine and freshwater argillaceous sediments. Geochim. Cosmochim. Acta, 27, 669—694. Oxford.
POWELL, T. G., DOUGLAS, A. G. and ALLAN, J. (1976): Variations in the type and distribution of organic matter in some carboniferous sediments from Northern England. Chem. Geology, 18, 137—148. Amsterdam.
POWELL, T. G. and MCKIRDY, D. M. (1973): Relationship between ratio of pristane to phytane, crude oil composition and geological environment in Australia. Nature, Phys. Sci., 243, 37—39. London.
PRASHNOWSKY, A. A. (1971a): Biogeochemische Untersuchungen an Gesteinen der Bohrung Münsterland I. (Teil I, II). N. Jb. Geol. Paläont. Abh., 138, 37—55, Stuttgart.
— (1971b): Biogeochemische Untersuchungen an Tonsteinen des Ruhrkarbons. Geol. Rundschau, 60, 744—812, Stuttgart.
PRASHNOWSKY, A. A. and BURGER, K. (1966): Petrographisch-biogeochemische Untersuchungen des Flözes Zollverein 5 im Ruhr-Karbon. N. Jb. Geol. Paläont. Abh., 126, 142—182, Stuttgart.
PRÁT, S. (1932): Fysiologie chemická a fysikálně chemická. Aventinum, 487 p., Praha.
PRÁT, S. and KOMÁREK, K. (1934): Vegetace u měděných dolů. Sborník MAP, 8, 8, 1—16, Praha.
PUCHELT, H. (1967): Zur Geochemie des Bariums in exogenem Zyklus. Sitz.-Ber. Heidelberg Akad. Wiss., Math.-nat. Kl., 4. Abh., Heidelberg.
QUINN, A. W. and GLASS, H. D. (1958): Rank of coal and metamorphic grade of rocks of the Narragansett basin of Rhode Island. Economic Geol., 53, 563—576.
RAKOVSKII, V. E. (1949): Obshchaya khimicheskaya tekhnologia torfa. Gos. Energ. Izdat., Moskva-Leningrad.
RAMAGE, H. (1927): Gallium in flue dust. Nature, 119, p. 783, London.
RANKAMA, K. and SAHAMA, T. G. (1950): Geochemistry. The Univ. of Chicago Press, 912 p., Chicago.
RAO, C. P. and GLUSKOTER, H. J. (1973): Occurrence and distribution of minerals in Illinois coals. Illinois State Geol. Surv. Circ., 476, 56.
RATAJCZAK, T. (1974): Caractéristique minéralogique et pétrographique des roches stériles du bassin Monilleure de Lublin. Polska Akad. Nauk, Prace Geol., 85, 1—81 (Polish, with French summary).
RATYNSKIĬ, V. M. (1943): Nakoplenie germaniya v uglyakh. Doklady AN SSSR, 40, 198—200; 223—227, Moskva.
— (1945): Sur la génèse du germanium dans la houille. Comptes rendus (Doklady) de l'Académie des Sciences de l'URSS, 49, 2, 119—122, Moskva.
— (1946): Germaniĭ v uglyakh. Trudy biogeokhimicheskoĭ lab. AN SSSR, 8, 181—223, Moskva.
RATYNSKIĬ, V. M. and ZHAROV, YU. H. (1976): O soderzhanii galliya v iskopaemykh uglyakh. Geokhimiya, 11, 1753—1755.
RAYNAUD, J. F. and ROBERT, P. (1976): Les méthodes d'étude optique de la matière organique. Bull. Centre Recherches Pau-SNPA, 10, 109—127, Pau.
RAZUMNAYA, E. G. (1957): Izucheniya raspredeleniya molibdena i vanadiya v uglisto-kremnistykh slantsakh metodom centrifugirovaniya. Sb. Sovrem. metody mineral. issledovaniya gorn. porod. rud i mineralov. Gosgeoltechizdat, p. 49—54, Moskva.

Retcofsky, H. L. and Friedel, R. A. (1976): Carbon-13, nuclear magnetic resonance spectrum of coal extract: a note on aromaticity of coal. Fuel, 55, p. 363. London.
Retcofsky, H. L., Stark. J. M. and Friedel, R. A. (1968): Electron spin resonance in American coals. Anal. Chem., 40, 1699—1704.
Reynolds, F. M. (1939): Note on the occurrence of barium in coal. J. Soc. Chem. Ind., 58, 64—66, London.
— (1948): Occurrence of vanadium, chromium and other unusual elements in certain coals. Jour. Soc. Chem. Ind., 67, 341—345, London.
Reynolds, R. C. (1965): The concentration of boron in Precambrian seas. Geochim. Cosmochim. Acta, 29, 1—16. Oxford.
— (1972): Boron: Element and geochemistry. In: R. W. Fairbridge (ed.): The Encyclopedia of Geochemistry and Environmental Sciences. Van Nostrand Reinhold Company, 88—90, New York.
Rice, G. S. and Hartmann, I. (1939): Coal mining in Europe. U. S. Bur. of Mines, Bull. 414, 1— —222.
Robinson, W. O., Bastron, H. and Murata, K. J. (1957): Biogeochemistry of the rare earths with particular reference to hickory trees. Abstract. Program 1957 Annual Meetings. Geol. Soc. of Am. Atlantic City, New Jersey, USA, 116—117.
Rogoff, M. H. (1959): Role of microbiological processes in the formation of peat and their relationship to coal genesis. Abstract. Program 1959 Annual Meetings, Geol. Soc. of Am. Pittsburgh, Pa., USA, 105 A—106 A.
Ronov, A. B. (1958): Organicheskiĭ uglerod v osadochnykh porodakh. Geokhimiya, 5, 409.
Rösler, H. J. and Lange, H. (1965): Geochemische Tabellen. Leipzig.
Rösler, R. and Zscherpe, G. (1971): Die Bestimmung der natürlichen Radioaktivität von Braunkohlen in der DDR. Neue Bergbautechn., 1, 340—345, Freiberg.
Rost, R. (1937): Minerály hořících hald na Kladensku, Rozpravy II. tř. Čes. akad., 47, 11, 1—19, Praha.
— (1940): O červených náletech realgaru a selenu z hald. Čas. NM, Praha, 114, 102—107.
— (1942a): Doplňky k mineralogii hořících hald na Kladensku. Rozpravy II. tř. Čes. akad., 52, 25, 1—4, Praha.
— (1942b): Mineralogické zprávy z Kladenska. Rozpravy II. tř. Čes. akad., 52, 26, 1—8, Praha.
— (1950): Zpráva o dílčím státním výzkumném úkolu VId: Geochemie kladensko-rakovnického permokarbonu. GGF KU, Praha. Unpublished.
Rotter, R. (1952): O výskytu germania v našem uhlí a popelovinách a o možnostech jeho isolace. Knihovna ÚÚG, 3, 1—63, Praha.
Rouir, E. V. (1954): Le germanium dans les charbons belges. Ann. (Bull.) Soc. Géol. Belgique, 77, B283— —B288, Bruxelles.
Ruch, R. R., Gluskoter, H. J. and Kennedy, E. J. (1971): Mercury content of Illinois coals. Environ. geol. Notes. Ill. State Geol. Surv., 43, 1—15, Washington.
Ruch, R. R., Gluskoter, H. J. and Shimp, N. F. (1974): Occurrence and distribution of potentially volatile trace elements in coals. Illinois State Geol. Surv. Environ. Geol. Notes, 72, 96.
Ruhemann, S. and Raud, H. (1932): Über die Harze der Braunkohle, I. Die Sterine des Harzbitumens. Brennstoff-Chem., 13, 341—345, Essen.
Ruhland, W. (1958): Handbuch der Pflanzenphysiologie, IV. Die mineralische Ernährung der Pflanze. Springer-Verlag, Berlin–Göttingen–Heidelberg.
Rukhin, L. B. (1962): Osnovy obshcheĭ paleogeografii. Gostoptechizdat, Moskva, 627 p.
Rushev, D. and Dragostinov, P. (1965): Differential thermal analysis and IR spectral analysis of petrographic ingredients of lignites. Erdoel und Kohle, 18, 372—375, Berlin, Hamburg.
Ryazanov, I. V. and Yudovich, Ya. E. (1974): K diffuzionnoĭ teorii redkometaľnogo obogashcheniya kontaktnykh zon ugoľnykh plastov. Litologiya i poleznye iskopayemye. AN SSSR, 4, 64—75, Moskva.
Řehák, J. (1966): Kontaktní metamorfóza uhelných slojí na Těšínsku. Věstník ÚÚG, 40, 341—346, Praha.

- (1968): Vzájemné stáří vrásových struktur a prouhelnění slojí na Ostravsku. Věstník ÚÚG, 43, 331—340, Praha.
- (1971): Vznik a prouhelnění slojí v československé části Hornoslezské pánve. Uhlí, 19, 25—28, Praha.

SACKETT, W. M., ECKELMANN, W. R., BENDER, M. L. and BE, A. W. (1965): Temperature dependence of carbon isotope composition in marine plankton and sediments. Science, 148, p. 235, Washington.

SAKANONE, M. (1960): Geochemical studies on the radioactive sediments, IV. Uranium and germanium in the carbonaceous materials found in the „Misasagroup" sedimentary beds of the Tertiary. J. Chem. Soc. Japan, Pure Chem. Sect, 81, 102, 1520.

SAPRYKIN, F. J. and SVENTIKHOVSKAYA, A. N. (1965): Rare metal mineralizations in recent peat marshes. Materialy Sov. Rabot. Labor. Geol. Organiz., 19, Leningrad, 7, 95—102.

SAROSIEK, J. and KLYS, B. (1962): The tin content in plants and soil in Sudety. Acta Soc. Botan. Polan., 31, 737; Chem. Abstr., 60, 16451h (1964), Easton.

SARROT-REYNAUD DE GRESSENEUIL, J. (1950): Radioactive properties of Alpine coals. Trav. Lab. Geol. Fac. Sci. Univ. Grenoble, 30, 43—54, Grenoble.

SAUKOV, A. A. (1950): Geokhimiya. Gos. izd. geol. lit., Moskva, 346 p.

SAVCHUK, M. G. (1967): Trace elements in peats of Novosibirsk and Tomsk regions. Biosfere Ikh. Primen. Seľskokhoz. Med., Sib. Daľnevost, 191—196.

SAVCHUK, S. V. (1971): Zakonomernosti raspredeleniya sery v uglyakh Donetskogo basseïna. Izv. Dnepropetr. gorn. inst., 54, 36—44, Dnepropetrovsk.

SAVEĽEV, V. F. (1964): O selenosoderzhashchikh obuglennykh rastiteľnykh ostatkakh iz verkhnemelovykh osadochnykh porod odnogo raïona Sredneï Azii. Zap. Uzbekist. otd. Vsesoyuzn. miner. obshch., 16, Tashkent.

SAVKEVICH, S. S. (1970): Jantar. Nedra, Leningrad.

SCHEJBAL, C., HONĚK, J., ADAMUS, B. and KÜHN, P. (1972): Petrograficko-geochemická charakteristika brouskového horizontu a přilehlých částí uhelných slojí na dole Zápotocký a Nejedlý II v Kladně. Sb. věd. prací VŠB, 18, 49—80, Ostrava.

SCHNITZER, M. and NEYROUD, J. A. (1975): Alkanes and fatty acids in humic substances. Fuel, 54, 17—19, London.

SCHOFIELD, A. and HASKIN, L. (1964): Rare-earth distribution patterns in eight terrestrial materials. Geochim. Cosmochim. Acta, 28, 437—446, Oxford.

SCHUHMACHER, J. P., HUNTJENS, F. J. and VAN KREVELEN, D. W. (1960): Chemical structure and properties of coal, 26. Studies on Artificial Coalification. Fuel, 39, 223—234, London.

SHACKLETTE, H. T. (1965): Element content of bryophytes. U. S. Geol. Surv. Bull. 1198-D, 1, Washington.

SHEIBLEY, D. W. (1975): Trace elements in coal. Advances in Chemistry Series 141, Edm. Babu, Washington.

SHIROKOV, A. Z., LAZEBNIK, P. V. and DOLGOPOLOV, V. M. (1971): O genezise germaniya v uglyakh. Izv. Dnepropetr. gorn. inst., 54, 50—53, Dnepropetrovsk.

SEKANINA, J. (1931): Rosickýit, die natürliche γ-Schwefelmodifikation. Zeit. Krist., 80, 174—189, Leipzig.

- (1932): Letovicit, ein neues Mineral und seine Begleiter. Zeit. Krist., 83, 117—122, Frankfurt a. M.
- (1948): Koktait, nový nerost ze skupiny syngenitové. Práce Moravskosl. akad. věd přír., 20. Spis 1, Brno, 1—26.

SENESI, N., GRIFFITH, S. M., SCHNITZER, M. and TOWNSEND, M. G. (1977): Binding of Fe^{3+} by humic materials. Geochim. Cosmochim. Acta, 41, 969—976.

SILBERMINZ, V. A. (1935): On the occurrence of vanadium in fossil coals. Doklady AN SSSR III (8), 3 (63), 117—120, Moskva.

- (1936): Germanium in den Kohlen des Donezbeckens. Miner. Syrje, 11, p. 16. Chem. Abstr., 30, 7306, Easton.

SILBERMINZ, V. A. and RUSANOV, A. K. (1936): The occurrence of beryllium in fossil coals. Doklady AN SSSR, 1, p. 27. Chem. Abstr., 30, 7072, Moskva.

SILBERMINZ, V. A., RUSANOV, A. K. and KOSTRIKIN, V. M. (1936): K voprosu o rasprostranenii germaniya v iskopaemykh uglyakh. Sbor. posvyashchennyĭ akad. V. I. Vernadskomu k 50-letiyu nauchnoĭ deyateľnosti. Izd. AN SSSR, Moskva, 1, 169–190, Moskva.

SILVERMAN, S. R. and EPSTEIN, S. (1959): Carbon isotopic compositions of petroleums and other sedimentary organic materials. Bull. Am. Ass. Petrol. Geol., 42, 988–1012, Tulsa.

SIMONSEN, J. L. and BARTON, D. H. R. (1952): The terpenes. Cambridge Univ. Press, 3, p. 337, Cambridge.

SKRIGAN, A. I. (1951): Composition of turpentine from a swamp resin 1,000 years old. Dokl. Akad. Nauk SSSR, 80, 607–609. Chem. Abstr., 46, 5337 (1952), Moskva.

SLAVÍK, F. (1925): Z mineralogie kladenského karbonu a jeho podloží. Čas. NM, 99, 113–120, Praha.

SMALES, A. A. and SALMON, L. (1955): Determination by radioactivation of small amounts of Rb and Cs in sea water. Limnol. Oceanog., 10, 226–232, Lawrence.

SMITH, J. W. and BATTS, B. D. (1974): The distribution and isotopic composition of sulfur in coal. Geochim. Cosmochim. Acta, 38, 121–133, Oxford.

SNIDER, J. L. (1953): Reconnaissance for uranium in coal and shale in southern West Virginia and southwestern Virginia. U. S. Geol. Survey TEI-409., 1–28. Oak Ridge, Tenn., USA.

SOKOL, R. (1919): Rudná žíla v Mantově u Chotěšova. Sb. Měst. musea v Plzni, 1–9, Plzeň.

SOMASEKAR, B. (1969): The petrology, physics and chemistry of the three seams of the Central Bohemian carboniferous formations, Kladno coalfield, Czechoslovakia. Ph. D. thesis, VŠB Ostrava.

— (1971): Boron in the main seam of the Kladno coalfield, Czechoslovakia. Sb. věd. prací VŠB, 17, 9–14, Ostrava.

SPEARS, D. A. (1974): Relationship between water-soluble cations and paleosalinity. Geochim. Cosmochim. Acta, 38, 567–575, Oxford.

SPEIGHT, J. G. (1971): The application of spectroscopic techniques to the structural analysis of coal and petroleum. Appl. Spectrosc. Rev., 5, p. 211.

STAATZ, M. H. and BAUER, H. L. JR. (1954): Gamma group, in: Radioactive deposits of Nevada by T. G. Lovering. U. S. Geol. Survey Bull. 1009-C, 76–77, Washington.

STACH, E., MACKOWSKY, M. TH., TEICHMÜLLER, M., TAYLOR, G. H., CHANDRA, D. and TEICHMÜLLER, R. (1975): Coal Petrology (transl. and English revision by D. G. Murchison, G. H. Taylor and F. Zieske). Berlin–Stuttgart (Gebrüder Borntraeger), 1–428.

STADNICHENKO, T., MURATA, K. J. and EXELROD, J. M. (1950): Germaniferous lignite from the District of Columbia and vicinity. Science, 112, p. 109, Washington.

STADNICHENKO, T., MURATA, K. J., ZUBOVIC, P. and HUFSCHMIDT, E. L. (1953): Concentration of germanium in the ash of American coals. A progress report. U. S. Geol. Survey Circ., 272, 1–24, Washington.

STADNICHENKO, T., ZUBOVIC, P. and SHEFFEY, N. B. (1956): Beryllium in the ash of American coals. Abstract. Geol. Soc. America, Bull., 67, p. 1735, New York.

— — — (1961): Beryllium content of American coals. U. S. Geol. Survey Bull., 1084-K, 253–295, Washington.

STAPLIN, F. (1969): Sedimentary organic matter. Organic metamorphism and oil and gas occurrence. Bull. Canad. Petrol. Geol., 17, 1, 47–66.

STOCK, A. and CUCUEL, F. (1934): Die Verbreitung des Quicksilbers. Naturwissenschaften, 22/24, 390–393, Berlin.

STOKLASA, J. and BAREŠ, J. (1926): O fysiologickém významu jodu v rostlinném organismu. Věstník Čes. akad. zeměd., 2, 1030–1038, Praha.

STRAKHOV, N. M. (1960): Osnovy teorii litogeneza. Izd. AN SSSR, Moskva, 547 p.

STREIBL, M. and HEROUT, V. (1969): Terpenoids — especially oxygenated mono-, sesqui-, di- and triterpenes. Organic Geochemistry (ed. Eglinton/Murphy). Springer-Verlag, Berlin, 401–424.

STREIBL, M., KRISTÍN, M., KRUPCÍK, J. and STRÁNSKÝ, K. (1972): Investigation of the chemical composition of Nováky brown coal. VI. On the chemical composition of brown coals. XII. Occurrence of a-phyllocladane (josene) in Slovakian lignite. An. Real Soc. Esp. Fís. Quím. QUIMICA 68, 879–882, Madrid.

STREIBL, M., VAŠÍČKOVÁ, S., HEROUT, V. and BOUŠKA V. (1976): Chemical composition of Cenomanian fossil resins from Moravia. Collection Czech. Chem. Comm., 41, 3138–3145, Praha.

STUTZER, O. (1940): Geology of coal. Chicago.

SUGGATE, R. P. (1974): Coal ranks in relation to depth and temperature in Australian and New Zealand oil and gas wells. New Zealand J. Geol. Geophys., 17, 149–167.

SVOBODA, J. V. and BENEŠ, K. (1955): Petrografie uhlí. NČSAV, Praha, 262 p.

SWAIN, F. M. (1961): Limnology and amino-acid content of some lake deposits in Minnesota, Montana, Nevada and Louisiana. Bull. Geol. Soc. Am., 72, 519–546, New York.

– (1970): Non-marine organic geochemistry. Cambridge Univ. Press, 445 p.

SWAIN, F. M., BLUMENTALS, A. and MILLERS, R. (1959): Stratigraphic distribution of amino acids in peats from Cedar Creek Bog, Minnesota and Dismal Swamp, Virginia. Limnol. Oceanogr., 4, 119––127, Lawrence.

SWAIN, F. M., BRATT, J. M. and KIRKWOOD, S. (1967): Carbohydrate components of some Paleozoic fossils. J. Paleont., 41, 1549–1554, Menasha.

– – – (1968): Possible biochemical evolution of carbohydrates of some Paleozoic plants. J. Paleont., 42, 1078–1082, Menasha.

– – – (1969): Carbohydrate components of Upper Carboniferous plant fossils from Radstock, England. J. Paleont., 43, 550–553, Menasha.

SWAINE, D. J. (1962): Trace elements in coal, II. Origin, mode of occurrence, and economic importance. C.S.I.R.O. Div. Coal Res. Tech. Commun., 45.

– (1967): Inorganic constituents in Australian coals. Mitt. Naturforsch. Ges. Bern (N.F.), 24, p. 49.

– (1971): Personal communication in: Wedepohl, K. H. (1969), II-2, Zinc.

SWANSON, V. E., FROST, T. C., RADER, L. F. JR. and HUFFMANN, C. JR. (1966): Metal sorption by northwest Florida humate. U. S. Geol. Survey Prof. Papers 550-C, p. 174, Washington.

SÝKORA, L. (1959): Rostliny v geologickém výzkumu. NČSAV, Praha, 322 p.

SZÁDECKY-KARDOSS, E. (1952): Rock metamorphism and coal. Acta Geol. Acad. Sci. Hungaricae, 1, 205–225, Budapest.

SZALAY, A. (1974): Accumulation of uranium and other micrometals in coal and organic shales and the role of humic acids in these geochemical enrichments. Archiv Min. Geol., 5, 23–36. Uppsala.

SZALAY, A. and SZILAGYI, M. (1969): Accumulation of microelements in peat humic acids and coal. In: Advances in Organic Geochemistry (1968) (edit. P. A. Schenk and I. Havenaar), 527 p., Pergamon Press.

SZALAY, S. (1954): The enrichment of uranium in some brown coals in Hungary. Magyar Tudom. Akad., Acta Geol., 2, 299–310, Budapest.

– (1957): The role of humus in the geochemical enrichment of U in coal and other bioliths. Acta Phys. Acad. Sci. Hung., 8, 25–35, Budapest.

ŠANTRŮČEK, P. (1958): Germanium v uhelných slojích chebské pánve. Věstník ÚÚG, 33, 367–370, Praha.

– (1961): Germanium v hnědouhelných slojích sokolovského revíru. Věstník ÚÚG, 36, 75–78, Praha.

ŠIMEK, B. G. (1940): Obsah germania v uhlí Ostravsko-karvinského revíru. Chem. listy, 34, 181–185, Praha.

ŠIMEK, B. G., COUFALÍK, F. and ŠTÁDLER, A. (1948): Obsah germania v uhlí Ostravsko-karvinského revíru. Zprávy Ústavu pro věd. výzkum uhlí, 167–174, Praha.

ŠINDELÁŘ, J. (1957): Minerální složení popelovin černých uhlí. Čas. pro mineral. geol., 2, 442–448, Praha.
ŠKUTA, K. (1972): Význam obsahu H_2O v uhelných slojích. Uhlí, 20, 109–113, Praha.
ŠPETL, F. (1956): Úprava uhlí. SNTL, 444 s., Praha.
– (1977): Personal communication.
ŠULCEK, Z. (1956): Zpráva o problému XX/2, další průzkum surovin obsahujících germanium. Geofond, Praha. Unpublished.
ŠVASTA, J., ZAHRADNÍK, L., ŠULCEK, Z., ŠŤOVÍK, M., BOUBERLE, M. and ROTTER, R. (1955): Výskyt germania v československém uhlí a jeho produktech. Geotechnica, 20, 1–142, Praha.
TARAKANOVA, E. I. (1968): Trace elements in the peat bogs of the Middle Ural and Central Transural regions. Litol. Polezn. Iskop. 2. 136–140.
TEICHMÜLLER, M. and TEICHMÜLLER, R. (1965): Die Inkohlung im Saar-lothringischen Karbon, Verglichen mit der im Ruhrkarbon. Z. Deut. Geol. Gess., 117, 243–279, Hannover.
TEICHMÜLLER, R. (1951): Zur Metamorphose der Kohle. Compt. Rend. 3e Cong. Strat. Geol. Carbonifère, p. 615–623, Heerlen.
TEICHMÜLLER, R. and TEICHMÜLLER, M. (1947): Problems of coalification in the Ruhr. Z. Deut. Geol. Ges., 99, 40–75, Hannover.
TEREBENINA, A. and ANGELOVA, G. (1962): Composition of the ashes of alcohol-benzene extracts of some Bulgarian coals. Compt. Rend. Acad. Bulg. Sci., 15, p. 495; Chem. Abstr., 58, 10005b (1963), Sofia.
THOMAS, B. R. (1966): The chemistry of the order Araucariales. Acta Chem. Scan., 20, 1074–1081, Copenhavn.
– (1970): Modern and fossil plants resins. In Phytochemical Phylogeny, ed. J. B. HARBOURNE, Acad. Press, p. 59–79, London.
TIMOFEEV, P. P. and BOGOLYUBOVA, L. I. (1971): Organicheskoe veshchestvo i ego izmenenie v protsesse ugleobrazovaniya In: „Osadkonakoplenie i genezis uglei karbona SSSR". Nauka, Moskva.
TINGEY, C. L. and MORREY, J. R. (1973): Coal structure and reactivity. Battelle Energy Program Report, 1973.
TKACHENKO, K. T. and TRETENKO, M. E. (1958): K voprosu o veshchestvennom sostave zol burykh uglei dneprovskogo basseina. Izv. Dnepropetr. gorn. inst., 29, 104–108, Dnepropetrovsk.
TKACHEV, YU. A., SKIBA, N. S. and BONDARENKO, C. F. (1965): Distributions of strontium and barium in coals of Kirgizia. Litol. Geokhim. i Polezn. Iskop. Osad. Obrazov. Tianshan, Akad. Nauk. Kirg. SSR, Inst. Geol., 24.
TOMAŇA, M. (1957): Příspěvek k poznání geochemie uranu v kladenském uhlí. Sb. VŠCHT v Praze. SPN, 191–197, Praha.
TOMŠÍK, J. (1959): Několik poznámek k salinitě sedimentačního prostředí siltovců a jílovců karbonu ostravsko-karvínské pánve. Sbor. prací konf. o geologii OKR. VŠB Ostrava, 129–131.
– (1963): Indikátory sedimentačního prostředí usazených hornin. Archív Geol. průzkumu, 1–58. (Manuscript). Ostrava.
TOYODA, S., SUGAWARA, S. and HONDA, H. (1966): Chemical structure and properties of heated coal in the early state of carbonization, V. Electron spin resonance of coal. J. Fuel Soc. Japan, 45, 876–883.
TRAVIN, A. B. (1957): O putyakh nakopleniya germaniya v uglyakh i nekotorye zadachi ego dal'neishikh issledovanii. Izv. vost. fil. AN SSSR, 1, 44–48, Moskva.
– (1960): Nekotorye zakonomernosti rasprostreneniya germaniya v uglyakh Zapadnoi Sibiri. Geol. i geof., 2, Novosibirsk.
TREIBS, A. (1936): Chlorophyll and hemin derivatives in organic mineral substances. Angew. Chem., 49, 682–686, Berlin, Leipzig.
TSCHAMLER, H. and DERUITER, E. (1966): A comparative study of exinite, vitrinite, and micrinite. In: Coal Science, Am. Chem. Soc. Adv. Chem. Ser., 55, 332–343, New York.

TYROLER, J. (1958): K otázce rozšíření germania v uhelných slojích v Týnci u Plzně a o metodice jeho stanovení. Sb. VŠCHT v Praze, 293—296, Praha.
TYROLEROVÁ, P. (1959): Geochemie germania v radnické pánvi. Sb. VŠCHT v Praze, 3, 353—363, Praha.
UZUNOV, J. (1965): O geokhimii vanadiya v nekotorykh ugoľnykh mestorozdeniyakh Bolgarii. Karpato-balkan. Geol. assoc. VII. kongr., Sofia, dokl. IV.
— (1967): Vanadiĭ v lignitakh Vostochno-Marshukogo basseĭna. Izv. Geol. In-ta Bolgarskoĭ Akad. nauk, 16.
VÁCL, J. and ČADEK, J. (1959): Základní geologický výzkum hrádecké části žitavské pánve. Unpublished. Geofond, Praha.
VAN KREVELEN, D. W. (1950): Graphical-statistical method for the study of structure and reaction processes of coal. Fuel, 29, 269—284, London.
— (1954): Unser derzeitiges physikalisches und chemisches Bild der Kohle, Brennst.-Chem., 35, p. 257, 289.
— (1961): Coal. Elsevier, Amsterdam, 514 p.
— (1963): Geochemistry of coal. In Organic Geochemistry (ed. I. A. Breger). Pergamon Press, New York, 183—247.
VAN KREVELEN, D. W. and SCHUYER, J. (1957): Coal sciences. Elsevier, Amsterdam, 352 p.
VAROSSIEAU, W. W. (1950): Ancient buried and decayed wood from a biological, chemical, and physicochemical points of view. Proc. Intern. Botan. Congr. Stockholm, 7, 567—569.
VAROSSIEAU, W. W. and BREGER, I. A. (1952): Compt. rend. 3e Congrès Strat. Géol. Carbonifère (Heerlen, 1951), Maastricht, p. 637.
VASIĽEV, S. F. and D'YAKOVA, M. K. (1958): Thermal solvent extraction as a method of processing peat into gas, chemical products and motor fuels. Novye metody rats. ispolz. mest. topliv. Trudy Soveshch. Riga, 37—46. Chem. Abstr., 56, 2666d.
VAVREČKA, P., LANG, I., ŠEBOR, G. and NĚMEC, J. (1978): Současné názory na chemickou strukturu uhlí. Uhlí, 26, 397—402. Praha.
VČELÁK, V. (1959): Chemie und Technologie des Montanwachses. Praha, 818 p.
— (1962): Hnědé uhlí, jeho vlastnosti, zpracování a využití. Praha, 177 p.
VELIEV, YU. YA., NACHADZHANOV, D. N. and ADAMCHUK, I. P. (1977): K geokhimii bora v yurskikh uglyakh Gissarskogo khrebta. Dokl. Akad. Nauk Tadzh. SSR, 20, 1, 57—60.
VELIKOWSKY, I. (1956): Earths in Upheaval. V. Gollanz, London.
VERNADSKIĬ, V. I. (1925): O chemickém složení živé hmoty v souvislosti s chemií kůry zemské. Sb. přír., Praha, I, 1—16.
— (1929): La Biosphère. Paris, 232 p.
VETEJŠKA, K., ZAHRADNÍK, L. and MAZÁČEK, J. (1967): Vzácné a stopové prvky. St. nakl. techn. lit., Praha, 149 p.
VIDAVSKIĬ, V. V. and PROKOPEC, E. I. (1932): Sbornik rabot po khemii ugleĭ. Kharkov.
VINE, J. D. (1953): Parts of Colorado, Utah, Idaho and Wyoming; in Search for and geology of radioactive deposits. Semiannual progress report, U. S. Geol. Survey TEI-330, p. 117, Oak Ridge, Tenn., USA.
— (1955): Uranium-bearing coal in the United States. Contr. to the Intern. Conf. on peaceful uses of atomic energy. Genève, Switzerland. U. S. Geol. Survey Paper, United Nations, No. P/55, Washington.
— (1957): Geology and uranium deposits in carbonaceous rocks of the Fall Creek Area, Bonnville County, Idaho. U. S. Geol. Survey Bull. 1055-I. Washington.
— (1959): Dopplerite from cretaceous rocks in Wyoming. Abstr. Program 1959 Annual Meetings. Geol. Soc. Am., p. 133A, Pittsburgh, Pa, USA.
VINE, J. D. and MOORE, G. W. (1952): Uranium-bearing coal and carbonaceous rocks in the Fall Creek Area, Bonnville County, Idaho. U. S. Geol. Survey Circ., 212, 1—10, Washington.

VINOGRADOV, A. P. (1965): Oligo-Elemente in biologischen Objekten (I, Se, Li, Al, Mn). Agrokhimiya 8, 20, Moskva.
VINOGRADOV, A. P. (1947): Vvedenie v geokhimiyu okeana. Moskva, 212 p.
VINOGRADOVA, KH. G. (1954): Molibden v rasteniyakh v svyazi s ikh sistematicheskim polozheniem. Trudy biogeokhimicheskoï lab. Inst. geokhimii i analit. khimii, AN SSSR, 10, 82 – 93, Moskva.
VLASOV, K. A. (1968): Geokhimiya redkykh elementov. Moskva.
VLÁŠEK, Z. (1957): Geochemické poměry jihočeských sedimentů se zvláštním zřetelem na stopové prvky. Research report. Geofond, Praha.
VNUKOV, A. V., PORTNOV, A. G. and PROKOPOV, N. S. (1971): O nekotorykh zakonomernostyakh kontsentratsii redkikh elementov v ugol'nykh plastakh. Izv. Zabajkal. fil. Geogr. o-va SSSR, 7, 36 – 41.
VOGT, T. and BRAADLIE, O. (1942): Geokjemisk og geobotanisk malmleting. IV. Plantevekst og jordbunn ved Rorosmaemene. Kgl. Norske Videnskap. Selskabs. Forh., 15, p. 25. Oslo.
VOLODARSKIĬ, I. KH., ZHAROV, YU. N. and RATYNSKIĬ, V. M. (1976): O kharaktere raspredeleniya galliya v iskopaemykh uglyakh i produktakh ikh ezzhiganiya. Khimiya tverdogo topliva, 6, 18 – 21. Moskva.
VOSKRESENSKAYA, N. T. (1968): Talliĭ v uglyakh, 2, 207 – 217, Moskva.
VOTAVOVÁ, Z. (1958): Výzkum germania v kladenském uhelném revíru. Výzkum surovinové základny a výroby GeO_2. Laboratoř anorganické chemie – Nerostné suroviny. ČSAV, Praha. Unpublished.
– (1959): Výskyt germania v rosicko-oslavanské kamenouhelné pánvi. Geofond, Praha. Unpublished.
VOTAVOVÁ, Z. and KRÁL, R. (1959): Distribuce germania v kladenské pánvi. Sb. VŠCHT, 3, 337 – – 352, Praha.
VRTAL, V. (1978): Světové zásoby uhlí. Uhlí, 26, 4, 169 – 171. SNTL Praha.
VYJÍDÁK, B. (1967): Occurrence of tungsten in the northern part of the Cheb Basin. Geochemie v Československu. VŠB v Ostravě, 183 – 191.
VYSLOUŽIL, J. (1927): Lučebná skladba soli kamenné z M. Ostravy. Příroda (Brno), 20, 7 – 8, 88.
WALKER, C. T. (1968): Evaluation of boroń as a paleosalinity indicator and its application to offshore prospects. Am. Assoc. Petrol. Geol. Bull., 52, 751 – 766.
– (1972): Boron geochemistry in marine environments. In: R. W. FAIRBRIDGE (ed.): The Encyclopedia of Geochemistry and Environmental Sciences. Van Nostrand Reinhold Company, New York, 90 – – 92.
WALKER, C. T. and PRICE, N. B. (1963): Departure curve for computing paleosalinities from boron in illites and shales. Bull. Amer. Ass. Petrol. Geol., 47, 833 – 841. Tulsa.
WANDLESS, A. M. (1954): The occurrence of sulphur in British coals. Colliery Guard., 189, 4887, 557 – – 560, London.
WASSOIEVITCH, N. B., KORGHAGINA, J. I., TOPATIN, N. V. and TCHERNITCHEV, V. V. (1969): Die Hauptphase der Erdölbildung. Z. angew. Geol., 15, 611 – 622.
WEDEPOHL, K. H. (ed.) (1969): Handbook of Geochemistry, I, II – 1, II – 2, II – 3, Springer-Verlag, Berlin, 442 p.
WEISS, G. (1976): K průběhu změn prouhelnění s hloubkou v čs. části hornoslezské pánve. Sb. Geol. průzkumu Ostrava, 11, 7 – 34. Ostrava.
WELCH, S. W. (1953): Radioactivity of coal and associated rocks in the anthracite fields of eastern Pennsylvania. U. S. Geol. Survey, TEI-348, 1 – 31. Oak Ridge, Tenn., USA.
WELTE, D. (1969): Organic geochemistry of carbon. In Handbook of Geochemistry (ed. K. H. Wedepohl, II-1, 6-L, 1 – 30).
WENINGER, M. (1965): Über Gehalte an Germanium, Zinn und einigen anderen Spurenelementen in ostalpinen Graphit- und Talkgesteinen. Tchermak Mineral. Petrog. Mitt., 10, 475 – 490, Wien.
WICKMAN, F. E. (1953): Was the isotopic constitution of carbon changed by coalification? Geochim. Cosmochim. Acta, 3, 244 – 252, Oxford.

WILKE, E. and RÖMERSPERGER, H. (1930): Über den Jodgehalt der Kohle. Chem. Zbl., 2, 1016—1017, Berlin.
WOLLRAB, V. and STREIBL, M. (1969): Earth waxes, peat, montan wax and other organic brown coal constituents. In Organic Geochemistry (ed. G.EGLINTON and M. T. J. MURPHY), Springer-Verlag, Berlin, 576—598.
YEN, T. F. and SPRANG, S. R. (1977): Contribution of ESR analysis toward diagenic mechanisms in bituminous deposits. Geochim. Cosmochim. Acta, 41, 1007—1018, Oxford.
YOKOKAWA, C. (1969): A further electron spin resonance study of solvent action on coal. Fuel, 48, 29—40. London.
YUDOVICH, YA. E. (1972): Geokhimiya ugolnykh vklyuchenii v osadochnykh porodakh. Izd. Nauka, Leningrad, 84 p.
— (1978): Geokhimiya iskopaemykh uglei. Nauka, Leningrad.
ZAHRADNÍK, L. (1959): Dílčí zpráva o úkolu XX/1. Další průzkum surovin obsahujících germanium. Geofond, Praha. Unpublished.
ZAHRADNÍK, L., FORMÁNEK, Z., ŠŤOVÍK, M., TYROLER, J. and VONDRÁKOVÁ, Z. (1959): Geochemicko- -prospekční průzkum germania a gallia v uhlí Vejprnického důlního pole. ÚNS Kutná Hora, pracoviště Praha, Geofond, Praha. Unpublished.
ZAHRADNÍK, L., TYROLER, J. and VONDRÁKOVÁ, Z. (1959a): Příspěvek ke geochemii germania ve středočeské pánvi. Geol. průzkum, 5, 140—141, Praha.
— — — (1959b): Studie o vazbě germania v uhlí plzeňské kamenouhelné pánve. Sborník prací Hornického ústavu ČSAV, 192—200, Praha.
— — — (1960a): Příspěvek ke geochemii germania v plzeňské kamenouhelné pánvi Věstník ÚÚG, 35, 459—468, Praha.
— — — (1960b): Obsah germania ve slojových pásmech plzeňské kamenouhelné pánve. Sb. VŠCHT, 4 (2), 267—275, Praha.
ZÁZVORKA, M. (1947): Mikroelementy — prvky sledované ve výživě. Chemický obzor, 66—71, Praha.
ZELENKA, O. (1972): Předběžné zhodnocení chemických rozborů popela hnědého uhlí z výzkumných vrtů v ústecké a teplické části severočeské hnědouhelné pánve. Uhlí, 20, 419—422, Praha.
— (1973): Chemicko-technologické ukazatele stupně prouhelnění uhelné sloje v mostecké, teplické a ústecké části severočeské hnědouhelné pánve. Uhlí, 21, 59—63, Praha.
— (1974): Vliv vosků a pryskyřic na chemicko-technologické vlastnosti uhlí v severočeské hnědouhelné pánvi. Uhlí, 22, 321—325, Praha.
ZELLER, H. D. (1955a): Preliminary geologic map of the Bar H area, Slim Buttes, Harding County, South Dakota. U. S. Geol. Survey Coal Inv. Map C 37, Washington.
— (1955b): Reconnaissance for uranium-bearing carbonaceous materials in southern Utah. U. S. Geol. Survey Circ., 349, p. 9, Washington.
— (1957): Results of exploratory drilling for uranium-bearing lignite deposits in Harding and Perkins Counties, South Dakota, and Bowman County, North Dakota. U. S. Geol. Survey Bull. 1055-C, Washington.
ZHEMCHUZHNIKOV, YU. A. (1948): Obshchaya geologiya iskopaemykh uglei. Ugletekhizdat, Moskva, 491 p.
— (1952): Ob uglefikatsii i metomorfizme uglei. Izv. AN SSSR, 1, 51—62, Moskva.
ZHOU, Y. P. (1974): Two types of germanium distribution in coal beds. Sci Geol. Sinica, 182—188. (Chinese with English abstract.)
ZUBOVIC, P. (1966a): Minor element distribution in coal samples of the Interior Coal Province. In Coal Science, Am. Chem. Soc., Adv. Chem. Ser., 55, 232—247, New York.
— (1966b): Physicochemical properties of certain minor elements as controlling factors in their distribution in coal. In „Coal Science", p. 221. Advances in chem., ser. 55, Ed. by R. F. Gloud, Washington.

ZUBOVIC, P., SHEFFEY, N. B. and STADNICHENKO, T. (1961a): U. S. Geol. Survey Prof. Paper, 424D, D 324–348, Washington.
– – – (1967): Distribution of minor elements in some coals in the Western and Southwestern Regions of the Interior Coal Province. U. S. Geol. Survey Bull. 1117-D, Washington. 3 p.
ZUBOVIC, P., STADNICHENKO, T. and SHEFFEY, N. B. (1961b): Geochemistry of minor elements in coals of the Northern Great Plains Coal Province. U. S. Geol. Survey Bull. 1117-A, Washington. 57 p.
– – – (1964): Distribution of minor elements in coal beds of the Eastern Interior Region. U. S. Geol. Survey Bull. 1117-B, Washington. 41 p.
– – – (1966): Distribution of minor elements in coals of the Appalachian Region. U. S. Geol. Survey Bull. 1117-C, Washington. 37 p.
ZÝKA, V. (1971): O významu stopových prvků pro živé organismy, I, II. Ústav nerostných surovin. Kutná Hora.
ŽÁK, L. (1961): Opál s germaniem z Března u Chomutova. Sb. ÚÚG, Praha, 18, 641–647.

INDEX

by M. Pačesová

Page numbers in bold type refer to the main information
of the elements' content in coals and plants

Absorption, 44, 54, 166, 184, 202, 203, 211, 218, 224
–, index, 116, 118
Acer, 42
Acetaldehyde, 77
Acetone, 77
Acid(s), abietic, 45, 48, 92
–, acetic, 77
–, agathene-dicarboxylic, 49
–, amino, 31, 33–35, 46, 55, 77, 79, 80, 83, 111, 135, 136, 213
–, ascorbic, 46
–, benzene hexacarbonic, 153
–, boric, 28
–, butyric, 77
–, carbonic, 135, 153
–, carboxyl, 80
–, citric, 46, 77
–, copalic, 49
–, cutinic, 47
–, cutininic, 47
–, diamino, 135
–, diterpenic carboxylic, 49
–, ellagic, 79
–, fatty, 46, 47, 75, 79, 136
–, formic, 46
–, fulvic, 83, 88, 211
–, giberellic, 50
–, glutamic, 46, 136
–, humic, 73–79, 81–83, 85, 86, 88, 91, 92, 96, 122, 126, 130–135, 187, 203, 205, 206, 208, 210, 211, 215

Acid, hydroxymonoamino carboxylic, 136
–, hymatomelanic, 83
–, inorganic, 53
–, ketonic, 112
–, lactic, 46, 77
–, linoleic, 46
–, linolenic, 46
–, α-lipoic, 46
–, malic, 46
–, mineral, 28, 78, 79
–, monoamino, 135
–, – dicarboxylic, 136
–, nicotinic, 151
–, nucleic, 46, 55
–, oleic, 46
–, organic, 46, 52, 60, 77, 80, 83
–, oxalic, 46, 77
–, palmitic, 46, 79
–, phellogenic, 48
–, phellonic, 48
–, phenolcarboxylic, 137
–, phloionic, 48
–, phloionolic, 48
–, phosphoric, 55, 60, 80
–, phytanic, 37
–, pyruvic, 77
–, resinoic, 48
–, ribonucleic, 59
–, sandaracopimaric, 49
–, silicic, 65, 196
–, stearic, 46
–, succinic, 58

Acid, sulphuric, 122, 158
–, terpenoic, 48
–, triterpentic carboxylic, 49
–, ursolic, 138
–, wax, 80
Acridine, 147
Actinolite, 109
Adsorption, 28, 187, 194, 203, 205, 215
Aeration, 81
Africa, 11, 161
Agathis australis, 49
Age, absolute, 12
–, radiometric, 11
Aglycone, 79
Agrostis alba, 61
Ak-Kinchungai, 13
Aksai, 184
Alabama, 169
Alanine, 35, 80
Alaska, 22
Albite, 105, 109, 156
Albona, 161
Alcohols, 47, 49, 69, 77, 80
Aldehydes, 49, 132, 133
Alder, 20, 24, 59
Alethopteris, 70
Aleuropelite(s), 160, 197, 209
Alfred, 79
Algae, 12, 16, 24, 42, 46, 47, 56, 63, 70, 73
–, oleaginous, 73
Algonkian, 156, 157
Alingite, 161
Alkalies, **58**, 73, **178**, 199, 230
Alkaline lyes, 77, 78
– carbonates, 78
Alkalinity, 73
Alkaloids, 48, 55, 179
n-alkanes, 36, 37, 47, 69, 136
Alluvium, sandy, 173
Almandine, 157
Alnus, 41
Alps, 163, 168
Altenbeichlingen, 180
Alteration, 103, 118, 216
Alum, 91, 94, 162
Alumina, 152, 154, 155, 231
Aluminium, 28, 52, 53, 56, **58**, 62, 92, 93, 135, 152, **164**, 165, 172, 205, 208, 213, 215
– succinate, 58
Alumosilicates, 172, 205

Alunogen, 158
Amber, 49, 64, 161, 162
–, Baltic, 49
Ambrite, 161
Amides, 46
Amines, 46, 55, 79
Ammonia, 55, 72–74, 79, 179
Amphiboles, 153, 155
Amsterdam Island, 22
Amylase, 78
Anaconda, 59
Analcime, 109
Andesite, 59, 100
Angiospermae, 12, 24, 38, 41, 42, 46, 49, 59, 71
Anhydrite, 160
Anisotropy, optic, 127
Ankerite, 153, 155, 157, 159, 160, 162, I/1, II/1, I/2, II/2
Annularia, 70
Anorthite, 156
Anthocyans, 50
Anthracene, 130, 137
Anthracite, 10, 13, 14, 39, 83, 96–98, 100, 105, 106, 108, 111–115, 117, 121, 123, 126–131, 135, 143, 153, 164–166, 168, 173, 177, 179, 180, 183, 185, 186, 188, 211, 230
Anthracitization, 108, 144
Antimonite, 156, 157
Antimony, 56, **65**, **182**, 189, 199, 200, 215
Apatite, 153, 155, 171
Arabans, 45
Arabinose, 45, 69, 136
Aragonite, 157
Araucariaceae, 49
Araucarixylon, 210
Archaeopteris, 39, 69
Argentine, 187
Argillite, 158, 183
–, bituminous, 186
–, coaly, 161, 186
Argentina, 187
Argon, 177
Aridization, 27, 31
Arizona, 185
Arkose(s), 160, 172, 205
Armeria hallerii, 67
Aromaticity, 147, 148, 150, 151
Aromatization, 111, 130
Arsculus hippocastanum, 60

Arsenic, 33, 56, **58**, 73, 85, 163, **165**, 189, 194, 198 to 200, 215, 229
Arsenopyrite, 153, 156, 157, 159, 165
Ascomycetes, 78
Ash, 20, 23, 32, 52, 53, 77, 86, 89, 95, 102, 105, 125, 128, 129, 132, 133, 154, 155, 169, 225, 227 to 229, 231
–, accumulation of trace elements, 189–216
– colour, 52
– matter, 52
– –, internal (primary, proper), 153, 189
– –, external (secondary, free), 153, 189, 206, 215, 228
– – –, postgenetic, 153
– – –, syngenetic, 153
Asia, 180, 183
Asidium marginale, 54
Asparagine, 35, 136
Aspergillus niger, 62
Asphalt, 10, 92, 94, 160
Asphaltenes, 151
Aspidium, 39
Asplenium viride, 64
Assam (India), 174
Assimilation, 50
Asterocalamites, 39
Astragalus, 65
Asturia (Spain), 175
Atmosphere, 11, 38, 96, 111, 112
Atropine, 48
Aurich, 92
Aussee, 92
Australia, 188
Austria, 168
Authigenesis, 28
Autunian, 160
Autunite, 163, 185, 210
Axerophtol, 50
Azurite, 160

Babice, 160
Bach, 161
Bacillus cellulosae methanicus, 76
– – *hydrogenicus*, 76
Bacteria, 16, 47, 55, 63, 65, 67, 70, 72, 73, 75, 76, 79, 89, 122, 182
–, soil, 55
Baggiatoa, 65
Baltic region, 49

Barite, 153, 156, 157, 159, 160, 162, 163, 166, 180, I/2
Barium, 32, 56, **59**, 88, 93, **166**, 167, 183, 189, 194, 197, 199, 200, 208, 212, 213, 215
Barley, 62
Barzas, 13, 39
Basidiomycetes, 78
Basin(s), 24, 25
–, Appalachian, 25, 188
–, Belgian, 24, 25
–, Brandov, 123
–, Central Asian, 184
–, – Bohemian Carboniferous, 24, 25, 30–32, 123, 203
–, Cheb, 123, 173, 187
–, Chelyabinsk, 25, 40
–, Chomutov–Ústí, 97
–, Chukotka, 40
–, coal-forming, 25
–, Czechoslovak, 198
–, Dnieper, 137, 170, 171, 179, 183, 187
–, Döhlen, 197, 209
–, Donets, 24, 40, 105, 107, 168, 171, 176, 181, 207
–, Eastern Maritsa, 164
–, East of the U.S.A., 40
–, Ebro, 184
–, European, 39
–, Gippsland, 107
–, Great Divide, 209
–, Intra-Sudetic, 198
–, Irkutsk, 24
–, Karaganda, 39
–, Karlovy Vary, 123
–, Kazakhstan, 203
–, Kiselovsk, 181, 187, 206
–, Kladno, 128, 152, 156, 157
–, Kladno–Rakovník, 177, 185, 210, 226, VIII/1
–, Kounov, 105
–, Kuznetsk, 13, 25, 39, 40, 103, 176, 183
–, Lena, 229
, Lenskiĭ, 176
–, limnic, 198
–, Ljubljana Coal, 231
–, Lower Silesian, 170, 173, 179, 183, 186, 197
–, Lowlands, 25
–, McAlester, 160
–, Minusinsk, 39
–, molassoid, 184
–, Moscow, 13, 25, 107, 116, 183
–, Narragansett, 105, 143

Basin, North Bohemian Brown Coal, 24, 25, 100, 107, 123, 128, 135, 137, 155, 162, 165, 168, 171, 181, 182, 224, 231
–, Nováky, 182
–, Ostrava, 152
–, Ostrava-Karviná, 24, 123, 180, 185
–, paralic, 198
–, Paris, 163
–, Pechora, 40, 184
–, Petřvald, 100, 104
–, Pkhelarovo Coal, 164
–, Plzeň, 128, 172, 173, 203, VI/1, VI/2, VI/3
–, Quangyen, 40
–, Radnice, 173
–, Rhenish, 24
–, Richmond, 40
–, Rosice, 152
–, Rosice-Oslavany, 24, 123, 173, 182
–, Ruhr, 24, 25, 105, 111, 134–136, 156, 170, 171, 174, 177–179, 182, 183, 188, 227, 231
–, Saar, 24, 103, 134, 135, 177
–, sedimentary, 27
–, Sokolov, 99, 112, 123, 127, 162, 165, 168, 173, 181, 182, 201
–, South Moravian, 162
–, South-Uralian, 137
–, Taimir, 40
–, Taranski, 107
–, Transcarpathian, 183
–, Třeboň, 177
–, Tunguzka, 40, 101
–, Upper Silesian, 17, 36
–, Vejvanov, 173
–, Volga-river, 23
–, Warwickshire, 184
–, Yakutsk, 176, 179
–, Žacléř-Svatoňovice, 123
–, Zirianski, 171
–, Žitava, 173
Batylykh series, 229
Bavaria, 184
Beech, 44, 52, 58, 59
Belgium, 156
Belorussian SSR, 85
Bennettitaceae, 12
1,2-benzanthracene, 137
Benzene, 15, 77, 78, 130, 137, 149
– ring, 145, 146
Benzol, 147
Benzolfluorene, 137

Benzophenylene oxide, 137
Berthelsdorf, 191
Beryllium, 56, **59**, 85, 93, **166**, 167, 169, 197, 199 to 201, 206, 208, 210, 212–215, 224, 230
Betula alba (verrucosa), 47
– *resinifera*, 65
Betulin, 47, 75
Beyrichite, 157
Bílina, 162
Bilinite, 162
Biolith(s), 10, 212
α-biotin, 55
Biotite, 95, 107, 108, 109, 155
Birch, 24, 47, 52, 59, 62, 63, 65, 66, 86
Birefringence, 127
Bismuth, 56, **60**, 85, **168**, 189, 199, 208, 213
Bitumen, 33, 70, 73, 80, 81, 83, 84, 112, 131, 133, 137, 156, 196
Bitumenization, 73, 97
Boehmite, 155
Boghead(s), 21, 42, 73, 112, 135, 136, 177, 196, 197, 230
Bog ores (iron ores), 91, 93
Bohemia, 123, 134, 172, 198, 229, 230, VII/2
Bohemian Massif, 185
Bombiceite, 138
Borate, 29
Borax, 29
Borkovice, 21, 92, 177
Borkovická blata, 20, 91, 93, 230
Bornite, 160, 197
Boron, 28–32, 35, 50, 56, **59**, 85, 93, 103, **166**, 189, 194, 198–200, 213–215, 230
Borosilicates, 166
Boskovice, 160
Boží Dar, 19
Brachyphyllum, 40
Branchite, 138
Brandov, 165, 170, 173, 178, 180, 183, 186, 188
Bráníšov, 173
Bravoite, 157, 179
Brazil-nut tree, 59
Březina, 161, 179, VII/1
Březno, 177
Brno, 160
Bromine, 56, 159, **177**
Brux, 163
Bryophyta, 12, 46
Buck wheat, 62
Bulgaria, 187

Buzzards Bay, 28
Bylomskoe (ASSR), 176
Byňov, 162
Bytom, 156
Bžany-Hradiště, 128

Cacoxene, 161
Cadmium, 56, **60**, 85, **168**, 188, 199, 200
Calamariaceae, 12
Calamites, 39, 69
Calcination, 160
Calcite, 93, 152, 153, 155–157, 159–163, 210
Calcium, 16, 17, 20, 27, 29, 52, 55, 56, **60**, 63, 73, 75, 88, 92, 93, 135, 152, 154, 155, 159, 164, **168**, 187, 189, 193, 194, 208, 213, 215
– oxalate, 60
California, 161
Callipteris, 40
Callixylon, 185
Cambrian, 12, 13, 38, 113, 184, 185
Camptopteris, 40
Campylium polygonum, 64
Canada, 41, 65
Cannel(s), 21, 42, 73, 112, 127, 135, 177, 196, 197, 207, 211, 227
Carbohydrates, 11, 81
Carbon, 10–12, 33, 36, 47, 48, 51, 53–55, 72, 73, 75, 77, 83, 86, 89, 96, 97, 103–105, 112–115, 123, 125, 128, 130–132, 134–137, 140–151, 189, 194, 195, 215
– dioxide, 11, 38, 43, 49, 67, 72–74, 77, 96, 111, 130, 131, 158, 177, 195
– –, atmospheric, 54
– –, in water, 54
– isotope, 54, 55
Carbonate(s), 28, 29, 52, 128, 130, 152, 153, 155 to 157, 159, 163, 171, 178, 198
Carboniferous, 12–14, 20, 21, 27, 28, 30, 32, 36, 38, 39, 42, 70, 73, 99, 104, 105, 107, 113, 134 to 136, 156, 158, 167, 175, 177, 191, 195, 200, 202, 230, VI/1, VI/2
Carbonization, 99, 135, 158
Carex, 86, 87
Carnotite, 210
Carotene, 50
Carotenoids, 50
Carpathians, 183
Carter County, 185
Carya alba, 66
Cassia County, 186

Caucasus, 176, 183
Caustobiolith(s), 10, 11, 21, 70, 72, 73, 89, 98, 106, 122, 136, 229
–, gaseous, 96
–, solid, 96
Cave Hills, 185
Cedar, 25
– Creek Bog, 79
Čejč, 162
Celestite, 66
Cellulose, 45, 46, 50–52, 54, 69, 70, 75–78, 80–84, 98, 112, 122, 131, 134
Cenomanian, 161, 177, 200, 201, 230
Ceratozamia mexicana, 64
Cerium, **187**, 213
Cesium, **178**
České Budějovice, 128, 156, 164, 165, 173, 186
Chalcedony, 156, 160, 163
Chalcocite, 160, 197
Chalcopyrite, 153, 156, 157, 159, 160
Chambéry, 89
Chamosite, 160
Characeae, 53
Chattanooga, 209
Chechum series, 229
Chelate(s), 50, 86, 213, 214, 217
Chesapeake Bay, 23
Chestnut, 24
Chinook mine, 153
Chitin, 45, 46, 48, 79, 80, 83
Chloride(s), 52, 152, 153
Chlorine, 53, **60**, 159, **168**
Chlorite, 93, 105, 153, 155, 107, 109, 161, 163
Chlorophyll, 49, 50, 55, 56, 61, 63, 67, 80, 84, 179, 213
Chloroplast, 46
Chlorosis, 60, 62
Chodov near Karlovy Vary, 168
Chomutov, 162, 177
Chotěšov, 173
Chotíkov, VI/1, VI/2
Chromium, 32, **61**, 74, 88, 93, **170**, 189, 197, 199, 200, 208, 210, 212, 213, 215, 227, 229
Chrysene, 137
Chrysocolla, 160
Churchill County, 186
Chvaleč, 180
Cladophlebis, 40
Cladorite, 94
Clarain, 21, 154, 185, 187, 188, 197, 204, 212

264

Clarain cuticle, 21
– megaspore, 21
– microspore, 21
Clay(s), 32, 74, 99, 155, 161, 165, 166, 170, 184, 203, 216, 221, 231
–, coaly, 166, 170, 179, 186, 230
– mineral(s), 28, 29, 92, 95, 103, 155, 162, 163, 170, 178, 196, 203, 205, 210, 231, VIII/1
–, saprocol, 74
–, sapropel, 74
Claysone(s), 27–29, 31–33, 108, 135, 136, 154, 161, 166, 171–173, 176, 184, 186, 187, 197, 199, 203 to 205, 208, VII/1
–, boghead, 230
–, coaly, 108, 180, 184–186, 212
–, freshwater, 32, 36
–, marine, 29, 36
–, sandy, 177, 186
Clematis vitalba, 59
Climate, 17, 24–27, 38, 40–42
Climatic oscillation, 33
Clinkers, coaly, 186
Club-mosses, 24, 39
Coal, amino acids content, 134
–, anisotropic, 100
–, anthracitic, 96, 100, 107, 123, 128, 135, 186, 207
–, bitumen, 136
–, bituminous, 10, 11, 13, 14, 39, 83, 89, 90, 96–98, 100, 101, 104, 105, 107, 108, 110, 111, 113–115, 117, 121–123, 126–129, 134–136, 137, 140, 143, 144, 147, 149, 151–155, 161, 163–167, 172, 177 to 186, 188, 198, 202, 211, 217
–, –, baking, 128, 129
–, –, coking, 107, 123, 128, 129
–, –, fat, 96, 107, 118
–, –, gas, 96, 123, 207
–, –, minerals, 155
–, brown, 10, 11, 13, 14, 21, 24, 83, 89, 90, 96–100, 107, 108, 111–115, 117, 118, 122, 123, 126–131, 134–138, 140, 143, 146, 152–154, 160, 161, 163 to 167, 170–172, 175, 177–188, 194, 198, 199, 207 to 211, 217, 223
–, –, earthy, 123
–, –, waxy, 123
–, cannel, 126
–, carbonaceous, 108
–, chemical elements content, 128, 129, 164–188
–, chemical structure, 141–151
–, coke, 127
–, collinitic, 209

Coal, contact-metamorphosed, 102
–, "diluvial", 89
–, gas, 118
–, humic, 10, 112
–, humitic, 197, 230
–, "paper", 134
–, pyritized, 178
–, reserve, 14
–, sapropelic, 24
–, sapropelitic, 112, 135, 230
– seam, 13, 14, 17, 25, 27, 32, 38, 39, 103
– – ashes, 62, 68, 86, 154
– – –, chemical elements content, 164–188
– – correlation, 225–231
– – identification, 225–231
– – mineralogy, 152–163
–, silitic, 177
–, structural parameters, 147
–, subbituminous, 96, 107, 108, 112, 118, 121, 123, 128, 129, 181, 211
–, waxy, 112, 136
–, xylitic, 137, 163, 166
Coalification, 42, 49, 72, 75, 76, 82, 83, 89, 90, 96 to 100, 103–105, 107, 108, 110–114, 116, 122–126, 128, 130, 131, 134, 141, 142, 144, 146, 147, 151, 166, 179, 181, 189, 198, 202, 203, 205, 209, 211, 215, 216, 223
– degree, 109, 114
– diagram, 116
– parameters, 117, 118
–, relation to mineral facies, 109
– scheme, 106
– stages, 90, 96, 121
Coalified log, VII/1
Coals, American, 155, 166, 170, 188, 208
–, Belgium, 174
–, Bohemian, 173, 177
–, British, 166, 171, 187
–, Bulgarian, 164
–, Colorado, 175
–, Columbia, 175
–, Czechoslovak, 172, 173, 178, 181
–, Donets, 168
–, Ekibastuzskoe, 171
–, England, 171, 172, 175
–, European, 192, 208
–, Garbagataiskoe, 171
–, G.D.R., 184
–, German, 164, 177, 184, 187
–, Illinois, 168, 175, 177, 181, 186

Coals, Indian, 187
-, Italy, 172
-, Japanese, 166, 172
-, Kladno, 170
-, Kentucky, 175
-, Montana, 175
-, Moravia, 173
-, Norway, 172
-, Ohio, 175
-, Ostrava-Karviná, 178
-, Peacock, 170
-, Pennsylvania, 153, 175
-, Polish, 184
-, Prince George County, 175
-, Ruhr, 165, 168, 178
-, South Africa, 165
-, Spanish, 180
-, Tennessee, 175
-, U.S.A., 168, 169, 171, 172, 177, 185, 214
-, U.S.S.R., 165, 182, 187
-, West Carpathians, 166
Cobalt, 33, 50, 56, **60**, 85, 86, 93, 158, **168**, 169, 179, 196, 197, 199, 200, 208, 210, 212, 213, 215, 227, 229, 230
Cocaine, 48
Coefficient of diffusion, 218, 219, 221, 223, 224
- - permeability, 220, 221
- - shrinkage, 107
Coffinite, 210
Coke, 15, 100, 103, 127, 211
Coking, 14
Colemanite, 29
Colinite, 118, 197, VI/2
Colloids, 76, 194, 196, 203
-, clay, 196
Colorado, 185, 186, 210
Colour, 124, 126, 127
Columbia, 194
Complex, alkaline, 210
-, alkaline-earth metal uranyl-carbonate, 210
-, metallo-organic, 213
-, organometallic, 182, 187, 210, 214, 217
-, organo-uranium, 210
-, uranium-organic, 185, 211
-, vanadium-porphyrin, 136
Compound(s), germano-organic, 205, 208
-, inorganic, 44
-, organic, 44, 45, 165, 171, 205
-, organometallic, 180, 194
-, uranium-organic, 185, 209

Compound(s), uranyl-organic, 210
Comptonia, 41
Concentration, 218, 219, 222–224, 227
Concretions, 156
-, calcareous, 156
-, ferruginous, IV/2
Condensation, 75, 111, 147
- index, 147, 148
Conductivity, 27
Conglomerates, 209
Coniferae, 12, 24, 39–42, 49, 52–54, 56, 58–65, 67, 68, 138, 200
Coniferyl alcohol, 46
Copiapite, 91, 92, 158, 160
Copper, 32, 33, 50, 52, 56, **61**, 74, 85, 86, 88, 93, 156, 160, **170**, 189, 194, 196, 197, 199, 200, 208, 212–215
Coprinus, 53
Cordaitaceae, 12, 40, 49, 69, 70
Correlation, stratigraphical, 27, 32, 33
Cotton grass, 17, 20
Coventry, 208
Cranston, 105, 143
Creosol, 136
Creston Ridge, 186, 227
Cretaceous, 12, 13, 38, 40, 41, 107, 133, 157, 161 to 163, 167, 170, 172–176, 179, 182, 185–187, 194, 195, 197, 202, 203, 217, 229, 230, IV/2
Cristobalite, 163
Crock, 191
Crustaceans, 73
Cryptogams, 53, 58, 64
Cryptohalite, 158
Cryptomeria, 138
Cupressaceae, 132
Cupressinoxylon durum, 132
- *hausruckianum*, 132
- *wardi*, 187
Cuticle, 10, 13, 24, 53, 75, 134
Cutins, 47, 51, 71, 72, 75, 80
Cyathea, 41, 53
- *caniculata*, 54
Cycadaceae, 12, 40, 41
Cyclization, 111, 131
Cyclostigma, 39
p-cymene, 47, 48, 138
Cypress, 25
Cysteine, 55
Cystine, 55, 136
Cytoplasm, 55

Czechoslovakia, 161, 178, 201
Czekanowskia, 40

Dachmoos, 92
Dacrydium, 138
Dakota, 165, 185, 186, 209
Danalopsis, 40
Darcy's law, 221
Darkov, 177
Darlington area, 185
Daun, 39
"Dead line", 118
Decarboxylation, 37, 80, 92, 122
Dehydration, 56, 92, 122
Dehydroabietane, 48
Dehydrogenation, 48, 137
Demethanation, 122
Denhardite, 161
Denmark, 187
Density, 124–126, 150, 184, 197, 212, 224
Deposit(s), allochthonous, 22
–, autochthonous, 20
–, bituminous, 20
–, brown coal, 20
–, caustobiolith, 16
–, Chinese coal, 25
–, clayey, 135
–, coal, 13, 14, 16, 23–26, 40, 41, 75
–, freshwater, 32
–, gaseous, 10
–, kukersite, 25
–, limnic, 24
–, liquid, 10
–, marine, 27, 32
–, mineral, 44
–, nickel, 60
–, oil, 14
–, peat, 16
–, silty, 135
–, solid, 10
Desorption, 223, 224
Desoxyphyllerythrin, 136
Dessulfovibrium, 65
Detritus, 32
Deuterium, 37, 55
Deuteroethio-porphyrin, 136
Devonian, 12, 13, 38, 39, 69, 70, 113, 175, 176, 185 to 187, 209
Devonshire, 24, 136
Diadochy, 28

Diagenesis, 28, 29, 32, 36, 48, 50, 97, 108, 109, 153, 183, 205, 217, 223, 226
Diamond, 130
Diaspore, 156
Diastrophism, 17, 22, 25, 26
Dictyophyllum, 40
Diffusion, 219–224
–, convective, 217, 220, 221, 223
–, molecular, 217
9,10-dihydroanthracene, 145, 146
Dihydrodiconiferyl alcohol, 46
Dihydrophenanthrene, 146
Dihydroporphyrin, 50
Diphenyl, 147
Diphenylene oxide, 137
Dipterocarpaceae, 49
Disaccharides, 45, 77, 83
Dismal Swamp, 23, 25, 79, 80
Dissimilation, 50
Disthene, 156
Distribution coefficient, 29
District, Almazno-Marevskii, 168
–, Bylym-Kabardinskoĭ, 178
–, Centralnyi, 168
–, Goose Creek, 186
–, Kladno, 180, 183, 186, 188
–, Lisikhanskii, 168
–, Žacléř-Svatoňovice, 181, 182
–, Stalinsko-Makeevski, 168
–, Zwickau-Oelsnitz, 168
Diterpenoids, 48, 138
Dithioether, 181
Doberlug, 191
Dölau, 191
Dolerite, 101, 102
Dolomite, 32, 128, 153, 155, 157, 160, 162, 163
Donbas, 177
Dopplerite, 91, 92, 163
Dorset, 175
Doubrava, 31, 172, 193
Dresden, 209
Dubňany, 162
Duchcov, 162, III/2
Durain, 21, 154, 188, 197, 204, 206–208, 212
Durham, 101, 163
Durite, 122, 127, 153
Duxite, 137, 162
"Dy", 76, 197
Dysprosium, 187
Dzhergalan, 184

East Midlands, 36
Ebersdorf. 165. 191
Eh, 24, 27, 32, 37, 44, 70, 72–75, 89, 96, 122, 195, 196, 216, 217, 223
Eisden, 167
Ekka, 103
Eksenyakh series, 229
Elaterite, 160
Elatides, 40
Electron spin resonance ESR, 118, 120–122
Elements, biogenic, 55, 56
–, biophyle, 58, 61, 171
– concentration, 193, 194
– distribution, 225–231
–, lithospheric, 56
–, organogenic, 55
–, stratigraphy purposes, 225–231
Energy, 11
–, chemical, 14
–, electrical, 14
–, geothermal, 14
–, mechanical, 14
–, solar, 14
–, thermal, 14
England, 25, 80, 154, 156, 161
Environment, 88
–, acidic, 224
–, aerobic, 48, 196
–, alkaline, 224
–, anaerobic, 48, 88, 196
–, aqueous, 197
–, brackish, 23, 27, 29, 32
–, coal-forming, 226
–, continental, 23, 24
–, depositional, 27
–, fluviatile, 24
–, freshwater, 23, 24, 27, 31–33, 35
–, lacustrine, 24, 36
–, marine, 23, 27, 32, 33, 35, 36, 86
–, oxidizing, 197
–, peaty, 80
–, primary, 226
–, residual, 29
–, saline, 32
–, sedimentary, 27, 32
–, subaqueous, 197
–, swampy, 80
–, terrestrial, 37
Enzymes, 50, 56, 72, 78
Eobacterium, 11

Eocene, 14, 24, 49, 50, 136, 186, 209
Epidote, 109, 153, 156, 229
Epigenesis, 97, 108, 109
Epizone, 109
Epsomite, 91, 94, 157, 158, 160, 162
Equisetaceae, 12, 40, 53, 69, 202
Equisetum arvense, 52, 59, 62, 64, 65
– *brongniarti*, 69
– *hyemale*, 54, 62
– *palustre*, 52, 59
– *silvaticum*, 66, 69
– *telmateia*, 65
Erbium, 187
Ergones, 49
Ericaceae, 79
Eriophorum vaginatum, 17
Erlandite, 168
Esters, 47, 49, 71, 80
–, glycerol, 47
–, *p*-hydroxyacetophenone, 79
–, phenolic, 79
Estonia, 23, 187
Ethers, 149
Ethyl alcohol, 77
Ethylene, 15
Europium, 187
Eusporangiales, 12
Evaporite, 29
Everglades, 23
Exines, 75, 122
Exinite, 113, 114, 118, 120, 122, 126, 128, 147–151

Fagus, 41, 42
Fallcreek, 186
Farnesane, 140
Fats, 7, 50, 51, 53, 54, 70, 79, 80, 85, 112, 136
Feldspar, 30, 95, 153, 156, 163
Fens, 76, 93, 94
–, sulpho-ferruginous, 94
Fermentation, 50, 72, 73, 78, 79
Ferns, 24, 39 41, 59, 61, 63, 64, 66, 67, 69
Ferrohydrite, 93
Fichtelgebirge, 92
Fichtelite, 48, 91, 92, 94, 138
Fick's law, 219
Ficus, 41
Fig Tree Formation, 11
Filtration, 217–221, 223
–, diffusive mechanism, 217
–, progressive, 218, 223

Fimmenite, 92
Fins, 184
Fir, 52
Flagstaffite, 48
Flavobacterium, 78
– *resinovarum*, 48
Flavoproteins, 50
Flöha, 191
Flora, 11, 38–40, 42, 50
–, cryptogamous, 39
–, dendriform, 41
–, fossil, 26
–, thermophilous, 42
Florida, 23
Fluoranthene, 137
Fluorapatite, 171
Fluorene, 137
Fluorine, 56, **61**, **171**, 177, 203
Fluorite, 160
Forces, van der Waals, 196
Formation, Arikaree, 185
–, Bochum, 136
–, Browns Park, 209
–, Essen, 136
–, Hellcreek, 185
–, Karviná, 31, 36, 103, 156, 173, 193, 230
–, Kladno, 30
–, Líně, 30, 31
–, Mississippi, 186
–, Ostrava, 31, 104, 155, 160, 178, 192, 193, 198, 230
–, Pennsylvanian, 186
–, Slaný, 30
–, Týnec, 30, 31
–, White River, 185
–, Žacléř, 170, 173
–, Zliv, 173
Fossilization, 161
Fracture, 124, 127
France, 184
Františkovy Lázně, 94
Freithal, 191, 209
Fremont County, 163
French Massif Central, 184
Friedelan, 138
Friedelin, 138
Fructose, 45, 136
Fuel, 14
–, gaseous, 14
–, heavy oil, 15

Fuel, liquid, 14, 15
–, nuclear, 14
–, solid, 14
Fulvenes, 183
Fungi, 12, 16, 46–48, 53, 63, 72, 75–80, 82
–, fiber, 24
– *imperfecti*, 78
Furane, 130
– chain, 149
Fusain, 137, 143, 154, 163, 168, 169, 173, 197, 202, 206, 207, 212, 216, 228
Fusinite, 113, 121, 150, 151, 211, VI/1
Fusinitisation, 72
Fusite, 127, 156
Fusitization, 144

Gadolinium, 187
Gagate, 187, 199
Galactans, 45, 52, 69, 70
Galactose, 46, 69, 70, 136
Galena, 64, 153, 155–157, 159–161, 180, 197
Gallium, 15, 32, **62**, 85, 93, **171**, 186, 189, 194, 197, 200, 205, 212–215, 223, 224, 227, 229
Gangamopteris, 40
Garnets, 153, 155, 160, 161, 163
Gas, 10, 15, 122
–, juvenile, 100
–, natural, 10, 14, 15, 98
Gasperian strip, 43
Gastrioceras subcrenatum, 36
G.D.R., 184, 188, 191, 207
Geisel, 50, 136
Gelification, 76, 98, 199, 217
Gels, 76, 89, 205
Geomicrite, 161
Geothermal gradient, 98–100, 103, 118
Germanates, 206
Germanium, 15, **62**, 85, 86, 170, **172**, 173–177, 185 to 187, 189, 192, 194, 195, 197–209, 212–215, 217, 222–224, 229, 230
Gesterwitz, 161
Gigantopteris, 40
Ginkgoaceae, 12, 40, 41
Gissarskogo khrebet, 166
Given's model, 146
Glauberite, 94
Glauconite, 28, 161
Glaucophane, 109
– schists, 109
Glossopteris, 40

Glucans, 70
Glucopyranose, 70
Glucosamine, 136
Glucose, 45, 46, 69, 78, 131, 136
Glutamine, 35
Glycerol, 79
Glycides, 44, 112
Glycine, 35, 36
Glycosides, 48, 79
Glyptostroboxylon tenerum, 132
Glyptostrobus, 41
Gmelinite, 157
Goethite, 93, 155, 160
Gold, 56, **59**, **165**, 215
Gonten, 92
Gramineae, 53
Granada, 184
Graphite, 99, 101, 107, 108, 116, 130, 142–144, 160, 211
Graphitization, 107, 108, 144
Grass, 25, 59, 61, 62, 64, 65
Great Plains (U.S.A.), 188
Gredneria, 41
Greenland, 40
Greenschists, 105, 108, 109
Greigite, 74
Gronningen, 100
Groups, alicyclic, 146
–, aliphatic, 146
–, aromatic, 146
–, –, dimensions, 147
–, carbonyl, 114, 115, 132, 133, 135, 137, 147
–, carboxyl, 71, 80, 86, 114, 115, 131, 135, 147
–, functional, 114, 115, 135
–, hydroaromatic, 151
–, hydroxyl, 86, 114, 115, 122, 131, 135, 146, 147, 149, 151, 164, 203
–, 4-hydroxy-3-methoxyphenyl, 213
–, methoxyl, 46, 78, 114, 115, 135
–, methyl, 146
–, methylene, 147
–, oxidic, 86
–, phenol, 164
–, phenol hydroxyl, 147
–, polar, 114
–, reactive, 116
–, thioether, 151
–, thiophenol, 151
Gulf Coast, 23
Gymnospermae, 39, 40, 42, 49, 69, 71, 136

Gypsum, 91, 92, 94, 153, 155, 157, 160–163, 210
Gyttja, 74, 197

Habartov, 112
Hafnium, **62**, 199
Hájek (Soos), 94, IV/1, V/1, V/2
Halite, 153, 156, 159
Halle, 136, 161
Halloysite, 155
Halogens, 177
Halotrichite, 158, 160–162
Hamstead, 154
Handlová, 100, 128, 162, 165, 166, 207
Hannover, 62
Harding County (South Dakota), 164, 185, 209
Hardness, 124–126
Hardpan, 91
Harrisburg, 186
Hartite, 138
Harz Mts., 64
Hatchetite, 138, 159, 160
Havírna, 162
Heath, 25, 62
Helchteren-Zolder, 167
Helium, 112, **177**
Hematite, 91, 93, 156, 160, 161
Hemicellulose, 45, 75, 77, 83, 84, 131, 134
Hemixylite, 21
Herzfeld's equation, 219
Heterotroph, anaerobic, 11
Heulandite, 109
Hexahydrite, 158, 162
Hexokinase, 56
Hexosans, 52
Hexoses, 45, 77, 136
Hickory tree, 66, 67
Highmoor bog, 16, 17, 19–21, 25, 83, 91–93, 230
– – "Jizerka", 18
– – "Mrtvý luh", 22
– – "Na Čihadle", 18
– – "V rezervaci", 19
Highmoor peat, 77, 84, 85
– –, chemical analyses, 84
Hilt's rule, 103, 104
Histidine, 35, 36
Hnojné, 207
Hofmanite, 138
Holmium, 187
Holtegoard, 92

Hornbeam, 59, 63
Horní Vernéřovice, 180, 186
Horse-tails, 39, 52, 56, 59, 62, 65, 69–71, 200
Horusice, 93, 177, 230
Hrádek, 173
Hřebeč, 161
Hronov, 160
Hrušov, 31, 178, 192, 230
Humates, 58, 81, 83, 89, 94, 155, 168, 171, 183, 198, 205, 224
Humboldtine, 162, 171
Humic component, 98
Humification, 74, 75, 81, 83, 122, 205
Huminitization, 98
Humins, 135
Humite, 10, 33
"Humolignins", 83
Humolith, 76–78, 80, 81, 83, 112
Humus, 16, 17, 20, 21, 73, 74, 79, 126, 193, 197, 212
Hungary, 86, 207
Hyderabat (India), 174
Hydrargyllite, 155
Hydrated aluminium oxide, 196
Hydration, 92
Hydrocarbonate, 158
Hydrocarbons, 36, 37, 47, 69, 80, 83, 92, 112, 118, 122, 134, 136–138, 147, 163
–, abiogenic, 11
–, diterpene, 49
–, monoterpene, 49
–, natural, 10
–, sesquiterpene, 49
Hydrogen, 10, 47, 50, 51, 53–56, 73, 75, 77, 80, 86, 89, 104, 112–114, 123, 128, 132, 134, 135, 138, 146, 147, 149–151, 189, 195, 210, 212, 215
Hydrogenation, 14
Hydrogen sulphide, 72–74, 95, 159, 181
Hydrogoethite, 93
Hydrolysis, 50, 71, 75
Hydrosilicates, 158
Hydrosols, 76, 196
Hydrosphere, 50
Hydroxide, 32, 94, 157, 171
Hydroxyls, 196
n-hydroxyphenol, 71
Hymenomycetales, 78
Hypergenesis, 122
Hyphae, 24, 48

Idaho, 186, 209
Idria, 161
Ilfeld, 191
Iller, 89
Illinois, 152, 169, 187
Illite, 28–30, 109, 152, 155, 231
Ilmenite, 105
Indiana, 134, 153, 169
Indium, **177**, 188, 200
Indole, 130, 131
Indonesia, 100
Inertinite, 108, 113, 114, 126, 128, 231
Inn, 89
Insect, 73
Inulin, 45
Iodine, 56, **63**, 159, **177**, 194, 198, 215
Ione Valley, 161
Ionic potential, 214
Ionite, 161
Ions ferric, ferrous, 74
Iosene, 138
Iron, 50, 52, 53, 55, 56, **61**, 88, 92–94, 135, 140, 152, 154–156, 158, **171**, 172, 181, 186, 194, 196, 208, 210, 212, 213, 215
Island, Batoan, 25
–, Sakhalin, 13
Isoprene, 48
Isoprenoids, 140
Isotope, 37, 58
Isotopic carbon composition, 130
Ivančice, 160

Jaklovec, 31, 178, 183, 192
Jarosite, 210
Jatulian, 13
Javorník, 136, 162
Jefferson County, 186
Jelenice, 30
Jeníkov, 128
Jet, 126
Jitra, 230
Jizerská louka, 85
Jurassic, 12, 13, 38, 133, 166, 167, 170, 176, 178, 179, 184, 195, 210, 217, 229

Kačice, VIII/2
Kaňk hill, 58, 62, 65, 67
Kaolin, 161, 162
Kaolinite, 28, 29, 92, 95, 109, 152, 155–157, 160, 163, 196, 210, 231

Kaolinization, 216
Kara-Kiche, 184
Karlovy Vary, 112, 136, 139
Karweil's diagram, 116
Katazone, 108
Kayakskoe, 102
Kazakhstan, 13
Kelčany, 162
Kenagaz, 13
Kentucky, 168, 169, 210
Keramohalite, 160–162
Kerchenite, 93
Kernite, 29
Kerogen, 11, 70, 73, 113, 118
Ketones, 80, 83, 137
Khibiny Mts., 60
Kittanning, 169
Kladno, 100, 156–158, 165, 166, 168, 173, 179, 181, 184, 188, 197, I/1, I/2, II/1, II/2, VIII/1, VIII/2
Kladnoite, 158
Kokand, 13, 39
Koktaite, 162
"Kolm" formation, 184
Kolozuby, 162
Konin, 163
Korozluky, 162
Kounov, 153, 156, 173, 186, 203, 204
Kozohrudky, 230
Kranzite, 138
Kratochvílite, 158
Krkonoše Mts., 156
Krušné hory Mts., 165, 182, 187, 209
Krypton, 177
Kukersite, 23, 187
Kutná Hora, 58, 62, 65, 67
Kyjov, 207
Kyzyl Bulak deposit, 13

Laajaniemi, 59
Lactose, 136
Lake Ponchartrain, 80
Laminaria digitata, 63
Laminaribiose, 70
Lampertice, 166
Langmuir's adsorption isotherm, 28
Lanthanides *see* Rare earth elements
Lanthanum, **187**, 189, 212, 213
Lapparentite, 158
Latrobe Valley, 24
Lauchagrund, 191

Laumontite, 109
Laurus, 41
Lavas, acid, 209
Lead, 33, 44, 50, 56, **64**, 74, 85, 88, 93, 156, **180**, 186, 189, 194, 197–200, 208, 210, 212, 213, 215, 229
Lee County, 186
Leguminosae, 49
Leicester, 153
Lematang, 100
Lenzites, 78
Leoben, 143
Lepidodendraceae, 12
Lepidodendron, 39, 70
Lepidodendropsis, 39
Lepidopteris ottonis, 40
Leptosporangiales, 12
Letovice, 160, 162
Letovicite, 158, 162
Leucoium aestivum, 66
Level of organic metamorphism LOM, 117, 118
Lhota p. Džbánem, 128
Lhotice, 105, 164, 165, 173, 185, 186, 230
Lias, 175, 208
Libušín, 158
Libverda, 84
Lichen, 62, 136
Lignin, 45, 46, 50–52, 54, 70, 71, 75–78, 80–84, 88, 98, 107, 111, 112, 122, 131, 133, 134, 205, 206, 213
–, decomposition products, 82
–, demethoxylized, 131
Lignites, 14, 42, 50, 96, 97, 108, 117, 121–123, 126 to 129, 131, 134, 163, 173, 175, 181, 182, 199, 207, 211, VII/2
–, chemical composition, 133
–, Japanese, 186
Lime, 47, 65
Limestone, 25, 33, 59, 61, 62, 128, 130
Limnic basin, 24
Limonite, 91–94, 155, 157, 160–162
Linnaeite, 156, 157
Lipids, 10, 47, 54, 55
Lipnice, 162
Liptobiolith(s), 10, 11, 13, 21, 75, 137, 197
Liptohumite, 10
Liquidambar, 24
Lithium, 32, 56, **58**, **178**, 189, 199, 210, 215, 230
Lituya Bay, 22
Lobatannularia, 40

Löbejün, 191
Lom, 162
Long Pine Hills, 185
Lowmoor bog, 16, 17, 19–21, 24, 25, 77, 83, 91, 92, 230
– peat, 77, 84, 85
Lublinite, 161
Lubná, 204
Lustre, 124, 126, 127
Lutetium, 187
Lužice, 162
Lužické hory Mts., 209
Lužná, 128
Lycopodiaceae, 12, 39, 53, 58, 66, 69–71
Lycopodium, 24, 53, 56, 58, 61
– *annotinum*, 58, 66
– *clavatum*, 58, 66
– *complanatum*, 54
– *dendroideum*, 54
Lyginodendron, 39

Macerals, 114, 118, 120–122, 147, 148
–, composition, 150
Mackinawite, 74
Maghemite, 93
Magnesite, 155, 163
Magnesium, 27, 29, 50, 52, 55, 56, **63**, 73, 88, 93, 152, 155, **178**, 189, 193, 199, 208, 210, 213–215
Magnetite, 93, 153, 156
Magnolia, 41
Malachite, 157, 160, 161
Malesice, 30
Maltase, 78
Maltose, 45, 136
Manebach, 191
Manganese, 43, 50, 52, 53, 56, **63**, 74, 80, 88, 93, **178**, 189, 194, 200, 208, 210, 212, 213, 215, 227
Mannans, 45, 70
Manning, 85
Mannose, 46, 69, 70, 136
Mantov, 173
Marcasite, 91, 94, 95, 103, 152, 153, 155–163, 180, 181, III/1
Mariánské Lázně, 84, 94
Mariopteris, 39
Marl, 66
Marsh, 16, 23
–, deltaic, 14
Maryland, 23, 194
Mascagnite, 158, 160, 162

Massachusetts, 28
Masugusbyn (Sweden), 86
Matières organiques dispersées MOD, 118
Mažice, 92
Měcholupy, 177
Meczek, 208
Medvezhi Islands, 13, 39
Meisdorf, 191
Meisdorf-Oppenrode, 174
Meissen, 178
Melandrium silvestre, 61
Melanterite, 91, 94, 157, 160–162
Mellite, 153, 162–164
Melnikovite, 152, 153, 163, 181
p-menthane, 47, 48, 138
Mercury, 50, 56, **62**, 88, **177**, 189, 200
Merlebach, 167
Mesozoic, 113, 167, 180, 185, 191, 195
Mesozone, 108
Messeix, 167
Meszko (Poland), 180
Meta-anthracite, 105, 107, 108, 117, 121, 126, 143
Metaautunite, 185, 210
Metabolism, 11, 50, 59, 83
Metagreywackes, 109
Metahalloysite, 163
Metamorphism, 29, 50
–, contact, 100, 101, 103, 198
–, critical, 99
–, dynamic, 105
–, progressive, 105
–, regional, 101, 109, 198
–, thermal, 100
Metauranocircite, 210
Metaxylite, 21
Metazeunerite, 210
Methane, 73, 74, 77, 79, 96, 110, 111, 118, 130, 177
Methanosareina, 77
Methine bridges, 140
Methionine, 136
Methods, electrodialysis, 206
–, electron spin resonance ESR, 88
–, gas chromatography, 161
–, infrared spectroscopy, 49, 137, 144, 146, 161
–, isotope analysis, 37
–, mass spectroscopy, 146, 161
–, Mössbauer spectrum, 88
–, nuclear magnetic resonance, 146
–, palynological, analysis, 229
–, paramagnetic resonance, 146

Methods, physical, 49
-, Rb/Sr, 11
-, silicate analyses, 229
-, spectrographic, 165, 168, 170, 183, 189, 192, 226 to 228
-, UV spectra, 149
-, X-ray, 30, 105, 142, 143, 146, 147
-, -, Debye-Scherrer, 142
Methylene bridge, 146
Methyl glyoxal, 77, 111
Methyl pheophorbide, 50, 136
Mezboltice, 162
Mica, 29, 30, 153
Micrinite, 113, 120, 147, 148, 150, 151
Microliths, 113, 147, 151
Microorganism, 35, 44, 48, 76–79, 82, 96, 97
Middletonite, 161
Midland Valley, 25
Mijngebied, 167
Milam County (Texas), 189
Millerite, 156, 157, 159, 160, 179, II/2
Mineralization, 198, 203, 211, 215, 231
Minerals, allogenic, 153
-, authigenic, 153
-, heavy, 229
-, light, 29
-, sulphidic, 165
Miocene, 25, 42, 100, 131, 133, 162, 165, 182, 185, 186, 227, 230
Mirabilite, 91, 94, 160, 162
Mirošov, 160, 173
Misasa group, 186
Mladějov, 173
Mofette, V/1, V/2
Mohria, 24
Molecular weight, 151
Molecule of coal, 144, 146
Molybdenum, 15, 33, 56, **63**, 64, 73, 85, 93, 170, **178**, 180, 186, 196–200, 208, 210, 212, 213, 215, 229
Monosaccharides, 45, 46, 69, 77, 83
Monoterpenoids, 48, 138
Montana, 185, 186, 195
Montgomery County, 186
Montmorillonite, 28, 73, 109, 155, 203, 231
Moravia, 123, 134, 181, 182, 230
Moravská Třebová, 160, 173
Morus alba, 60
Mosses, 19, 25, 40, 58–65, 67, 68
Most, 107, 162, 173, 207

Mouldering, 72, 74, 76, 165
Moulds, 62
Muckite, 161
Mülsengrund, 191
Münden, 178
Münsterland, 135
Muscovite, 95, 105, 109, 155, 156, 163, 178
Mustard, 62
Mycobacteria, 76
Mydlovary, 177, 230, VII/2

Nacrite, 156, 157, 159
Namurian, 31, 36
Naphthalene, 130, 137, 147
Naphthene, 149
Naphthol, 136
Neodymium, 187
Neogene, 162
Neon, 177
Neudorfite, 161
Neuropteris, 39
Neutralization, 75
Nevada, 186
New Brunswick, 39
New Mexico, 92, 161, 185
New Siberian Islands, 22
New Zealand, 29, 107, 161, 166, 194
Niacin, 50
Nickel, 32, 50, 56, 60, **64**, 73, 85, 86, 88, 93, 140, 158, 170, **179**, 185, 196, 197, 199, 200, 208, 210, 212–215, 227, 229, 230
Nilssonia, 40
Niobium, **179**, 199, 200, 212, 229
Nitrates, 55, 63
Nitrite, 55
Nitrogen, 33, 46, 50, 51, 53–55, 63, 73, 75–77, 79, 81, 89, 104, 111–113, 123, 128, 131, 132, 134, 135, 146, 150, 151, 177, **179**, 181, 189, 194, 195, 210, 213–215
North Carolina, 25
North Staffordshire, 137, 153
Northumberland, 36, 175, 183, 227
Nováky, 137, 163, 165
Nova Scotia, 39, 174
Nová Ves, 161, 207
Novosibirsk, 86
Nowa Ruda, 178
N.S. Wales (Australia), 165, 174, 182, 188
Nucleoids, diphosphopyridin, 55

Nucleoids, triphosphopyridin, 55
Nucleoproteins, 46, 80
Nymphaea, 88
– *orontium*, 88
Nýřany, 30, 156, 160, 180, 203, 204

Oak, 24, 52, 58–61, 63, 64, 86
Oat, 62
Obbergen,
Obbürg, 62
Ochre, clayey earthy, 94
–, ferric, 93, 94
n-octacosane, 69
n-octacosanol, 69
Oelsnitz, 174, 177, 191, 228
Ohio, 185
Ohnič, 101
Ohrenkammer, 191
Oil(s), 10, 14, 15, 47, 48, 50, 214
–, crude, 179
– fields, 118
– genesis, 118
–, synthetic, 15
Okefenokee Swamp (Georgia), 86, 88
Oldenburg, 92
Old Leyden, 186
Olefins, 15, 137
Oligocene, 49, 133, 174, 185
Ontario, 79
Onychiopsis, 41
Opal, 65, 92, 157, 162, 177
Opium, 67
Optical constants, 116
Ordering, turbostratic, 141, 142
–, two-dimensional, 141, 142
Ordovician, 12
Oregon, 25
Orestovia, 13
Organic matter, 32, 33, 36, 37, 44, 118, 166, 170, 208, 211, 215, 226, 228
Orijärvi, 60, 61, 68
Orites excelsa, 58
Orpiment, 162, 163, 165
Orthoclase, 155
Orthoquinone, 115
Oslavany, 59, 64, 160
Ostankino, 64
Osterwald, 183
Ostrava-Karviná coalfield (coal district), 17, 23, 28, 31, 34–36, 100, 104, 105, 128, 155, 156, 159, 160, 164–166, 168, 170, 172, 173, 177–183, 186, 188, 192, 193, 198, 207, 228, 230, 231
Otruby, 30
Oxalite, 162
Oxidation, 37, 75, 79, 82, 100, 103, 111, 122, 124, 131, 135, 157, 158, 160, 182, 195
Oxides, 52, 152, 154, 155, 158, 171, 178, 198, 203
Oxiquinones, 112
n-oxybenzaldehyde, 132
Oxygen, 10, 11, 37, 38, 47, 51, 53–55, 61, 72–75, 89, 104, 110–115, 121–123, 128, 131, 134, 135, 137, 147, 149–151, 189, 213–215
Ozocerite, 92, 160

Padochov, 160
Pakistan, 181
Palaeogene, 133, 174
Palaeosalinity, 27, 29, 30, 35–37, 135
Palaeozoic, 11, 37–39, 69–71, 228
Palladium, 56, 165, **180**
Palliardite, 94
Palms, 24, 41
Paraffin, 81, 137, 138
– chain, 80
Paragonite, 105
Park County, 134
Pawtucket, 105, 143
Peat, 10, 17, 19, 21, 48, 61, 73–76, 79–84, 86, 88–90, 93, 94, 96–99, 107, 108, 113, 117, 121, 123, 128 to 131, 134, 135, 137, 138, 164, 168, 177, 188, 193, 196, 208, 211, 223, 230, IV/1
– bog, 14, 16, 17, 20, 21, 23, 24, 46, 58, 74–76, 79, 84, 88, 89, 91, 93, 153, 158, 163, 181, 196, 230
– –, eutrophic, 24
– –, Finnish, 86
– – –, element content, 87
– –, fossil, 25, 39
– –, mineralogical characteristic, 91–95
– –, oligotrophic, 24
– –, recent, 39
– –, transitional, 16, 19–21
–, element analysis, 85
–, sapropelic, 17
–, stage, 75
–, structural substances, 85
–, woody, 17, 21, 77, 79, 196
Peat-clays, 74
Peatification, 72, 74–76, 78, 82, 83, 85, 86, 89, 97, 108, 211
Pechora, 176

Pechora Valley, 176, 199
Péclet's hydrodynamic criterion, 220
Pecopteris, 39, 40
Pectins, 50, 70, 75, 83
Pelite, 30
Pell, 167
Pelosiderite, 94, 152, 156, 158–161, I/1, II/1, I/2, II/2
Penninite, 156
Pennsylvania, 39, 103, 143, 186, 187
Pentosans, 50
Pentoses, 45, 77, 136
Peptization, 76
Permeability, 218, 223
Permian, 12, 13, 38–40, 103, 105, 113, 156, 160, 167, 176, 191, 198, 199, 209, 227–230
Permocarboniferous, 185
Perylene, 80
Petřkovice, 31, 178, 230
Petrovice, 177, 192
pH, 20, 21, 24, 27, 37, 44, 54, 58, 73–76, 84, 89, 159, 164, 185, 187, 195, 196, 208, 210–213, 216, 223
Phase, biochemical, 96–98, 108, 109
–, first critical, 11
–, geochemical, 97, 98, 108, 109
–, second critical, 11
–, vapour, 37
Phellogen, 48
Phenanthrene, 137, 147
Phenols, 75, 78, 79, 82, 83, 115, 136
Phenylpropane, 46
Pheophorbide, 136
Philippines, 25
Phlebopteris, 40
Phlobaphenes, 79
Phosphates, 52, 55, 64, 93, 152, 153, 187
Phosphorus, 55, **64**, 77, 92, 153, 171, **179**, 180, 187, 189, 194, 208, 215, 230
Photoassimilation, 67
Photocatalyst, 11
Photosynthesis, 11, 46, 63, 195
α-phyllocladene, 138
Phyllocladus, 138
Phylloretine, 92, 138
Phyllotheca, 40
Phytane, 140
Phytin, 56
Phytol, 37, 48, 50
Piauzite, 161

Picea alba, 65
– *excelsa*, 62
Picene, 137
Pickeringite, 158, 162
Piesky, 61
Pigments, 11, 50, 61, 80, 214
–, flavinoid, 84
–, organic, 47
Pila, 112, 136, 139
Pinaceae, 47–49, 52, 54, 59, 63, 93, 132
α-pinene, 47, 48, 138
Pinus australis, 92
– *nigra*, 63
– *silvestris*, 58, 59, 64, 65, 67, 92, 132
– *strobus*, 63
– *uliginosa*, 92, 94
Pinuxylon paxii, 132
Pitchblende, 210
Plagioclase, 155
Plankton, 16, 35, 42, 73, 80
Plants, 11, 14, 37–39, 43–45, 47, 49, 130, 166, 200, 202, 215
–, aquatic, 50
–, caryophyllaceous, 61
– composition, 54
–, continental, 50
–, dry-land, 17
– elements content, 58–68
–, fossil, composition, 69
–, herbaceous, 39
–, leguminous, 63, 64
– material, 23
– – accumulation, 16, 25, 26
– – decomposition, 25
– microorganism, 10
– organism, 10
–, recent, composition, 69
–, rubber, 41
–, sporiferous, 70
–, terrestrial, 57
–, –, content of elements, 57–68
–, –, ash, content of elements, 57–68
Platanus, 41
Platinum, 56, 165, **180**
Plaur, 17
Pleistocene, 194
Pleuromeia, 40
Plevno, 174, 187
Pliocene, 23, 24, 100
Plötz, 191

Plzeň, 160, 204
Podocarpoxylon severzovii, 133
Podocarpus, 138
Podzolization, 91
Polička, 177
Pollen (grains), 24, 50, 55, 64, 73, 80, 92, 229
Polyamantanes, 149
Polymerization, 50, 75, 82, 115, 122, 136, 161, 184
Polymers, 131
Polypeptides, 79
Polyporus, 78
Polystictus versicolor, 82
Polysaccharides, 45, 69, 70, 77, 78, 101
–, nitrogenous, 46, 111
Polyterpenoids, 140
Poplar, 25, 44, 52, 59, 62
Populus, 41, 42
– *tremula*, 66
Porosity, 103, 205, 216, 218, 219, 222–224
Porphyrin(s), 50, 140, 214
– rings, 56
Portsmouth, 105, 143
Poruba, 31, 104, 193
Post-peatification stage, 86, 89, 97, 108
Potassium, 16, 27–30, 52, 53, 55, 56, **58**, 60, 77, 88, 93, 152, 155, 164, 170, **178**–180, 208, 213, 215, 230, 231
Praseodymium, 187
Precambrian, 11–13, 37
– sea, 29
Prehnite, 109
Pressure, 76, 89, 98, 99, 103–107, 111, 122, 205
Pristane, 37
Process, aerobic, 11
–, biochemical, 89, 98, 111
–, – decay, 39
–, biogeochemical, 93, 100
–, biological, 11
–, coal-forming, 39
–, endothermic, 89
–, fermentative, 11
–, geothermal, 11
–, irreversible, 96
–, metabolic, 38
–, peatifying, 92
–, post-sedimentary, 32
–, supergenic, 196
Propylene, 15
Propylidenecyclohexanone, 138

Proteins, 10, 46, 50, 51, 55, 56, 65, 70, 72, 73, 76, 77, 79–83, 85, 112, 136, 179, 181
Protoplasm, 44, 56
Province(s), Angara, 40
–, Eastern coal, U.S.A., 169
–, Euroamerican, 40
–, Gondwana, 40
–, Interior coal, U.S.A., 169, 188
–, Katasian, 40
–, Rocky Mountains, 169
–, Siberian, 40, 41
–, Vastergotland, 184
Pseudolaricixylon firmoides, 132
Pseudomonas, 78
Psilomelane, 93, 160, 161
Psilophytales, 12, 13, 39, 69
Protochlorite, 156
Pteridopsida, 39
Pteridospermae, 39, 40, 70
Pulsatilla patens, 60, 64
Pumpellyite, 109
Purines, 46, 48
Putrefaction, 72–74, 76
Pyran chain, 149
Pyrene, 137
Pyridine, 130, 131
Pyridoxin, 50
Pyrimidine, 50
Pyrite, 33, 74, 86, 91, 94, 95, 122, 152, 153, 155–163, 165, 170, 177–184, 188, 231
Pyritization, 74
Pyrolusite, 93
Pyrophosphate, 48
–, dimethylallyl, 48
–, geranyl, 48
–, isopentenyl, 48
Pyrophyllite, 109
Pyropissite, 161, 207
Pyroretinite, 161, 162
Pyroxenes, 153, 155
Pyrrhotite, 156, 157, 160
Pyrrole, 130, 131, 214
– ring, 50, 140

Quartz, 95, 152, 153, 155–157, 159–163, 210, 231
–, smoky, 160
Quartzite, 11
Quaternary, 12
Quebec, 131
Queensland (Australia), 188

Quinoline, 130, 131

Radium, 112, **180**
Radnice, 30, 156, 160, 173, 203, 204
Radstock, 70
Radvanice, 160, 170, 173, 179, 180, 184, 186–188, 197
Rájov near Kynžvart, 85
Rakovník, 186
Ranihandzhi, 103
Rare earth elements, 66, **67**, 186, **187**, 199, 200, 229
– gases, 177
Ratio, Ag/Pb, 164, 228, 230
– As/S, 165
– Ba/Sr, 228, 230
– Ca/Sr, 228
– $^{13}C/^{12}C$, 128–130
– C/H, 131
– $(C\text{-}H_{arom})/(C\text{-}H_{aliph})$, 146
– $CH_3:CH_2:CH:C$, 146
– C/(H + O), 112
– C/N, 33, 36, 55, 73, 74, 83, 125, 134
– Co/Ni, 60, 74, 170, 197, 228, 230
– Cr/Ni, 74
– Fe/Mn, 228
– Ga/V, 228, 230
– glycine + alanine/glutamine + asparagine, 33
– H/C, 112, 113, 125, 134, 150
– K/Rb, 58
– Mg/Ca, 24
– Na/K, 230
– $^{15}N/^{14}N$, 179
– O/C, 112, 113, 125, 150
– O/H, 47
– Pb/Ag see Ag/Pb
– pristane/phytane, 37
– resin/wax, 137
– Si/Al, 231
– SiO_2/Al_2O_3, 231
– Th/U, 32, 33
– V/Cr, 74
– V/Ni, 74
– V/Mo, 74
Ratíškovice, 128, 162
Realgar, 158, 159, 162, 163, 165
Recent, 37, 229
Red Desert, 186
Reduction, 75, 131, 153, 181, 194, 195, 215, 217
–, bacterial, 181
Redwitz, 92

Reed, 16, 20, 24
Reflectance (R_{im}), 108, 116–118, 120, 121, 124, 150
Refractive index, 116, 118, 124, 131, 150
Resins, 10, 11, 45, 48–52, 54, 64, 70, 72, 75, 76, 80, 81, 85, 98, 112, 122, 134, 136–138, 147, 157, 161 to 163, 196
–, fossil, 161
–, liptobiolithic, 136
–, sapropelitic, 136
Retene, 48, 92, 138
Retinite, 94, 160–162
Rhacomitrium lanuginosum, 60, 61, 68
Rhamnose, 45, 69, 136
Rhenium, **64**, **180**
Rhetenite, 138
Rhine area, 156
– valley, 156
Rhizoclonium, 54
Rhodium, 180
Rhodochrosite, 93
Rhodope, W., 174
Rhomboclase, 158
Riboflavin, 50
Ribose, 45, 136
Riley's model, 145
Riss-Würmian, 89
River, Amazon, 22
–, Lena, 22
–, Mississippi, 23
–, Radbuza, 173
–, Volga, 23
Rochlederite, 161
Rock, argillitic volcanogenic, 158, 185
–, basic, 61
–, calcareous, 67
–, clastic, 177
–, clayey, 29, 32, 103, 172, 177, 186, 203
–, –, brackish, 29
–, –, freshwater, 32, 33
–, –, marine, 29, 33
– crystal, 160, 161
–, granitic, 216
–, – kaolinized, 216
–, – lateritized, 216
–, humolithic, 77
–, igneous, 130
–, illitic, 29
–, metamorphosed, 29
–, organic, 10
–, organogenic, 96

Rock, pelitic-psammitic, 88
-, phytogenic, 10
-, picritic, 103
-, sandy, 203
-, sedimentary, 10–12, 28, 29, 105, 178, 179, 183, 185, 217
-, – marine, 36
-, siliceous, 62, 64
-, silty, 103
-, tuffaceous, 185
-, tuffogenic, 31, 103
-, tuffogenous, 186
-, ultrabasic, 60–62, 64, 67, 68
-, volcanic, 100, 198, 209, 216
-, volcanodetrital, 216
-, zoogenic, 10
Rokycany, 160
Ronchamps, 184
Rosice, 160
Rosickyite, 162
Rosenhein, 92
Rosthornite, 161
Rostite, 158
Rostov, 89
Rotting, 72
Rubidium, 32, **58**, **178**, 189, 199
Ruda peat bog, 17, 93, 230
Rush, 17, 63
Russian Platform, 12
Rutile, 153, 156, 163

Sabal, 41
Saccharides, 45–47, 50, 51, 53, 55, 56, 60, 69, 70, 73, 75–77, 79, 83, 85, 136, 195
-, polymers, 46
Saccharose, 45
Sackville, 85
Sage Lincoln County, 185
Salinity, 28, 32, 33, 37, 135, 166
Salix capna, 59
– *viminalis*, 64
Salmiac, 158–160
Salsola hali, 60
Salt Range, 153, 181
Salts, 29, 31, 58, 152, 153, 159, 193, 198
-, alkali, 193
Salzkohle, 159, 181
Samarium, 187
Sand, 28, 29, 221, 229
Sandaracopimarinol, 49

Sandstone, 28, 135, 159, 160, 172, 173, 176, 182, 185, 186, 199, 203–205, 209, 216, 221, VI/3
-, clayey, 177
-, coaly, 160
-, porous, 185
-, tuffaceous, 209
Sangara, 137, 176, 229
Sapanthracon, 73
Saprocol, 73, 74, 89
Saprodil, 73
Sapromyxite, 13
Sapropel, 16, 17, 73, 74, 89, 97, 126
Sapropelite(s), 11, 21, 73, 196, 197
Saprophytes, 72
Saratov, 177
Sarcocaulon, 75
Saxony, 177
Scandium, **65**, 85, **182**, 187, 199, 200, 212, 213
Scheererite, 138, 161
Scheuchzeria palustris, 92
Schizoneura, 40
Schönfeld, 191
Schraufite, 161
Schroeckingerite, 163, 209
Schürmann rule, 103
Schwarzenberský pond, 93, 230
Scleretinite, 161
Sclerotia, 24, 48
Scopolamine, 48
Scotland, 39, 80, 138
Seam, Dickebank, 170, 178, 182, 227
-, Hartley, 172
-, Katharina, 170, 171, 179, 180, 183, 188, 227
-, Peacock, 170, 179, 183, 187
-, Prokop, 17
-, Yard (U.K.), 175
Sea-weeds, 63
Secale cereale, 62
Sedge, 17, 20
Sedimentation, 183, 209
-, cyclic, 24, 25
-, littoral, 35
-, marine, 28, 32
-, organogenic, 100
Sediments, 45, 50
-, bituminous, 11, 168
-, brackish, 27, 29
-, clastic, 89
-, clayey, 28, 37, 98, 155, 158
-, – marine, 32

279

Sediments, coaly, 161, 185
–, continental, 24, 32, 36, 39, 198
–, deltaic, 89
–, freshwater, 25, 27, 29, 33, 54, 135
–, interglacial, 89
–, lacustrine, 135
–, lagoonal, 39
–, marine, 24, 25, 27, 29, 32, 33, 35, 54, 135
–, organic, 195
–, pelitic, 189, 190
–, – freshwater, 28
–, – marine, 28, 30
–, sandy, 98, 229
–, sapropelic, 73, 74
–, sapropelitic, 197
–, silty, 98
Seeds, 53, 55, 56, 63, 64
Seehestadt, 23
Sejkov, 163
Selaginellaceae, 12
Selenides, 196
Selenium, **65**, **182**, 196
– (mineral), 158
Semi-anthracite, 121, 126, 186
Semigraphite, 107, 108
Sequoia reichenbachi, 41
Sericite, 160
Serine, 136
Sesquiterpenoids, 48, 138
Shale, 36, 184, 186, 187
–, bituminous, 187
–, black, 184, 186
–, clayey, 183
–, coal, 36
–, – marine, 36
–, – non-marine, 36
–, coaly, 186
–, combustible, 23
–, marine, 29, 36
–, micaceous, 161
–, – graphitic, 160
–, non-marine, 36
–, oil, 214
Shungite, 13
Siberia, 40, 41, 66, 176
Siderite, 93, 153, 155, 157, 159, 160, 171
Siegenite, 157, 158, 179
Sigillariaceae, 12, 39, 160
Silene inflata, 67
Silesia, 178

Silica, 59, 65, 152, 154–156, 200, 231
Silicates, 52, 53, 157, 166, 172, 178, 200, 203, 205, 206
Silicification, 156, 163
Silicon, 28, 56, **65**, 93, 152, **182**, 189, 194, 200, 208, 213, 215
Silt, 29
Siltstone(s), 28, 30, 36, 108, 197
–, clayey, 186
Silurian, 11–13, 38, 39
Silver, 33, **58**, 73, 85, 93, **164**, 189, 194, 197, 199, 200, 212, 213, 215
Simonelite, 138
Sinapyl alcohol, 46
β-sitostanol, 83, 140
β-sitosterol, 83, 84, 140
Skutičko, 161
Slaný, 177, 181
Slatina, 207
Slide Butte, 186
Sluggan Bog, 92
Sobuty Kavak, 184
Soda, 91, 94
Sodium, 27–29, 52, 56, **58**, 88, 93, 152, 155, 159, **178**, 189, 208, 213, 215
– carbonate, 195
– chloride, 60, 168
Soil, 32, 43, 44, 48, 56, 58, 60, 62–64, 66, 67, 73, 86, 194, 200
–, fossil, 32
–, podzol, 61, 64
–, residual, 63
–, seleniferous, 65
Sokolov, 162, 207
Sorption, 28, 44, 52, 89, 178, 196, 202, 203, 205, 207, 208, 211, 215–220, 223
–, selective, of ions, 56
– surface, 44
Sor Range, 181
Špania Dolina, 61
Specific gravity, 173, 197, 212, 218, 219
Spectra, infrared, 118–120
–, ultraviolet, 118
Sphagnetalia, 19
Sphagnum sp., 19, 20, 25, 60, 63, 71, 79, 86, 87, 89
Sphalerite, 60, 153, 155–157, 159–161, 188, 197, 203
Sphene, 155
Sphenophyllaceae, 12
Sphenophyllum thoni, 40
Sphenopsida, 39

Spherosiderite, 160, 162
Spilite, 157
Spin number N_s, 121
Spitsbergen, 22, 39
Spores, 10, 24, 38, 46, 50, 53, 64, 73, 75, 80, 229, VI/2
Sporinite, 118
Sporopollenins, 50, 51, 75, 80
Spruce, 46, 63, 64
Sri Lanca, 25
Staffordshire, 175
Stage, alteration, 96
-, bituminous-coal, 98
-, brown-coal, 98
-, diagenetic, 35
-, Gehren, 191
-, interglacial, 25
-, Manebach, 191
Starch, 45, 55, 69, 75, 78
Staurolite, 155
St. Eloy, 167
Stephanian, 30, 31, 40, 156, 160, 191
Stereocaulon pascale, 62
Steroids, 47, 140
Sterols, 83, 140
St. Gallen, 161
St. Hippolyte, 184
Stockheim-Neuhaus, 191
Streak, 124, 126, 127
Stříbro, 156
Strontium, 32, 56, **66**, 93, 166, **183**, 187, 189, 197, 199, 200, 208, 210, 213, 215, 227, 229
Suberin, 10, 48, 51, 72, 75, 80
Subhumite, 89
Subsaprocol, 89
Subsapropel, 89
Substances, bituminous, 73, 136, 156
-, clayey, 153, 156
-, fatty, 47
-, heterocyclic, 50
-, humic, 10, 74, 76, 86, 112, 122, 130, 131, 134, 135, 154, 156, 196, 200, 202, 204, 215, 216
-, huminitized, 98
-, hydrophilic, 45
-, inorganic, 212
-, -, in plants, 53
-, lipid, 47
-, lipoidal, 83
-, nitrogenous, 46, 73, 77
-, oil, 196

Substances, organic, 73, 200, 212, 214, 223
-, protobituminous, 83
-, protohumic, 78
-, resinous, 30, 161, 181
-, sapropelic, 74
-, siliceous, 153
-, waxy, 47
Succinate, 58
Succinite, 161
"Succinosis", 49
Suchá, 31, 193
Sulphates, 52, 55, 92–95, 152, 153, 157–159, 161, 162, 180–182, IV/1
Sulphides, 95, 100, 153, 155–159, 171, 172, 179–182, 194, 196, 197, 200, 201, 214, 215
-, black, 74
-, ferrous, 74
Sulphidization, 158, 159
Sulphur, 51, 54, 55, **65**, 77, 86, 88, 112, 123, 128, 134, 136, 150, 151, 158, 165, **180**, 181, 182, 189, 194, 196, 210, 214, 215, 224
- dioxide, 95
-, elementary, 181
-, inorganic, 181
-, isotopic composition in coal, 182
- (mineral), 91, 94, 95, 153, 157–162
-, organic, 180–182
-, "pyritic", 181
-, sulphate, 181, 182
-, sulphidic, 181, 182
Sumatra, 25
Šumava Mts., 22
Suppuvaara, 58
Surface extent, 216
Svatoňovice, 128, 160, 170, 173, 179, 186
Světec, 162
Svět near Třeboň, 85
Swamp iris, 88
Swamp(s), 14, 23–25, 40, 158, 214
-, tropical, 25
Sweden, 184
Sweetwater County (Wyoming), 169, 186, 209
Switzerland, 184
Sydney coalfield (Canada), 174

Taimir, 103
Tannins, 48, 52, 73, 79, 81
Tantalum, 66
Tashkent, 13, 39
Tashkumyr, 184

Tasmanite, 161
Taxodiacea, 25, 41, 88, 132
Taxodioxylon anthratoxoides, 132
– *ishikuraense*, 132
– *sequoianum*, 132
– *taxodii*, 132
Taxodium distichum, 69
Tea plant, 61
Tekoretin, 138
Telinite, 187, 197
Tellurides, 196
Tellurium, 158, 196
Temperature, 28, 33, 38, 40, 52, 76, 77, 81, 89, 98, 99, 101, 103, 105–108, 110, 118, 122, 181, 205, 218, 219, 222, 223, 228
Tennessee, 169, 185
Teplice, 209
Terbium, **66**, **187**
Terpene(s), 47–49, 75
Terpenoids, 47–49, 80, 138
Tertiary, 12–14, 20, 21, 38, 39, 41, 42, 73, 113, 122, 133, 137, 167, 175, 183, 185, 186, 196, 230
Těšín, 101
Tetrahedrite, 157, 160
2,6,10,14-tetramethyl hexadecane, 140
Texas, 23
Thalassia, 23
Thallium, 56, **66**, **183**, 184
Thenardite, 160
Thénard's blue, 158
Theocretine, 92
Thermal alteration index T.A.I., 117, 118
– mineral, 198
– springs, 166, 194, 198
Thiamine, 50
Thiazol, 50
Thioether, 181
Thiothrix, 65
Thlaspi calaminarium, 67
Thorium, 32, **66**, 112, **183**, 184, 194, 212, 215
Threonine, 35, 136
Thulium, **66**, **187**
Timan, 176
Time, 28, 108, 216–224, 226, 229
Tin, 56, **65**, 73, **183**, 187, 189, 199, 200, 208, 210, 212, 213, 215
Titanium, 56, **66**, 93, 152, **183**, 198–200, 208, 210, 212, 213, 215, 227, 229, 231
Tocopherols, 50
Toluene, 15

Tomsk, 86
Tonstein, 136, 155
Topaz, 156
Tortella tortuosa, 59, 61, 67
Tourmaline, 28, 29, 155
–, authigenic, 28
–, detrital, 28, 30
–, syngenetic, 28
Touškov, 173
Trace elements, 29, 32, 33, 73, 76, 190–193, 226
– – enrichment, 217–224
– – in coal, 189–216
– – – – ashes, 189–216
– – – –, concentration, 189–216
– – – –, distribution, 189–216
Tracheophytes, 65
Transbaikalia, 197
Transitional bog, 91, 230
Třeboň, 93, 177, 230
Triassic, 12, 13, 40, 69, 133, 167, 209, 210, 217
Tricyclo-alkanes, 149
Trifail, 161
2,6,10-trimethyldodecane, 140
Trinkerite, 161
Tripticene, 145
Trisaccharides, 45, 77
Triterpenoids, 48, 84, 138, 139
–, functional groups, 139
–, melting point, 139
Troctolite, 103
Tschermigite, 158–160, 162, III/2
Tuff(s), 59, 209
Tyne, 153
Týnec near Plzeň, 173
Tungsten, **67**, **187**, 197, 199, 212
Typha, 16
Tyulenevski Mine, 64

Uhelná, 136, 162
Ulmus, 41
United States, 65, 209
Urals, 13, 60, 64, 85, 86, 176, 227, 228
Ural Valley, 176
Uraninite, 103
Uranium, 15, 32, 64, **66**, 74, 85, 86, 163, 183, **184** to 186, 194, 196, 198–200, 208–213, 215, 217, 227
Uranyl, 210
– fulvates, 211
– humates, 211
Úsilné, 128, 156, 230

Utah, 186
Utricularia, 88
Uzbekistan, 182
Uznach, 161

Vacciniaceae, 79
Vaccinium myrtillus, 60
Valaite, 160
Valchov, 161
Vanadium, 15, 32, 33, 56, **67**, 73, 86, 140, 170, 171, 178, 185, **186**, 194, 196–200, 208, 210, 212–215, 217, 227, 229
– phenols, 214
– porphyrins, 214
Vanadyl, 214
Vancouver, 14
Vanilline, 132
Van Krevelen's model, 149
Vejprnice, 172
Velence Mts., 208
Vermiculite, 109
Veselí nad Lužnicí, 21, 230, IV/2
Victoria (Australia), 107, 134, 165, 188
Vimperka, 93, 230
Vinařická hora, 100, 157
Vintířov, 99, III/1
Viola calaminaria, 67
Virginia, 23, 25, 79, 165, 169, 172, 182, 185, 188
Viséan, 191
Vísky, 162
Vitamins, 50
Vitrain, 21, 121, 133, 154, 168–170, 173, 176, 178, 179, 183, 187, 188, 197, 201, 204, 206–208, 212, 216, 227, 229, 231
Vitrinite, 113, 114, 116, 118, 120–122, 124–128, 144 to 151, 164, 185, 206, 211
Vittinkii, 60, 65, 68
Vivianite, 91, 92, 94, 161, 162
Volatile matter (V^{daf}), 96, 102, 103, 104, 106–108, 110, 112, 114, 115, 117, 118, 123, 128, 135, 137, 146, 149–151, 182
Volcanogenic material, 29, VIII/2
Volchepolskoe, 174
Voltzia, 40, 69
– *brongniarti*, 69
Volume, 98

Wakkesdorf, 184
Walchia, 40
Walchowite, 161

Wales, 25, 152, 153
Warrior coalfield, 180
Warwickshire coalfield, 208
Washington, 25
Water, 11, 14, 43, 45, 47, 52, 55, 56, 72, 73, 76, 77, 79, 84, 86, 88, 89, 93, 97, 98, 102, 105, 110, 111, 114, 122, 123, 128, 152, 159, 177, 181, 182, 186, 187, 189, 193–196, 198, 203, 205, 208, 216
–, atmospheric, 19
– basin, 16, 17, 20
–, circulating, 203, 216
–, content in plants, 53
–, continental, 28
–, fresh, 37
–, ground, 19, 185, 200, 203, 208–211, 216, 220, 221, 223
–, heavy, 55
–, hydrothermal, 216
–, metamorphic origin, 216
–, ocean (sea), 28, 29, 37, 54
–, percolating, 209, 216, 226
–, porous, 37
–, post-volcanic, 216
–, saline, 37
–, sorbed, 37
–, surface, 216
–, swamp, 58
–, vapour, 14, 74, 158
Waterschei, 167
Waxes, 10, 37, 47, 50–52, 54, 69, 70, 72, 75, 76, 80, 81, 85, 98, 112, 122, 134, 137, 162, 196
–, liptobiolithic, 136
–, montan, 80, 136, 138, 139
–, peat, 136
–, sapropelitic, 136
Weichselia, 41
Wendel, 167
Weston County, 185
Westphalia, 156
Westphalian, 30, 31, 36, 39, 40, 136, 156, 160, 191
Wettin, 191
Wheelerite, 161
Whewellite, 157, 159, 162, 163, 168, I/2
Wigan, 161
Willow, 20, 59, 64
Witherite, 156, 163
Wood(s), 107, 113, 123, 127, 131, 173, 202, 211, 230
–, chemical composition, 132
–, coalified, 133, 172, 173, 185, 187, 199, 200, 203, 214, 220

Wood(s), fossil, 207
—, mineralized, 220
—, sideritized, 160
—, silicified, 162
Wyoming, 163, 185, 186

Xantophylls, 50
Xylans, 45, 70
Xylene, 15
Xylenite, 164
Xylenol, 136
Xylite(s), 21, 162, 196, 207, 209, 216
Xylose, 45, 46, 69, 70, 136

Yakutsk, 164, 199
Yew, 58, 60, 61, 63–65
Yorkshire, 120, 137, 153
Ytterbium, 93, **187**, 208, 213
Yttrium, **67**, 93, **187**, 208, 212, 213
Yubari, 137

Žacléř, 160, 166, 168, 173, 179, 183, 186, 197, 207, 230
Zálezly, 162
Zaratite, 157
Zastávka near Rosice, 128, 160
Zaunhaus, 191
Zbýšov, 160
Želénky, III/2
Zeolites, 157
Žeravice, 162
Zeunerite, 185, 210
Zinc, 50, 56, **67**, 74, 85, 88, 89, 93, 156, **188**, 189, 194, 196, 197, 199, 200, 212, 213, 215
Zircon, 103, 153, 155, 163, 215
Zirconium, **68**, 93, **188**, 189, 199, 200, 210, 212, 213, 215, 227
Žíšov, IV/2
Zwartberg, 167
Zwickau, 174, 177, 179, 191, 228